Pulsed Electrochemical Detection in High-Performance Liquid Chromatography

TECHNIQUES IN ANALYTICAL CHEMISTRY SERIES

BAKER • CAPILLARY ELECTROPHORESIS

CUNNINGHAM • INTRODUCTION TO BIOSENSORS

LACOURSE • PULSED ELECTROCHEMICAL DETECTION IN HIGH-PERFORMANCE LIQUID CHROMATOGRAPHY

MCNAIR and MILLER • BASIC GAS CHROMATOGRAPHY

METCALF • APPLIED pH AND CONDUCTIVITY MEASUREMENTS

MIRABELLA • MODERN TECHNIQUES OF APPLIED MOLECULAR SPECTROSCOPY

SCHULMAN and SHARMA • INTRODUCTION TO MOLECULAR LUMINESCENCE SPECTROSCOPY

STEWART and EBEL • CHEMICAL MEASUREMENTS IN BIOLOGICAL SYSTEMS

TAYLOR • SUPERCRITICAL FLUID EXTRACTION

Pulsed Electrochemical Detection in High-Performance Liquid Chromatography

WILLIAM R. LaCOURSE, PhD
University of Maryland Baltimore County
Baltimore, Maryland

A Wiley-Interscience Publication
JOHN WILEY & SONS, INC.
New York • Chichester • Weinheim • Brisbane • Singapore • Toronto

CHEM

Library of Congress Cataloging in Publication Data:

LaCourse, William R., 1957–
    Pulsed electrochemical detection in high-performance liquid
chromatography / William R. LaCourse.
        p.   cm. — (Techniques in analytical chemistry series)
    "A Wiley-Interscience publication."
    Includes bibliographical references and index.
    ISBN 0-471-11914-8 (cloth : alk. paper)
    1. High performance liquid chromatography.   2. Electrochemical
sensors.   I. Title.   II. Series.
    QD79.C454L324   1997
    547'.7046—dc21   97-548

This work is dedicated to my *Sweet-Ps*.
They have a talent for turning dreams into reality.

Contents

Foreword   **xi**
Series Preface   **xiii**
Preface   **xv**
About the Author   **xvii**

1   Historical Perspective   **1**

Electrode Reactivation   4
Origins of Pulsed Potential Cleaning   6
History of Pulsed Electrochemical Detection   7
Impact of PED Techniques   8
This Book   10

2   Electrochemical Fundamentals   **13**

Basic Concepts   13
Electrochemistry   14
Electrode Potentials   19
Potentiometry   27
Electrolysis   32
The Electrode Interface and Mass Transfer   39
Voltammetry   46
Electroanalytical Methods   58

3 Amperometric Detection in HPLC   **60**

Fundamental Principles of Amperometry   62
Electrolysis Cell Designs   63
Other Cell Considerations   69
The Origins of Current   74
Optimization of Applied Potential   75
Selectivity: Multiple Electrodes   79
Quantitative Aspects   82
Applications: A Brief Overview   82

4 Pulsed Amperometric Detection in HPLC   **86**

The Noble-Metal Electrode   90
Concepts of Pulsed Electrochemical Detection   105
Pulsed Amperometric Detection   112

**5 Integrated Pulsed Amperometric Detection
and Other Advanced Potential–Time Waveforms**   **122**

Oxide Formation and Dissolution Kinetics   123
Role of the Oxide in PED Mechanisms   130
Reverse Pulsed Amperometric Detection   132
Activated Pulsed Amperometric Detection   135
Integrated Pulsed Amperometric Detection (IPAD)   135
Multicycle Waveforms   144

6 Pulsed Voltammetry: Waveform Optimization   **149**

Pulsed Voltammetry   153
PAD Waveform Optimization   155
RPAD and APAD Waveform Optimization   176
IPAD Waveform Optimization   176
Other System Considerations   177
Pulsed Voltammetry (PV): A Quantitative and Mechanistic Tool   180
Conclusions   181

7 Applications of PED   **182**

Carbohydrates and Alditols   182
Aliphatic Alcohols and Polyalcohols   198
Aminoalcohols, Aminosugars, and Aminoglycosides   205

Amines, Aminoacids, Peptides, and Proteins   213
Sulfur-Containing Compounds   218
Miscellaneous Applications and Reviews   223

8   Instrumental Considerations   **229**

The Instrumental Setup   230
Postcolumn Reagent Addition   233
Maintaining HPLC–PED Systems   236
Troubleshooting HPLC–PED Systems   242

9   Future Aspects of PED   **245**

Microelectrode Applications in PED   245
Indirect Detection Methods   254
Pulsed Voltammetric Detection (PVD)   257
Conclusions   260

Appendix A   Pulsed Voltammetry Program   **262**

Appendix B   Tables of Applications   **293**

Table B.1.   Applications Related to Carbohydrates and Alditols   293
Table B.2.   Applications Related to Aliphatic Alcohols   304
Table B.3.   Applications Related to Aminoalcohols, Aminosugars,
and Aminoglycosides   304
Table B.4.   Applications Related to Amines, Aminoacids, Peptides,
and Proteins   305
Table B.5.   Applications Related to Sulfur-Containing Compounds   306
Table B.6.   Inorganic, Electroinactive, and Miscellaneous
Applications   307
Table B.7.   Instrumental Developments and PED Reviews   308

Index   **317**

Foreword

The popularity of pulsed electrochemical detection (PED) has steadily increased over the last decade for the sensitive and reliable detection of many polar aliphatic compounds in aqueous samples following their separations by high-performance liquid chromatography (HPLC). PED is based on multistep potential waveforms applied at solid electrodes to alternate the operations of anodic detection with oxidative cleaning and reductive reactivation of the electrode surfaces. Compounds detected by these multistep waveforms at gold and platinum electrodes include all alcohols, alditols, and carbohydrates, as well as many organic amines and sulfur compounds. The development of HPLC–PED represents a successful marriage of two analytical technologies. Furthermore, it is an example of a successful cooperative research effort by scientists in academia (Iowa State University) and industry (Dionex Corp.) that can serve as a template for the pursuit of future academic–industrial partnerships.

Because the principles of HPLC have been summarized in numerous monographs, the author has made the appropriate choice to emphasize the electrochemical principles on which PED technology is based. Also presented are numerous interesting examples of HPLC–PED applications for the analysis of complex samples. Also included is a valuable appendix listing all published applications of HPLC–PED to date. Furthermore, operational guidelines are included to assist with equipment setup and procedural optimization. This text is written in a very readable style that is especially appropriate for graduate students and other newcomers to this analytical methodology. However, experts in the separate disciplines of *liquid chromatography* and *electroanalysis* also will benefit from the extensive coverage given to the subject area. This text can be highly recommended especially for use in graduate-level special topics courses and industrial short courses. Furthermore, it is must reading for persons interested in the analytical challenges offered by polar aliphatic compounds that lack strong chromophores and are, therefore, not amenable to direct detection by conventional uv–vis (ultraviolet–visible) photometric techniques.

DENNIS C. JOHNSON

Series Preface

Titles in the *Techniques in Analytical Chemistry Series* address current techniques in general use by analytical laboratories. The series intends to serve a broad audience of both practitioners and clients of chemical analysis. This audience includes not only analytical chemists but also professionals in other areas of chemistry and in other disciplines relying on information derived from chemical analysis. Thus, the series should be useful to both laboratory and management personnel.

Written for readers with varying levels of education and laboratory expertise, titles in the series do not presume prior knowledge of the subject, and guide the reader step by step through each technique. Numerous applications and diagrams emphasize a practical, applied approach to chemical analysis.

The specific objectives of the series are

- To provide the reader with overviews of methods of analysis that include a basic introduction to principles but emphasize such practical issues as technique selection, sample preparation, measurement procedures, data analysis, quality control, and quality assurance.

- To give the reader a sense of the capabilities and limitations of each technique, and a feel for its applicability to specific problems.

- To cover the wide range of useful techniques, from mature ones to newer methods that are coming into common use.

- To communicate practical information in a readable, comprehensible style. Readers from the technician through the Ph.D. scientist or laboratory manager should come away with ease and confidence about the use of the techniques.

Books in the *Techniques in Analytical Chemistry Series* cover a variety of techniques including capillary electrophoresis, biosensors, supercritical fluid extraction, measurements in biological systems, inductively coupled plasma–mass spectrometry, gas chromatography–mass spectrometry, Fourier transform infrared spectroscopy, and other significant topics. The editors welcome your comments and suggestions regarding current and future titles, and hope you find the series useful.

FRANK A. SETTLE

Lexington, VA

Preface

The separation and sensitive detection of polar aliphatic compounds, which have poor detection properties by traditional analytical techniques, has always been a significant analytical challenge. The growth of biotechnology elevated the need to be able to determine bioactive compounds such as carbohydrates, oligosaccharides, amines, peptides, glycopeptides, and thiocompounds in complex mixtures. Over the past 15 years, pulsed electrochemical detection (PED) following high-performance liquid chromatography (HPLC) has been extensively used for the determination of carbohydrates and other polar aliphatic compounds, and as a consequence a large number of methods have been developed to enable the analysis of a wide variety of samples. During this time, a plethora of articles, anecdotal information, and misinformation has been disseminated in the scientific community. I wrote this book in order to assemble a comprehensive and relevant review of PED and its applications. Since my intent was to write a book for anyone interested in PED, it was important to write this text in a style that would be readable and useful to both the expert and novice.

The book is divided into three major parts. The first part focuses on the background material needed to understand the principles and relevance of PED. Hence, Chapter 1 both documents the history behind the conception and development of PED and establishes the need for PED in a world of numerous analytical techniques. Chapters 2 and 3 were written specifically with the nonelectrochemist in mind. These chapters offer a concise overview of all the fundamentals of electrochemistry and dc (direct-current) amperometry needed to fully comprehend PED.

The second part begins with an in-depth discussion of PED (Chapter 4) using voltammetry and other electroanalytical techniques to promote understanding and insight. The advantages and applicability of all existing PED waveforms are discussed in detail (Chapter 5) including their optimization (Chapter 6). A pulsed voltammetry (PV) program, which was specifically written to optimize pulsed amperometric detection (PAD) waveforms, is included in Appendix A.

The third part covers the practical aspects of HPLC–PED, which begins with a summary and overview of the major applications (Chapter 7). For handy reference, all the known applications are categorized and listed in tabular form in Appendix B. Since any application is dependent on a properly operating system, Chapter 8 gives practical insight and information uniquely concerning instrumental setup, optimization, and troubleshooting for HPLC–PED. Finally, Chapter 9 summarizes the developments on the horizon for this technique of interest to both the end user and researcher.

This book is intended for use by analytical chemists, biochemists, carbohydrate chemists, separation scientists, biologists, biotechnologists, and others working in academia and industries that include biochemical, medical, pharmaceutical, and food industries. In addition, this book can be a useful reference for the experienced analyst and newcomer alike, as well as a specialty text for undergraduates, graduates, postdoctorals, and so on.

I would like to express my deepest appreciation to Dennis C. Johnson for all his guidance, discussions, and collaborations. He is an excellent mentor. I would also like to acknowledge the support of numerous Dionex personnel who have been courteous, professional, and most helpful throughout the years.

I am very grateful to Christine M. Zook for her invaluable assistance in the preparation of this manuscript. Her zeal and dedication have not gone unnoticed. I am also indebted to Catherine O. Dasenbrock and past and present graduate and undergraduate students in my research laboratory. Finally, I warmly appreciate the unwavering support, encouragement, and inspiration of my wife, Katy, and of my daughter, Maggie.

WILLIAM R. LaCOURSE

About the Author

William R. LaCourse received his Ph.D. in analytical chemistry from Northeastern University in 1987. After completing a postdoctoral appointment at Ames Laboratory, U.S.D.O.E., he held the position of Scientist at Iowa State University/Ames Laboratory, working with Dennis C. Johnson. In 1992, he joined the faculty at the University of Maryland Baltimore County. In addition, he has 5 years of industrial experience in the pharmaceutical industry on the development of product assays of both human and veterinarian formulations. Dr. LaCourse's research interests include basic and applied research on hydrodynamic electroanalytical techniques in liquid chromatography and capillary electrophoresis, adsorption phenomena at noble-metal electrodes, and advanced sample preparation techniques (e.g., microdialysis and pressurized fluid extraction). Presently, the thrust of his efforts is to further understand and expand the limits of pulsed electrochemical detection (PED) techniques for the determination of polar aliphatic compounds.

Pulsed Electrochemical Detection in High-Performance Liquid Chromatography

1 Historical Perspective

That's one small pulse for electrochemistry, one giant leap for analysis.

It wasn't that long ago that the mere mention of electroanalytical chemistry would conjure up visions of dropping mercury electrodes, magical polishing techniques, and tricky instrumentation. The high sensitivity at moderately low cost afforded by electroanalytical techniques was often overshadowed by limited voltammetric selectivity. Hence, prior to the early 1970s, assays using electroanalytical techniques centered on "simple" mixtures of metal ions and/or organic compounds. Applications involving complex biological samples were practically nonexistent—that is, nonexistent only until *electrochemical detection* (ED) in flow-through cells and *high-performance liquid chromatography* (HPLC) were combined to form one of the most powerful bioanalytical techniques available today.

Since the first commercially available HPLC electrochemical detector was introduced in 1974, an overwhelming number of journal articles on design, performance, theory, and applications have appeared in the scientific literature. In addition, the number of papers presented at scientific meetings and symposia continues to grow at a breathtaking pace. Electrochemical detection is highly sensitive, whereas HPLC, in its numerous forms, is far superior to electroanalytical voltammetry in resolving power for analysis of complex mixtures. Unlike the combination of gas chromatography (i.e., high pressure) and mass spectrometry (i.e., vacuum), which operate at opposing pressure conditions, ED and HPLC are mutually compatible. Both techniques are heterogeneous in nature, in that a

1

liquid phase is in contact with a solid phase or a phase boundary. In ED, a supporting electrolyte is in contact with the electrode surface, and in chromatography, the mobile phase is in contact with the stationary phase. Both techniques rely on diffusion to deliver the analyte of interest from the bulk liquid phase to the phase interface. Figure 1.1 illustrates the basic similarities of ED and HPLC. Electrochemical detection is predominantly a water-based system, and it has been estimated that 70–80% of all liquid chromatographic separations are also performed under aqueous conditions.

Amperometric detection is typically performed at a solid anode (e.g., Au, Pt, and glassy carbon) under constant applied potential, which is typically denoted as *dc amperometry*. Numerous aromatic compounds are detected easily by anodic reactions. These include phenols, aminophenols, catecholamines, and other metabolic amines. Electronic resonance in aromatic molecules functions to stabilize free-radical intermediate products of anodic oxidations, and, as a consequence, the activation barrier for electrochemical reaction is lowered significantly, see Figure 1.2A. In dc amperometry, the compounds are self-stabilizing, and the electrode is considered to be inert. The electrode functions to accept and donate electrons to the analyte of interest. Any chemical involvement of the electrode (i.e., adsorption) with the electrochemical process often leads to fouling of the electrode surface. Fouling of the electrode surface is detrimental to the sensitivity and reproducibility of the HPLC–ED system. Thus, the goal is to use the most inert electrode material available, which accounts for the popularity of glassy-carbon electrodes in dc

ELECTROCHEMISTRY   CHROMATOGRAPHY

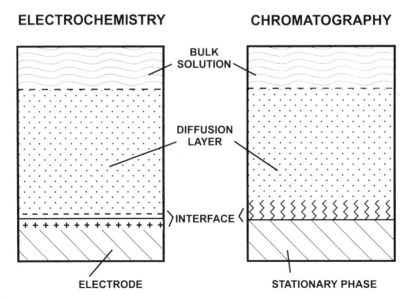

Figure 1.1. Fundamental similarities of electrochemical detection and high-performance liquid chromatography.

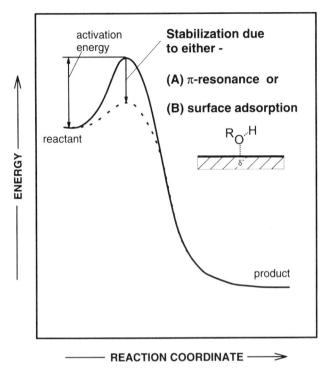

Figure 1.2. Reaction profiles of electrochemical reactions with (*A*) self-stabilized and (*B*) surface-stabilized intermediates. Since both mechanisms of stabilization achieve the same overall effect, the same plot can be used to visualize each approach.

amperometry. Glassy carbon as an electrode is very resistant to fouling. The term *resistant* does not connote impervious, and fouling does occur. Hence, the daily ritual of polishing the electrode surface or some other method of electrode reactivation must be undertaken to please the redox gods.

As expected, the absence of π-resonance for aliphatic compounds results in very low oxidation rates at an inert electrode, although the reactions may be favored thermodynamically. What is needed is an alternative way to stabilize the free-radical intermediates formed in the detection process or to *catalyze* the electrochemical reaction. Hence, stabilization can be achieved with the use of catalytic electrode surfaces, which have unsaturated *d*-orbitals. In other words, noble-metal electrodes can be used to stabilize free-radical products via adsorption, and, as a consequence, the activation barrier of the electrochemical reaction is lowered significantly (see Fig. 1.2*B*). Unfortunately, exploiting adsorption to the electrode to assist in the electrochemical detection process does not come without a price. In contrast to inert electrodes, which may suffer from fouling over a period of hours or even days, the adsorption of organic molecules and free radicals to catalyze detection results in fouling of the electrode in a matter of seconds. Thus, the historical perspective of nonreactivity for polar aliphatic organic compounds at noble metals can be attributed to rapid fouling or occlusion of the electrode surface, and the irony is that this fouling is a result of high, but transient, catalytic activity.

ELECTRODE REACTIVATION

From a fundamental point of view, many electrochemical systems suffer from the same lament: namely, that the electrode is in direct contact with the test solution at which a heterogeneous reaction is carried out. This arrangement often leads to fouling of the electrode surface. As noted above, this effect is drastically enhanced for detections using electrocatalytic surfaces. As a consequence, the electrode must undergo some form of treatment prior to performing an analytical measurement. A quick perusal of the literature shows a wide range of techniques and methods for cleaning electrode surfaces. Stulik [1] has summarized many of these techniques and grouped them into four different categories.

  1. Mechanical Grinding and Polishing.   Mechanical polishing is usually the first step in the preparation of the surface of a solid anode. Common metallographic procedures are employed to polish the electrode to a mirrorlike finish. Emery wheels, papers, cloths, and polishing suspensions of alumina and diamond are used with gradually decreasing grain size. The polished electrode is then washed with soaps and organic solvents, and thoroughly rinsed with water. Mechanical polishing is harsh in nature and typically results in both physical changes and chemical modifications of the electrode surface. Mechanical polishing is often the only recourse for a severely fouled electrode surface. As expected, electrodes with flat surfaces (e.g., disks, rings, and ring–disk electrodes) are most suitable for mechanical polishing, and it is for this reason that planar electrodes have been so popular in HPLC–ED. In addition, mechanical polishing can be done only external to electroanalytical systems.

  2. Thermal Methods.   At the heart of all the energy methods of electrode reactivation is the liberation and/or utilization of heat energy to induce mechanical and chemical changes at the electrode surface. These methods include placing the electrode in a flame [2,3], subjecting the electrode to a radio-frequency (rf) discharge [4–6], or irradiating it with laser radiation [7–9]. Typically a very thin layer of adsorbed particles and electrode material is removed from the electrode surface. These treatments often lead to highly active electrode surfaces, which decay rather rapidly with use. Although the use of laser pulses has been proposed for periodic cleaning for electrochemical detection in HPLC [10,11], these methods are generally awkward to apply for routine measurements.

  3. Chemical Pretreatment.   The major objectives of most electrode reactivation procedures are to remove passivating layers and preserve the electrode surface in a state of high sensitivity. Often metal electrodes have a passivating oxide layer, which inhibits electrode activity. Strong mineral acids (e.g., chromic acid or hot nitric acid) can be used to remove passivating layers to leave the electrode in a "clean" state. Oxidized electrode surfaces can be formed with solutions containing ferric or ceric ions and bromine water, whereas reduced surfaces can be formed by treatment with ferrous or arsenious ions [12]. Hence, chemical treatment can be used to essentially reduce or oxidize the electrode surface. Sometimes chemical methods are combined with electrochemical methods. Obviously, these methods cannot be used "on-line" in a flow-through system.

4. Electrochemical Pretreatment. Probably the most common form of electrode cleaning and reactivation is with the use of polarizations under suitable solvent conditions. Oxidation and reduction of the electrode surface removes passivating layers and chemisorbed impurities. This approach is particularly attractive because it can be carried out in the electrochemical cell using the same instrumentation used in the experiment. In addition, the procedure can be readily incorporated either at the onset or periodically throughout the experiment. It is this aspect of electrochemical reactivation that germinated the origins of pulsed electrochemical detection.

These reactivation methods are used either singly or in combination. Often mechanical polishing is the first step and the step of last resort of any electrode pretreatment. Electrochemical pretreatment of the electrode surface is often combined with submersion of the electrode into a chemical solution (e.g., cold aqua regia) to remove inert oxides. Because of the great variety of electroanalytical systems and types of measurements, no one system is the best system. Generally, if the electrochemical cell can be taken apart readily at a convenient time, as in batch experiments, any of the treatments described can be used to obtain electroanalytical measurements.

When it comes to flow systems (e.g., HPLC), electrode deactivation phenomena are much less pronounced. This observation is often attributable to the electrode being continuously washed with the mobile phase, or carrier solution, and the electrode is only occasionally exposed to the sample, as a chromatographic band. Fouling often occurs over a period of hours or days, and it is generally classified as long-term, which complements the short-term nature of chromatographic experiments. Hence, in dc amperometry using a glassy-carbon electrode, the electrode need be polished only at the beginning of each workday. Unfortunately, for electrocatalytic detections the rate of fouling is considerably more rapid, and complete passivation of the electrode can take place within seconds. In order to perform chromatographic experiments, the electrode must be continuously reactivated, or "cleaned," on-line. Of all the possible cleaning pretreatments, only electrochemical reactivation can be performed efficiently online and exploit the same electrochemical instrumentation to perform detection and reactivation.

ORIGINS OF PULSED POTENTIAL CLEANING

The application of alternated positive and negative potential pulses as an effective means of reactivating noble-metal electrodes that have become fouled by adsorption of organic impurities from solutions of supporting electrolyte has been employed since the early part of this century. In 1924, Hammett [13] used pulsed waveforms to reactivate Pt electrodes during studies of the anodic oxidation of H_2. Armstrong et al. [14] in 1934 similarly used potential pulse reactivation in studies of the cathodic reduction of O_2. Over the next four decades, the use of pulsed waveforms for reactivation of fouled noble-metal electrodes was used extensively in research directed toward the development of hydrocarbon fuel cells [15–18]. It is interesting to note that almost every publication on voltammetric data at noble-metal electrodes briefly describes a procedure for maintaining electrode activity and reproducibility. These procedures typically include the application of repeated cyclic scans or alternated anodic and cathodic polarizations. Other than fuel cell research, interest in pulsed waveforms for electroanalytical techniques was in a lull.

The use of pulsed potential pretreatment for electroanalytical systems can be traced to a classic paper by Kolthoff and Tanaka [12] , in which they studied the polarization curves of platinum electrodes in different supporting electrolytes. The explosive growth of HPLC in the 1960s and 1970s, which stimulated the application of electrochemical detectors in flow systems, also promoted the use of pulsed waveforms to facilitate electroanalytical detection at noble-metal electrodes. Many of these procedures were based on the work of Kolthoff and Tanaka. A brief literature survey reveals the benefit of electrode reactivation by alternated positive- and negative-potential excursions [15–19]. Clark et al. [20] reported greater reproducibility for anodic oxidation of ethylene at Pt when cleaning pulses were applied, and MacDonald and Duke [21] offered a similar report relating to the detection of *p*-aminophenol. The benefits of pulsed potential cleaning were extended to inorganic species by Stulik and Hora [22] in 1976 for the cathodic detection of Fe(III) and Cu(II) at Pt electrodes in order to achieve greater reproducibility.

Sensitivity and reproducibility enhancements have also been reported for pulsed potential cleaning at carbon electrodes by several researchers, including Fleet and Little [23], van Rooijen and Poppe [24], Ewing et al. [25], Berger [26], and Tenygl [27]. The oxidation of carbon electrodes leads to many oxygen-containing groups (e.g., phenols and carboxylates) on the surface, which exhibit acid–base properties. Pulsed potential activation of carbon electrodes has been shown to lower the detection limits of epinephrine, norepinephrine, and dopamine [28] and, in certain instances, increase the selectivity of the electrode for dopamine over ascorbic acid [29]. It should be noted that pulsed waveforms at carbon electrodes do not enable the satisfactory detection of a large variety of polar aliphatic compounds. This is attributable to the absence of appropriate catalytic properties of carbon surfaces, which must exist to support the anodic reaction mechanisms of aliphatic compounds.

HISTORY OF PULSED ELECTROCHEMICAL DETECTION

The most prominent example of pulsed potential cleaning and/or reactivation of solid anodes is represented by the advent of pulsed electrochemical detection (PED) for HPLC. In fact, its prominence has led to this book. The importance of PED warrants a historical survey of its beginning.

Professor Dennis C. Johnson, an academic descendant of Kolthoff, had been studying anodic oxygen-transfer reactions at noble-metal electrodes for a number of years. He had pioneered a number of different areas of electroanalytical chemistry involving mechanisms, flow systems, photoelectrochemistry, and other phenomena. In the late 1970s, a student under Johnson had noticed that a significant response resulted for an aliphatic alcohol, which was being used to dissolve an inorganic sample, at a platinum electrode. The response for the alcohol was reproducible as long as the electrode was continuously reactivated with positive- and negative-potential excursions. Research efforts focused on pulsed waveforms at noble-metal electrodes were initiated in 1981 by the Johnson group at Iowa State University, and pulsed amperometric detection (PAD) was introduced in 1981 for the detection of simple alcohols at Pt electrodes in flow-injection systems [30,31].

Dionex Corporation (Sunnyvale, CA) was working on alkaline-tolerant polymer-based columns to resolve carbohydrates as anions. The obstacle for carbohydrate analysis was narrowed down to a problem with detection. PAD was able to detect alcohols and, therefore, carbohydrates under alkaline conditions. It may have been a confluence of celestial orbs, or a crossing of destinies, which resulted in the Johnson group and Dionex working together to put forth the first commercial electrochemical detector for carbohydrates following *high-performance anion-exchange chromatography* (HPAEC) [32,33]. Electrochemical detection at noble-metal electrodes using triple-step potential waveforms quickly became known as *pulsed amperometric detection* (PAD) [32–36]. A variety of techniques were soon to follow, including *pulsed coulometric detection* (PCD) [37,38], which involves electronic integration of the amperometric signal, and *potential sweep–pulsed coulometric detection* (PS–PCD) [38,39], which expands PCD to incorporate a triangular potential sweep in the detection step. The technique of PS–PCD is now known as *integrated pulsed amperometric detection* (IPAD) [40]. We now use the generic title of *pulsed electrochemical detection* (PED), which includes the techniques PAD and IPAD, for all detection strategies based on the application of multistep waveforms regardless of the specific form of signal measurement [41]. The analytical significance of this technique is probably best reflected in the fact that PED instrumentation has become available from at least seven other commercial manufacturers.

PED has spawned a number of advanced waveforms, such as *reverse PAD* (RPAD) [42] and *activated PAD* (APAD) [43]. These waveforms are used to overcome the effects of large surface oxide formation signals. Other pulsed waveforms, which were applied to noble-metal electrodes for the purpose of studying PAD response mechanisms at rotating-disk electrodes, are *pulsed voltammetry* (PV) [37,44–47] and *pulsed voltammetric detection* (PVD) [43]. Neuberger and Johnson applied PV in 1987 to study the PAD response of carbohydrates [37,44,45], and similar studies from the Johnson group focused on characterizing the PAD response of sulfur-containing compounds at noble-metal electrodes [42,46]. Ewing et al. [48] used PV to obtain voltammetric information at Pt microelectrodes in static biological microenvironments. More recently, LaCourse and Johnson [49] developed PV as a definitive method for the optimization of PAD waveforms to be applied in HPLC.

IMPACT OF PED TECHNIQUES

The popularity and importance of PED as an analytical technique has gained in prominence over the past decade. This popularity is reflected in the growing number of publications to appear since its inception (see Fig. 1.3). The onset in this distribution reflects the flurry of academic publications put forth to introduce and explain PED. The dip in the early years (i.e., 1984) probably reflects the reluctance of industry and the scientific community to accept this technique, which was saying that polar aliphatic compounds could be detected with high sensitivity. I have both witnessed and experienced audiences who stood up and expressed their disbelief. This reticence to accept just any scientific claim as the immediate truth is good for science; it weeds out mistakes without altering the course of orthodox science that has been garnered through decades of research and challenge. I know of very few people who purchased cold-fusion stocks. Needless to say, the evidence substantiating PED was overwhelming, and it is now an accepted analytical technique.

PED is best known for its application to the direct and sensitive determination of carbohydrates. Since the majority of carbohydrates have poor spectroscopic properties, the most sensitive methods of carbohydrate detection in a flow system required some form of pre- or postcolumn derivatization. Refractive-index (RI) detection can be used for the direct determination of carbohydrates, but the sensitivity is often limited. In addition, RI is a bulk property detector, which means that the chemical nature of the solute is irrelevant, and, as a consequence, chromatograms can often be complex even for simple sample matrices. Therefore, imagine the introduction of a system that allows for the separation of a wide variety of carbohydrates and glycoconjugates and affords the analyst with direct

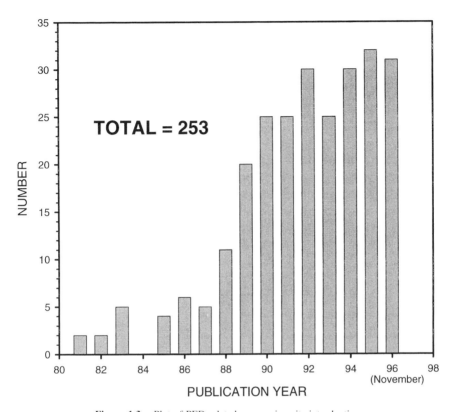

Figure 1.3. Plot of PED-related papers since its introduction.

detection with high sensitivity. Professor Johnson related to me the story about the time when he was visiting an industrial laboratory, and a scientist asked if he could detect monosaccharides in aqueous sample down to 1%. He suggested that the best results would be obtained if the samples were diluted below 1%. The scientist was astounded! Presently, subpicomole limits of detection for neutral monosaccharides is readily achieved.

PED has revolutionized the way we look at carbohydrates. The presence of glycosylated compounds in biosystems was always known. In fact, the ubiquitous nature of carbohydrates relegated them to the role of lubricants or structural components. The limelight had been on the peptides and proteins. Basic aminoacid building blocks had been observed to combine together to make singular strands with form and function. Aminoacid analysis, peptide/protein sequencing, and peptide synthesis have been developed to the point of automation. On the other hand, carbohydrate analysis, sequencing, and synthesis had

been anything but routine. The advent of PED also brought forth the possibility
of glycosequencing. Basic monosaccharide building blocks combine together to
make not only singular strands, but branched structures (biantennary, triantenn-
ary, etc.). Glycocompounds have been determined to play roles in intercellular
communication, immunoresponse, bacterial adhesion, and inflammation. Glycoproteins are markers for diseases, such as cancer. PED following HPLC has
facilitated research and discoveries in these areas, and every other application
where carbohydrate analysis is important.

Although PED is best known for carbohydrate analysis, the technique is also
applicable to any compounds containing an amine group that is not quaternary or
a sulfur moiety that is not fully oxidized. Hence, it is expected that PED will
make even more contributions to the advancement of science and analytical
research. An aminoacid analyzer based on PED may be just around the corner.
PED may be the ideal detection methodology for capillary electrophoresis. Not
only are electrochemical systems easily miniaturized, but PED is most amenable
to the determination of nonchromophoric compounds, which are exceedingly
difficult to detect without derivatization. Recently, we have seen the detection of
carbohydrates using CE–PAD [50] with mass detection limits for glucose of the
order of femtomoles.

The impressive accomplishments in HPLC–PED thus far only accentuate the
need for novel chromatographic separations for polar aliphatic compounds, a
deeper understanding of PED and its limits, and application of this technology to
real-world problems of critical significance.

THIS BOOK

Figure 1.4 is a schematic drawing of a typical HPLC–PED setup. The PED
unit outputs a precise, reproducible potential–time waveform to the working
electrode, the current is then sampled only during the detection step for a specified length of time, and the average signal is plotted versus time on an output
device (chart recorder, integrator, etc.). Each of these steps is illustrated in
Figure 1.4. In the following chapters, the basics of electrochemistry and electrochemical detection in liquid chromatographic systems will be reviewed to
prepare the reader to fully understand the fundamental tenets of PED. Pulsed
electrochemical detection will be reviewed from its most basic principles to the
latest and most advanced waveforms, including the optimization of PED waveforms. Published applications are reviewed on the basis of compound classes.
In addition, practical and anecdotal information has been compiled into a single
chapter to let the reader in on all the "ins and outs" of working with PED
systems. At this point, readers should be well equipped to initiate their own
PED assays.

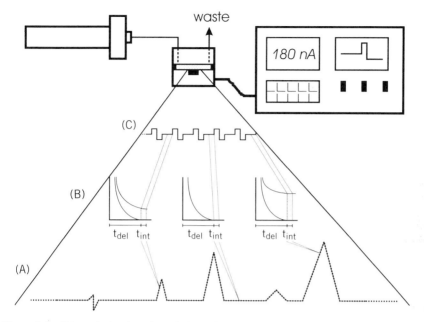

Figure 1.4. Schematic drawing of a typical HPLC–PED setup showing how a (*A*) chromatogram is produced from the (*B*) sampled current response derived from the application of a (*C*) potential–time waveform to the working electrode of an electrochemical cell. The current is integrated typically for a fixed interval of time (t_{int}) after a delay period (t_{del}), which allows capacitive current to dissipate.

REFERENCES

1. K. Stulik, *Electroanalysis* **4**, 829 (1992).

2. S. Gilman, in *Electroanalytical Chemistry,* Vol. 2, A. J. Bard, ed., Marcel Dekker, New York, 1967, p. 111.

3. E. L. Goldstein and M. R. Van de Mark, *Electrochim. Acta* **27**, 1079 (1982).

4. N. Oyama and F. C. Anson, *J. Electroanal. Chem.* **88**, 289 (1978).

5. J. F. Evans and T. Kuwana, *Anal. Chem.* **51**, 358 (1979).

6. C. W. Miller, D. H. Karweik, and T. Kuwana, *Anal. Chem.* **53**, 2319 (1981).

7. E. Hershenhart, R. L. McCreery, and R. D. Knight, *Anal. Chem.* **56**, 2256 (1984).

8. M. Poon and R. L. McCreery, *Anal. Chem.* **58**, 2745 (1986).

9. K. Stulik, D. Brabcova, and L. Kavan, *J. Electroanal. Chem.* **250**, 173 (1988).

10. M. Poon and R. L. McCreery, *Anal. Chem.* **59**, 1615 (1987).

11. K. Sternitzke, R. L. McCreery, C. S. Bruntlett, and P. T. Kissinger, *Anal. Chem.* **61**, 1989 (1989).

12. I. M. Kolthoff and N. Tanaka, *Anal. Chem.* **26**, 632 (1954).

13. L. P. Hammett, *J. Am. Chem. Soc.* **46**, 7 (1924).

14. G. Armstrong, F. R. Himsworth, and J. A. V. Butler, *Proc. Roy. Soc. London* (A) **143**, 89 (1934).

15. S. Gilman, *J. Phys. Chem.* **67**, 78 (1963).

16. M. W. Breiter, *Electrochim. Acta* **8**, 973 (1963).

17. J. Giner, *Electrochim. Acta* **9**, 63 (1964).

18. S. Gilman, in *Electroanalytical Chemistry*, Vol. 2, A. J. Bard, ed., Marcel Dekker, New York, 1967, pp. 111–192.

19. R. Woods, in *Electroanalytical Chemistry*, **9** A. J. Bard, ed., Marcel Dekker, New York, 1976, pp. 20–27.

20. D. Clark, M. Fleishman, and D. Pletcher, *J. Electroanal. Chem.* **36**, 137 (1972).

21. A. MacDonald and P. D. Duke, *J. Chromatogr.* **83**, 331 (1973).

22. W. Stulik and V. Hora, *J. Electroanal. Chem.* **70**, 253 (1976).

23. B. Fleet and C. J. Little, *J. Chromatogr. Sci.* **12**, 747 (1974).

24. H. W. van Rooijen and H. Poppe, *Anal. Chim Acta* **130**, 9 (1981).

25. A. G. Ewing, M. A. Dayton, and R. M. Wightman, *Anal. Chem.* **53**, 1842 (1981).

26. T. A. Berger (Hewlett-Packard Corp., Avondale, PA), U.S. Patent 4,496,454 (Jan. 29, 1985).

27. J. Tenygl, in *Electrochemical Detectors* (Proc. Symp., 1981), T. H. Ryan, ed., Plenum Press, New York, 1984, pp. 89–103.

28. J. Mattusch, K. H. Hallmeier, K. Stulik, and V. Pacakova, *Electroanalysis* **1**, 405 (1989).

29. G. Mamantov, D. B. Freeman, F. J. Miller, and H. E. Zittel, *J. Electroanal. Chem.* **9**, 305 (1965).

30. S. Hughes, P. L. Meschi, and D. C. Johnson, *Anal Chim. Acta* **132**, 1–10 (1981).

31. S. Hughes and D. C. Johnson, *Anal. Chim Acta* **132**, 11–22 (1981).

32. P. Edwards and K. K. Haak, *Am. Lab.* 78–87 (April 1983).

33. R. D. Rocklin and C. A. Pohl, *J. Liq. Chromatogr.* **6**, 1577–1590 (1983).

34. S. Hughes and D. C. Johnson, *J. Agric. Food Chem.* **30**, 712–714 (1982).

35. S. Hughes and D. C. Johnson, *Anal. Chim. Acta* **149**, 1–10 (1983).

36. D. C. Johnson and W. R. LaCourse, *Anal. Chem.* **62**, 589A–597A (1990).

37. G. C. Neuberger and D. C. Johnson, *Anal. Chim. Acta* **192**, 205–213 (1987).

38. G. C. Neuberger and D. C. Johnson, *Anal. Chem.* **60**, 2288–2293 (1988).

39. L. E. Welch, W. R. LaCourse, D. A. Mead, Jr., and D. C. Johnson, *Anal. Chem.* **61**, 555–559 (1989).

40. W. R. LaCourse and D. C. Johnson, in *Advances in Ion Chromatography*, Vol. 2, P. Jandik and R. M. Cassidy, eds., Century International, Medfield, MA, 1990, pp. 353–372.

41. D. C. Johnson and W. R. LaCourse, *Electroanalysis* **4**, 367–380 (1992).

42. T. Z. Polta and D. C. Johnson, *J. Electroanal. Chem.* **209**, 159 (1986).

43. D. G. Williams and D. C. Johnson, *Anal. Chem.* **64**, 1785 (1992).

44. G. C. Neuberger and D. C. Johnson, *Anal. Chem.* **59**, 150–154 (1987).

45. G. C. Neuberger and D. C. Johnson, *Anal. Chem.* **59**, 203–204 (1987).

46. P. J. Vanderberg, J. L. Kowagoe, and D. C. Johnson, *Anal. Chim. Acta.* **260**, 1–11 (1992).

47. L. A. Larew and D. C. Johnson, *J. Electroanal. Chem.* **262**, 167–182 (1989).

48. T. K. Chem, Y. Y. Lau, D. K. Y. Wong, and A. G. Ewing, *Anal. Chem.* **64**, 1264–1268 (1992).

49. W. R. LaCourse and D. C. Johnson, *Anal. Chem.* **65**, 50–55 (1993).

50. T. M. O'Shea, S. M. Lunte, and W. R. LaCourse, *Anal. Chem.* **65**, 948–951 (1993).

2 Electrochemical Fundamentals

And GOD said, 'Let there be electrons': and there was electrochemistry.

The goal of this chapter is to acquaint the reader with the fundamentals of electrochemistry. Electrochemistry covers a wide range of techniques, and, as a consequence, the focus of this chapter will be limited to those principles that are necessary for the understanding of electrochemical detection, and in particular PED, as it applies to high-performance liquid chromatography. For a more in-depth treatment of the topics introduced in this chapter or electrochemistry in general, the reader is directed to any of the books listed in the bibliography at the end of this chapter.

BASIC CONCEPTS

If there is to be a beginning to understanding electrochemistry, one must first start with the electron. The *electron* (e) has a mass of 9.11×10^{-28} g and a charge of 1.6×10^{-19} C. The *coulomb* (C) is the Systeme International (SI) unit for electrical *charge* (Q), and one coulomb represents 6.25×10^{18} electrons. It is important to note that charge is a quantity and that the electron is arbitrarily given a negative charge.

The valence-band electrons of certain substances (i.e., typically metals) are not confined to any one particular atom, and, as a consequence, the electrons are able to migrate under the influence of an electric field. Since these substances conduct electricity, they are classified as *conductors*. The electron flow is known to us as *current* (i), and if one coulomb flows past a given point in one second, we say the current equals one *ampere*, or amp. Effects that impede the flow of charge (current) are classified as *resistance* (R), and the basic unit of resistance is the *ohm* (Ω). Substances that strongly oppose electron flow are called *insulators*.

The difference in charge between two points is defined as the difference in *electrical potential* (E), or *voltage*, or *electromotive force*. The voltage between two points represents the energy needed to move a unit of charge from the more negative point (lower potential) to the more positive point (higher potential). The concept of electrical potential difference is similar to that of a concentration difference, which is the driving force for ions and molecules to migrate from regions of higher concentration to lower concentration. In the case of electricity, the species that moves is the electron. Voltage is measured in units of *volts* (V). One volt will push one ampere of current through one ohm of resistance.

The most fundamental relationship in the electrical sciences is that of Ohm's law:

$$E = iR \tag{2.1}$$

ELECTROCHEMISTRY

Electrochemistry deals with the relationships that exist between electricity and chemical reactions. In electrochemical techniques, an electrical charge, or a quantity of electrons, is transferred between compounds and/or electrodes. The quantity of electricity that contains one mole of electrons (6.023×10^{23} e mol^{-1}) is 96,485 C, or one *faraday* (F).

When a substance gains/accepts electrons, it is *reduced*, and the process is called *reduction*. When a substance loses or donates electrons, it is *oxidized*, and the process is called *oxidation*. Oxidation and reduction must occur simultaneously; there cannot be one without the other. In an oxidation–reduction (*redox*) reaction, the *oxidizing agent*, or *oxidant*, is reduced, and the *reducing agent*, or *reductant*, is oxidized. Hence, for the reaction

$$\underset{\substack{\text{reducing} \\ \text{agent}}}{Ni^0} + \underset{\substack{\text{oxidizing} \\ \text{agent}}}{Cu^{2+}} \;\rightleftharpoons\; Ni^{2+} + Cu^0 \tag{2.2}$$

Ni^0, or Ni(s), causes reduction of Cu^{2+} and is, in turn, oxidized; and Cu^{2+} causes oxidation of Ni^0 and is, in turn, reduced. One could say, "it's all a matter of perspective as to which species is inducing the redox reaction to proceed." What is important to remember is that electrons are transferred from one species to another. In electroanalysis, either the donor or the receiver is typically an electrode.

It is important that redox reactions be balanced in both mass and charge. A redox reaction can be envisioned as occurring in two steps, called *half-reactions*. A half-reaction is simply a depiction of the oxidation or reduction process of the reaction alone. Hence, Reaction (2.2) can be viewed as consisting of the following two half-reactions:

$$\text{Reduction:} \quad Cu^{2+} + 2e^- \rightleftharpoons Cu^0 \tag{2.3}$$

$$\text{Oxidation:} \quad Ni^0 \rightleftharpoons Ni^{2+} + 2e^- \tag{2.4}$$

The sum of the half-reactions should be the overall redox reaction. One of the more popular methods of balancing redox reactions uses the concept of half-reactions on the premise that it is easier to balance half the reaction at a time than the entire reaction at once. A detailed method of balancing redox reactions by half-reactions can be found in almost any undergraduate chemistry text.

Electrochemical Cells

Physical separation of the oxidation and reduction processes of a redox reaction can be accomplished by constructing a device such that the reactants of the redox reaction do not come into direct contact with each other, and, as a consequence, the transferred electrons can be made to flow through an external circuit. The development of such a system begins with the construction of a *half-cell*. A half-cell can be as simple as the placement of an electrical conductor, or *electrode*, in contact with a solution of its own ions. For Reaction (2.4), a Ni wire is dipped into a solution of $NiNO_3$. Under these conditions, the metal wire can undergo oxidation to metal ions [Reaction (2.5)]:

$$M^0 \dashrightarrow M^{n+} + n_{e^-} \tag{2.5}$$

$$M^{n+} + n_{e^-} \dashrightarrow M^0 \tag{2.6}$$

$$\overline{M^{n+} + n_{e^-} \rightleftharpoons M^0} \tag{2.7}$$

and leave electrons behind on the metal-wire electrode; or the solution ions of the metal can undergo reduction [Reaction (2.6)] and pick up electrons from the metal-wire electrode to form metal atoms. Since this reaction can proceed in either direction, it is said to be *reversible*. These two reactions form a *redox couple*, which is in dynamic equilibrium at the electrode surface [Reaction (2.7)]. Reaction 2.7 is an overall expression of the redox reaction and not the mathematical summation of Reactions 2.5 and 2.6. If either the forward or reverse reaction is favored, the electrode will become positively or negatively charged,

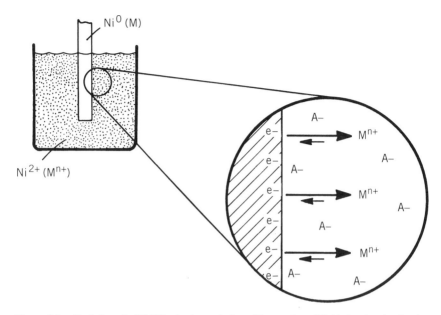

Figure 2.1. Depiction of a Ni (M) wire in a solution of its own ions (M^{n+}) showing the development of excess charge at the electrode surface. A^- represents generic counterions.

respectively, in relation to the solution. For Ni, the forward reaction, or oxidation, is favored, and, as a result, the electrode potential will be negative because of an excess of electrons at the electrode surface (Fig. 2.1). Hence, a potential difference, or *electrode potential (E)*, is established across the electrode solution interface. The relationship between the electrode and the metal-ion activity in also known as the *electrode response*, which may be exploited in potentiometry.

Another redox couple is shown in Reaction (2.8). Note that both forms of iron are ions

$$Fe^{3+} + e^- \rightleftarrows Fe^{2+} \tag{2.8}$$

and, as a consequence, both are present in the half-cell in solution form. If an *inert* electrode (e.g., platinum wire or carbon rod) is placed in the solution, both the forward and reverse reactions will occur at the inert electrode. Inert electrodes act as electron sinks to either receive or donate electrons to and from the substances in the solution. Inert electrodes do not play a role in the redox reaction. Since there is a plethora of redox couples, a more generalized equation is expressed as follows:

$$oxd + ne^- \rightleftarrows red \tag{2.9}$$

where *oxd* and *red* denote the oxidized and reduced species of the couple, respectively, ne^- is the number of electrons involved in the charge transfer, and the double arrows indicate an equilibrium.

The combination of two half-cells is denoted as an *electrochemical cell*, which is further classified as either a *galvanic* or an *electrolytic cell*. A *galvanic cell* generates a potential difference between the electrodes. The redox couples of the half-cells are spontaneous, and we create a source of practical electricity. Galvanic cells are commonly called *batteries*. In contrast, in an electrolytic cell, the redox couples of the half-cells are nonspontaneous, and a potential difference must be imposed on the cell from an external source in order for the overall redox reaction to proceed. *Electroplating*, which entails the deposition of a thin layer of metal (e.g., silver or gold) on a substrate, involves an electrolytic cell. Each electrode of an electrochemical cell is defined in terms of the reaction that occurs at its surface. The electrode at which reduction occurs is the *cathode*, and the electrode at which oxidation occurs is the *anode*. These definitions apply to both galvanic and electrolytic cells. A simple way to memorize which process occurs at which electrode is depicted in Figure 2.2. Simply ask yourself whether oxidation occurs at the anode or the cathode. The answer will be obvious from the question. It is extremely important to note that electrodes are *not* defined in terms of electrical polarity (i.e., positive or negative). In a galvanic cell, the cathode is positive with respect to the anode; in an electrolytic cell, the cathode is negative with respect to the anode. Hence, never judge an electrode by its polarity.

Figure 2.3 shows an electrochemical cell that utilizes Reaction (2.2) to generate electricity. Each half-cell is connected through a wire to a *meter*, which is used to monitor the potential difference between the two electrodes. When a nickel atom leaves the electrode and enters the solution as a nickel ion, two electrons are left on the Ni electrode, rendering it electron-rich and of negative $(-)$ polarity. At the copper electrode, reduction is occurring, resulting in an electron-deficient electrode, making it of positive $(+)$ polarity. If electrons are to flow through the wire, the charge imbalance in the solutions caused by ions that are generated or consumed in the half-cells must be compensated for in order to maintain electroneutrality. Hence, a *salt bridge*, which contains a source of positive and negative ions, is needed to complete the electrochemical cell. A salt bridge is designed as a source of positive and negative ions, which can flow into and out of the half-cell solutions in order to maintain electroneutrality. The positive ions are *cations*, and the negative ions are *anions*. It is desirable that the

Figure 2.2. Mnemonic to remember the anode from the cathode.

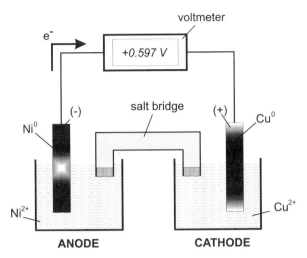

Figure 2.3. Simple electrochemical cell used to generate electricity.

cations and anions constituting the electrolyte in the salt bridge have about the same mobility (i.e., rates of diffusion and migration) in the solution. Potassium chloride satisfies these requirements, and it is commonly used in most salt bridges. With the electrochemical cell complete, electrons will flow through the electrical wire from the negative electrode to the positive electrode.

It would be a bit cumbersome to draw an electrochemical cell as in Figure 2.3 each time a cell was referred to, and, as a matter of convenience, chemists have evolved a shorthand notation for representing electrochemical cells. As represented by

$$\text{Ni(s)} \mid \text{Ni(NO}_3\text{)}_2\text{(conc)} \| \text{Cu(NO}_3\text{)}_2 \text{ (conc)} \mid \text{Cu(s)}$$

the chemical constituents of the cell are arranged in order from left to right depicting the solution circuit from anode to cathode. A single vertical line denotes a phase boundary across which an electrode potential exists. Double vertical lines indicate a salt bridge. A redox couple in the solution phase, which occurs at an inert electrode, as in Reaction (2.8), is represented by

$$\text{Pt} \mid \text{Fe(NO}_3\text{)}_2 \text{ (conc), Fe(NO}_3\text{)}_3 \text{ (conc)}$$

This half-cell is depicted as the anode. Since the electrode potential is concentration-dependent, concentrations of the cell constituents are denoted parenthetically.

ELECTRODE POTENTIALS

Unless electrode potentials can be quantified in some way, most of what we know as electrochemistry would truly fall under the heading of an "art form" rather than a science. Fortunately for us, the strengths of oxidants and reductants can be determined, *relatively* speaking. Electrochemistry has evolved from several different disciplines, and this diversity has been reflected, historically, in a diversity of conventions relating to cells, polarities, and electrode reactions. Formerly half-reactions were written as oxidations, but this book will follow IUPAC (International Union of Pure and Applied Chemistry) convention of writing all half-reactions as reductions.

When a silver wire is placed in contact with a solution of silver ions, as in Reaction (2.10), the metal wire, or electrode,

$$Ag^+ + e^- \rightleftarrows Ag^0 \qquad (2.10)$$

tends to acquire a charge, which is a result of the equilibrium between the forward and reverse reactions. In the case of Reaction (2.10), the forward reaction is favored, and the electrode develops a positive charge. According to LeChatelier's principle, if a system at equilibrium is disturbed by a change in concentration of one of its components, the system will tend to shift its equilibrium position so as to counteract the effect of the disturbance. Hence, if the concentration of Ag^+ ions is increased, the reduction of Ag^+ to Ag^0 will be favored, and the electrode potential will become more positive. On the other hand, if the concentration of Ag^+ ions is decreased, the reverse reaction, or oxidation, will be favored, and the electrode potential will become more negative. For a uniform ranking of half-reactions, a standard set of conditions must be chosen under which all electrode potentials are compared. Hence, by convention, when all components of the half-reaction are at standard-state concentrations, the electrode is defined to be at its *standard potential*, or $E°$. For dilute solutions, the activity of a solute is approximately equal to its concentration. In addition, pure substances (i.e., solid or liquid) are defined as unit activity, and the activity of a gas is equivalent to its partial pressure in atmospheres.

If we take a reading of the meter shown in Figure 2.3, we measure the difference in the electrode potentials of the two half-cells. The *difference* between the two potentials (ΔE) is the cell's *voltage*. The measured difference in potential is given by

$$\Delta E_{cell} = E_{cathode} - E_{anode} \qquad (2.11)$$

If the half-reactions in the half-cells are at standard-state concentrations, the standard potentials can be used as in Equation (2.11), and the meter in Figure 2.3 will read $+0.597$ V.

There is *no method in existence* for measuring the absolute value of the potential at a single electrode. Thus, it is not possible to rank half-reactions and half-cells in order of absolute electrode potentials. But if we chose a particular half-cell to act as a common reference, we would then be able to measure and rank half-cell potentials relative to one another. The entire chemical community has accepted the *standard hydrogen electrode* (SHE), or the *normal hydrogen electrode* (NHE), as the ultimate reference electrode against which all electrode potentials are measured. Figure 2.4 depicts the construction of a typical hydrogen electrode. The electrode is composed of platinum foil, which has been coated with a finely divided layer of platinum (platinum black). The platinum black provides a large surface area for the redox couple shown in the following reaction:

$$2H^+ + 2e^- \rightleftharpoons H_2(g) \tag{2.12}$$

The coated electrode is immersed in an acidic solution of H^+ activity of 1.0 M and through which hydrogen gas at a partial pressure of 1.0 atm (unit activity) is bubbled. The platinum electrode develops a potential based on the equilibrium of Reaction (2.12). The absolute potential of the SHE is *not known*, but it is *assigned* a value of exactly zero at all temperatures. Unfortunately, not all decisions in life are this easy.

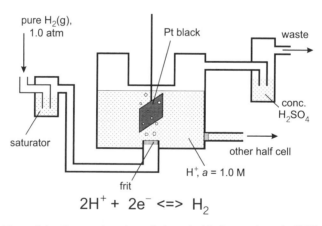

Figure 2.4. Construction of a typical standard hydrogen electrode (SHE).

At this point, if we desire to know the relative potential of an unknown half-cell, we simply measure the electrochemical cell potential using the SHE as the anode. As demonstrated by the following equations

$$\Delta E_{cell} = E_{cathode} - E_{SHE} \tag{2.13}$$

$$\Delta E_{cell} = E_{cathode} - 0 \tag{2.14}$$

$$\Delta E_{cell} = E_{cathode} \tag{2.15}$$

the half-cell potential is equivalent to the potential difference of the overall cell when the SHE is used as the anode.

Thus, for any half-cell, the relative half-cell potential can be determined in relation to the SHE, which is an *arbitrarily* chosen reference half-cell. Table 2.1 shows a list a $E°$ values for half-cells. As with IUPAC convention, all half-cells are listed as reductions. As the $E°$ value becomes more positive, the tendency for the half-cell reaction to proceed in the forward direction, which signifies reduction, is favored. Permanganate ion (MnO_4^-) is a well-known oxidizing agent, and in a redox reaction it would be reduced. From Table 2.1, the strong tendency for MnO_4^- to undergo reduction is reflected in its very positive $E°$ value for the MnO_4^-/Mn^{2+} couple. On the other hand, Zn^0 is known as a strong reducing agent. In fact, a Jones reductor column, which is commonly used in sample preparation to reduce metal ions to a lower valence state, is simply a column containing Zn metal as an amalgam with Hg^0. From Table 2.1, the strong tendency of Zn to reduce other species is denoted by the very negative $E°$ value (-0.762 V) of the Zn^{2+}/Zn^0 redox couple. With a negative value corresponding to the desire to undergo reduction, the reverse reaction, or oxidation, is then strongly favored.

TABLE 2.1   Standard Electrode Potentials ($E°$) for Selected Half-Cell Reactions

Reaction	$E°$ at 298 K, V	
	SHE	Ag/AgCl
F_2 (g) + 2 e$^-$ $\rightleftarrows$ 2 F$^-$	+3.06	+2.84
H_2O_2 + 2 H$^+$ + 2e$^-$ $\rightleftarrows$ 2 H$_2$O	+1.77	+1.55
Ce^{4+} + e$^-$ $\rightleftarrows$ Ce^{3+}	+1.61	+1.39
MnO_4^- + 8 H$^+$ + 5 e$^-$ $\rightleftarrows$ Mn^{2+} + 4 H$_2$O	+1.51	+1.29
Cl$_2$ + 2 e$^-$ $\rightleftarrows$ 2 Cl$^-$	+1.36	+1.14
$Cr_2O_7^{2-}$ + 14 H$^+$ + 6 e$^-$ $\rightleftarrows$ 2 Cr^{3+} + 7 H$_2$O	+1.33	+1.11
O_2 (g) + 4 H$^+$ + 4 e$^-$ $\rightleftarrows$ 2 H$_2$O	+1.23	+1.01
Br$_2$ (aq) + 2 e$^-$ $\rightleftarrows$ 2 Br$^-$	+1.087	+0.865
Br$_2$ (l) + 2 e$^-$ $\rightleftarrows$ 2 Br$^-$	+1.065	+0.843
Ag$^+$ + e$^-$ $\rightleftarrows$ Ag	+0.800	+0.577
Hg$_2^{2+}$ + 2 e$^-$ $\rightleftarrows$ 2 Hg	+0.79	+0.568
Fe^{3+} + e$^-$ $\rightleftarrows$ Fe^{2+}	+0.771	+0.549
O_2 + 2 H$^+$ + 2 e$^-$ $\rightleftarrows$ H$_2$O$_2$	+0.682	+0.460
I$_3^-$ + 2 e$^-$ $\rightleftarrows$ 3 I$^-$	+0.545	+0.323
Cu^{2+} + 2 e$^-$ $\rightleftarrows$ Cu	+0.337	+0.115
Hg$_2$Cl$_2$ (s) + 2 e$^-$ $\rightleftarrows$ 2 Hg + 2 Cl$^-$	+0.2676	+0.0453
AgCl (s) + e$^-$ $\rightleftarrows$ Ag + Cl$^-$	+0.2223	+0.0000
2H$^+$ + 2 e$^-$ $\rightleftarrows$ H$_2$ (g)	+0.0000	-0.2223
PbSO$_4$ + 2 e$^-$ $\rightleftarrows$ Pb + SO$_4^{2-}$	-0.356	-0.578
Cd^{2+} + 2 e$^-$ $\rightleftarrows$ Cd	-0.403	-0.625
Zn^{2+} + 2 e$^-$ $\rightleftarrows$ Zn	-0.763	-0.985
Li$^+$ + e$^-$ $\rightleftarrows$ Li	-3.045	-3.267

Source: Data were taken from *Lange's Handbook of Chemistry*, 13th edition.

This same logic can be extended to the spontaneity of electrochemical reactions. If we place silver metal (Ag^0) in an acidic solution, it is predicted that no reaction will occur. This is because the Ag^+/Ag^0 half-cell reaction ($+0.800$ V) has very little drive to undergo oxidation in relation to the H^+/H_2 redox couple (0.000 V). On the other hand, if Zn metal (-0.762 V) is placed in a solution containing H^+ ions, as confirmed in our undergraduate freshman chemistry laboratory experiments, H_2 would be produced. The metal readily undergoes oxidation and H^+ is reduced to H_2 gas. Thus, by simply comparing the standard electrode potentials, we can determine relative oxidizing and reducing strengths, and probable reaction spontaneity.

It is important to remember that the SHE is an arbitrarily chosen reference half-cell reaction. We could just as easily have chosen another reaction, such as

$$AgCl(s) + e^- \rightleftharpoons Ag^0 + Cl^- \qquad (2.16)$$

Hence, this reaction would have been given the E° value of 0.000 V, and all other half-cell reactions would have been compared relative to it. Table 2.1 shows comparative standard potentials of other half-cell reactions using Reaction (2.16) and SHE as references. Notice that *relative order* of the reactions is no different from that using a SHE. The SHE electrode was chosen since it is extremely reproducible and produces a potential very near to that which is predicted theoretically.

The Nernst Equation

The E° values in Table 2.1 are all at standard-state conditions, or unit activities. As discussed earlier, electrode potentials are dependent on the concentrations of reactants and products in the redox couple. Therefore, since most real-world laboratory conditions would be something other than standard conditions, it is important to understand the quantitative relationship between the concentration of substances constituting a half-reaction and the electrode potential of the half-cell.

For the general half-reaction below written as a reduction under standard conditions, the *equilibrium constant*, or K, is given by Equation (2.18):

$$aA + bB + ne^- \rightleftharpoons cC + dD \qquad (2.17)$$

$$K = \frac{C^c D^d}{A^a B^b} \qquad (2.18)$$

where A–D are the activities of the reactants and products of the half-reaction and a–d are the respective reaction coefficients. The Gibbs free energy (ΔG°) of the reaction can be related to the equilibrium constant, or K, via the following expressions:

$$\Delta G° = - RT \cdot \ln K = - RT \cdot \ln \frac{C^c D^d}{A^a B^b} \tag{2.19}$$

where R is the molar gas constant (8.31441 J mol^{-1} K^{-1}) and T is the absolute temperature. The Gibbs free energy of any reaction is equivalent to the work (w_e) that can be derived from the electrochemical reaction [Eq. (2.20)]. The maximum electrical work available can also be expressed in terms of electrode potential as in Equation (2.21), and, therefore, the change in free energy associated with the reaction is proportional to its half-cell potential [Eq. (2.22)]:

$$\Delta G° = -w_e \tag{2.20}$$

$$w_e = nFE° \tag{2.21}$$

$$\Delta G° = -nFE° \tag{2.22}$$

Hence, one can obtain values of thermodynamic state functions from electrochemical measurements or calculations.

If Equation (2.19) is made equivalent to Equation (2.22), we find that the standard electrode potential is related to the equilibrium constant, as follows:

$$E° = \frac{-RT}{nF} \ln K = \frac{-2.303 \, RT}{nF} \log \frac{C^c D^d}{A^a B^b} \tag{2.23}$$

Thus far, we have discussed the relationships between free energy, the electrode potential, and the equilibrium constant under standard conditions. What about other conditions and/or concentrations? For Reaction (2.17) under nonequilibrium conditions, we can calculate the reaction quotient, or Q, via the expression

$$Q = \frac{C^c D^d}{A^a B^b} \tag{2.24}$$

The general relationship between free-energy change under standard conditions, $\Delta G°$, and the free-energy change under any other conditions, ΔG, is given by the following relationship:

$$\Delta G = \Delta G° + 2.303 RT \log Q \tag{2.25}$$

If Equation (2.23) is substituted into Equation (2.25), then

$$\Delta G = -2.303 RT \ln K + 2.303 RT \log Q \tag{2.26}$$

This equation implies that the magnitude of the free energy for the system depends on how far the system is from the equilibrium state. Since the change in free energy is related to the half-cell potential via the expression

$$\Delta G = -nFE \quad (2.27)$$

substituting Equation (2.22) into Equation (2.25) and making Equation (2.27) equivalent to (2.25) yields the Nernst equation

$$E = E^\circ - \frac{2.303\ RT}{nF} \log \frac{C^c D^d}{A^a B^b} \quad (2.28)$$

on rearrangement. The most common form of the Nernst equation is

$$E = E^\circ - \frac{0.0592}{n} \log \frac{C^c D^d}{A^a B^b} \quad (2.29)$$

where the coefficient of the log term is simplified for the reaction occurring at 298 K. When a system is at equilibrium, ΔG must be zero, and the reaction quotient is then equal to K [Eq. (2.23)].

Because of the importance of the Nernst equation, we have taken the time to derive it via a thermodynamic pathway. If one desires, a kinetic argument can be employed to derive the Nernst equation as well. With the Nernst equation, chemists can calculate potentials of half-reactions and half-cells in which the participating substances are not at standard-state concentrations. Such a voltage is a quantitative measure of the driving force behind a reaction.

Limitations of Electrode Potentials

Electrode potentials, at either standard or nonequilibrium conditions, are extremely important to understanding electroanalytical processes. It is important to note that these calculated potentials, or thermodynamic potentials, are only theoretical, and often calculated potentials do not agree with experimentally observed potentials. There are certain inherent limitations that one should always keep in mind.

In calculating electrode potentials, molar concentrations are typically substituted for activities, where the activity, a_x, of the species is given by

$$a_x = f_x[X] \quad (2.30)$$

Here, f_x is the activity coefficient of the solute X, and the bracketed term is the molar concentration of X. Activity coefficient data for ions in solutions of the kinds commonly encountered in electroanalytical work are limited, and, as a consequence, molar concentrations are substituted. This may lead to substantial errors. The assumption that molar concentration is equivalent to activity of a solute is valid only in dilute solutions.

The reaction species listed for a half-cell seldom exist as simple ions or molecules; in fact, often they are involved in competing equilibria. This effect is often noted in that the electrode potential can be quite different depending on the supporting electrolyte. For instance, the Ag^+/Ag^0 couple has an $E°$ of $+0.800$ V. If Cl^- is used for the supporting electrolyte, Ag^+ and Cl^- will combine to form AgCl at the electrode surface, and the $E°$ value for the Ag/AgCl couple will be $+0.222$ V. The Cl^- serves to remove any Ag^+ from the solution as it is formed and, as reflected in the standard potential values, reduction is less favored. In terms of competing equilibria, the solution near the Ag electrode contains the many silver species [i.e., Ag^+, AgCl(s), AgCl(aq), $AgCl_2^-$, $AgCl_3^{2-}$, $AgCl_4^{3-}$], and the net effect is to reduce the equilibrium concentration of the free metal ion, which, in turn, affects the electrode potential. The use of a chloride salt as part of the supporting electrolyte for a Ag^+/Ag^0 electrode is an extreme example. Frequently the competing equilibria effects are more subtle and unknown. Only if the competing equilibria are known and constants for the processes are available can these effects be compensated for. More often than not, this information is lacking, and the chemist is then forced to neglect these differences.

As a result of these discrepancies, a second system has been proposed substituting a quantity called the formal potential, or E', in place of the $E°$ in oxidation–reduction calculations. The formal potential of a system is the potential of the half-cell with respect to the SHE when concentrations of reactants and products are 1 M and all other concentrations of cell constituents are carefully noted. It is important to remember that formal reduction potentials are for a specified set of experimental conditions, and are only pertinent to that particular system.

The use of formal potentials can compensate for activity differences and competing equilibria, but there are additional reasons why the calculated potentials may be different from the observed potentials, which are beyond the intention of formal potentials. These reasons include liquid junction potentials, reversibility of electrode reactions, and effects of current.

In the majority of electrochemical cells, solutions of varying composition are in contact with each other. Hence, liquid junctions or boundaries are formed at the interface of these solutions. Differences in the diffusion rates of ions migrating across the liquid junction will result in a *junction potential*. Consider 0.1 M HCl in contact with 0.01 M HCl (Fig. 2.5). The natural tendency of this system is for the ions to diffuse from zones of high concentration to zones of low concentration. In the process, H^+ and Cl^- will diffuse, each according to its own physical characteristics (i.e., weight, size, shape). The lighter H^+ will diffuse at a much faster rate than the larger and heavier Cl^-. As shown in Figure 2.5, a charge imbalance will occur with an excess of positive charges due to excess H^+ on one end of the boundary and an excess of negative charge due to Cl^- on the high concentration side of the boundary. A separation of charges results in a potential difference, and this is generally called a *liquid junction potential*. Junction potentials will occur at all solution interfaces, such as at the ends of salt bridges and at membranes separating anode and cathode compartments of an electrochemical cell. These potentials are dependent on both the nature of the ions (i.e., rate of diffusion) and the magnitude of the concentration gradient.

Liquid junction potentials are almost impossible to eliminate, and, instead,

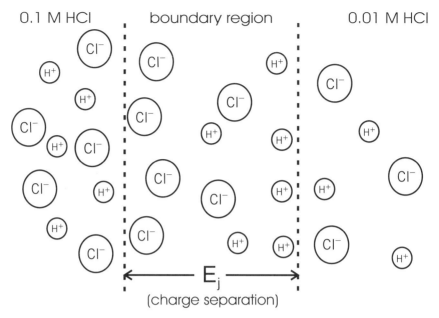

Figure 2.5. Illustration of how a junction potential develops across a boundary region due to a concentration gradient.

every attempt is made to minimize their effects. Hence, a saturated solution of potassium chloride is typically used in a salt bridge because the diffusion rates of K^+ and Cl^- differ by ~4%, and they generally produce junction potentials of <5 mV.

In the discussion of the Nernst equation and cell potentials, we have treated all the redox couples as if the reduction and oxidation reactions were favored equally and occurred rapidly. In other words, the redox reaction is completely reversible. In reality, many half-cell reactions do not behave in a completely reversible manner. Half-cell reactions represent the overall process and not the details of the reduction or oxidation mechanisms involved in the process. Hence, changing the concentration of reductant or oxidant (red or oxd) in a redox reaction may only indirectly affect the potential if an intermediate is involved in the reduction–oxidation mechanism. For example, the MnO_4^-/Mn^{2+} half-cell proceeds through an intermediate (MnO_2) in the redox mechanism. Changing the concentrations of MnO_4^- and Mn^{2+} will lead to a potential change, but the response will not follow the Nernst equation theory. In fact, many of the data found in tables of standard electrode potentials are based on calculations that are, in turn, based on equilibrium or thermal measurements in lieu of the actual measurement of the potential of an electrode based on the redox reaction. The reason is that no suitable electrode can be prepared. For example, the reaction

$$2CO_2(g) + 2H^+ + 2e^- \rightleftharpoons H_2C_2O_4 \qquad (2.31)$$

does not obey the Nernst equation because the rate at which CO_2 combines to produce oxalic acid is extremely slow. Hence, the electrode reaction is not reversible, and no electrode system can be produced that will produce the expected $E°$ value. This does not mean that knowing the standard electrode potential is not important, for the potential is useful for computational purposes.

In our discussion of electrode potentials and the Nernst equation, we have assumed the current to be zero or negligible. With current, reactions occur at the electrode and alter the concentrations of reactants and products from which the electrode response in determined. Hence, current perturbs the system that we are attempting to measure. As will be discussed later in this chapter, current will lead to a variety of effects that are deleterious to the intent of determining the electrode potential.

It is important to remember that electrode potentials are thermodynamic properties of redox substances, and, therefore, they are predictive only of whether a given reaction will or can occur. Electrode potentials offer *absolutely no* insight into the rate at which a reaction proceeds.

POTENTIOMETRY

Techniques based on the measurement of potentials of electrochemical cells come under the heading of *potentiometry*. In addition, these measurements are performed with little or no current (i.e., $i = 0$ or at equilibrium). Earlier we discussed how the electrode potential of a metal electrode in solution is a function of the ionic activities of the constituents composing the half-cell. Hence, if we chose the components to which an electrode is sensitive as analytes, the concentration or activity of these analytes can be monitored via measurement of the electrode potential. Potentiometric methods require an indicator electrode, a reference electrode, and a device for measuring a potential. The indicator electrode is also known as the *working electrode*. Using a pH meter with its glass membrane electrode (indicator) to measure H^+ activity is a potentiometric technique. Essentially, we are determining the potential of a galvanic cell with the indicator electrode reacting via a potential change to the concentration of a particular analyte.

Potentiometry is the most widely used analytical technique because of its simplicity, versatility, and low cost. It is used to measure analyte concentrations in titrations, process streams, biological fluids, and a plethora of other situations. However, potentiometry plays a minor role in the understanding of hydrodynamic electroanalytical methods for HPLC. Since reference electrodes are important to all electroanalytical techniques, including potentiometry, we will describe the common reference electrodes, and will give only a cursory review of indicator electrodes.

Since the potential of a single half-cell cannot be measured, all indicator electrode potentials must be measured against a reference electrode. The ideal reference electrode should meet the following criteria: (1) be reversible and follow the Nernst equation, (2) be prepared from readily available materials, (3) maintain its potential after exposure to finite currents, and (4) should not suffer thermal hysteresis. Also, the electrode response should be stable over time. In practice, only a few reference electrodes are commonly used.

Reference Electrodes

Standard Hydrogen Electrode

As described earlier in this chapter, the SHE is the reference electrode against which all half-reactions are measured. This electrode is highly reproducible and shows nearly ideal Nernstian response and behavior. With all these positive attributes, the SHE would theoretically be the electrode of choice. Unfortunately, the SHE is quite fragile in construction, and it is inconvenient to set up and operate for routine electroanalytical experimentation. The calomel and Ag/AgCl electrodes are considered to be more convenient and rugged, and, as a consequence, they are frequently used as reference electrodes.

The Calomel Electrode

Calomel reference electrodes are the most commonly used reference electrodes. The half-cell reaction for this electrode is

$$Hg_2Cl_2(s) + 2e^- \rightleftharpoons 2Hg(l) + 2Cl^- \qquad (2.32)$$

and its potential is given by

$$E = E^\circ - \frac{0.0592}{2} \log [Cl^-]^2 \qquad (2.33)$$

Note that the electrode potential is dependent on the chloride concentration. If the solution is saturated with KCl, the electrode is called the *saturated calomel electrode* (SCE). Other concentrations of chloride are also used, and Table 2.2 lists the most common forms of the calomel reference electrode and the corresponding electrode potentials.

The calomel electrode is typically composed of an inner tube that is filled with Hg/Hg_2Cl_2 paste and plugged with glass wool. The inner tube is placed inside an outer tube filled with a particular concentration of KCl. The SCE will typically show some KCl crystals along the wall of the outer tube. The outer tube is connected to the outside world via a small asbestos fiber, porous glass frit, or porcelain plug. A generalized rendition of an SCE is shown in Figure 2.6A.

The calomel electrode cannot be used at temperatures greater than 80°C

TABLE 2.2 Common Reference Electrodes and Their Electrode Potentials at 298 K

Reference Electrode	Fill Solution	Potential (mV) versus SHE[a] (Liquid Junction Potential Included)
Calomel	0.1 M KCl	335.6
	1.0 M KCl	283.0
	3.5 M KCl	250.1
	Saturated KCl	244.53
Ag/AgCl	0.1 M KCl	288
	1.0 M KCl	222.34
	Saturated KCl	198
Ag/AgBr	1.0 M KBr	71.06
Ag/AgI	1.0 M KI	−152.44
Hg/HgO	0.1 M NaOH	165
	1.0 M NaOH	140
Hg/Hg$_2$SO$_4$	Saturated KCl	655
	Saturated K$_2$SO$_4$ (295 K)	658

[a]Standard hydrogen electrode - 0 mV.

Source: Data were taken from *Lange's Handbook of Chemistry*, 13th edition.

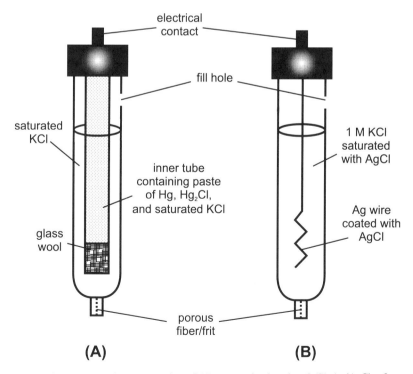

Figure 2.6. Schematic representation of (*A*) saturated calomel and (*B*) Ag/AgCl reference electrodes.

because of disproportionation of Hg(I) into Hg and Hg(II) ion. In addition, this electrode tends to equilibrate slowly to temperature changes.

Silver–Silver Chloride Electrode

Figure 2.6*B* shows the simple construction of this widely marketed reference electrode. In the Ag/AgCl reference electrode, a Ag wire coated with AgCl is immersed in a solution of KCl, and, as with the SCE, the electrode is connected to the analyte solution via a porous frit. The redox reaction that occurs in the Ag/AgCl electrode is

$$Ag^+ + e^- \rightleftarrows Ag^0 \qquad E° = +0.800 \ V \qquad (2.34)$$

The Nernst equation for Reaction (2.34) is

$$E = +0.800 - \frac{0.0592}{1} \log \frac{1}{[Ag^+]} \qquad (2.35)$$

In this half-cell, the potential is determined by the Ag^+ concentration, but the Ag^+ concentration is dependent on the solubility product equilibrium of AgCl:

$$AgCl(s) \rightleftarrows Ag^+ + Cl^- \qquad (2.36)$$

$$K_{sp} = [Ag^+][Cl^-] \qquad K_{sp} = 1.8 \times 10^{-10} \qquad (2.37)$$

By substituting Equation (2.37) into Equation (2.35), we see that the half-cell

$$E = +0.800 - \frac{0.0592}{1} \log \frac{[Cl^-]}{K_{sp}} \qquad (2.38)$$

potential is dependent on the chloride concentration. Since the K_{sp} of AgCl is known, the Nernst equation can be further simplified to

$$E = +0.222 - \frac{0.0592}{1} \log [Cl^-] \qquad (2.39)$$

(2.39)

$$AgCl + e^- \rightleftarrows Ag° + Cl^- \qquad E° = +0.222 \ V \qquad (2.40)$$

Note that on simplification the $E°$ value of Equation (2.39) corresponds to the $E°$ value for the AgCl/Ag redox couple noted in Reaction (2.40).

This electrode is easily constructed, can be operated over a wider range of temperatures than the SCE, and can be used in nonaqueous solutions. The Ag/AgCl reference electrode is easily miniaturized.

Indicator Electrodes

The ideal indicator electrode should respond rapidly, reproducibly, and selectively to changes in the concentration of the analyte of interest. Indicator electrodes are classified according to the mechanism by which the electrode potential is produced. There are two basic types of indicator electrodes: metallic and membrane. Metallic indicator electrodes develop a potential based on the equilibrium position of a redox half-reaction at the electrode surface. Metallic indicator electrodes are further classified as electrodes of the first kind, electrodes of the second kind, electrodes of the third kind, and redox or inert electrodes. Membrane indicator electrodes, or ion-selective electrodes (ISEs), develop a potential based on the difference in concentration of a particular ion on two sides of a membrane. Table 2.3 lists a brief summary of metallic and membrane indicator electrodes with an example of each.

Measuring Potentials

With only an electronic voltmeter, the potential of an indicator electrode versus a reference electrode can be easily measured. The beauty of this technique is that the chemist can get fast, simple, and economical determinations of concentrations of analytes by a simple comparison of the potential developed by an

TABLE 2.3 Summary of Indicator Electrodes Used in Potentiometry

Type	Description	Example
First kind	Metal electrode in contact with its own cations	$Ag\|Ag^+$; $Ag^+ + e^- \rightleftarrows Ag^0$
Second kind	Slightly soluble salt of a metal in contact with precipitating anion	$Ag\|AgCl(s)\|Cl^-$; $AgCl + e^- \rightleftarrows Ag^0 + Cl^-$
Third kind	Electrode responds to a alternate cation	$Tl^+\|TlCl, AgCl(s), Ag$; $TlCl + e^- \rightleftarrows Tl^+ + Cl^-$ $AgCl \rightleftarrows Ag^+ + Cl^-$ $K_{sp,TlCl} <<<< K_{sp,AgCl}$
Redox	Inert electrode in solution of redox couple	$Fe^{3+}, Fe^{2+}\|Pt$; $Fe^{3+} + e^- \rightleftarrows Fe^{2+}$
Membrane	Charge separation at an interface	A. Crystalline 1. Single crystal (LaF_3 for F^-) 2. Mixed crystal (Ag_2S for S^{2-} and Ag^+) B. Noncrystalline 1.Glass (silicate glasses for H^+) 2. Liquid (liquid ion exchangers for Ca^{2+}) 3. Immobilized polymer (ionophore in plastic matrix for K^+) 4. Enzyme (immobilized enzymes for specific substrates)

indicator electrode immersed in a test solution versus its potential when immersed in a standard solution of analyte. These analytes consist of anions, cations, and neutral molecules. A major aspect of potentiometry is that current in the electrochemical cell is appreciably small, and, as a consequence, the technique is nondestructive. It is important to remember that the presence of current is reflected in a reduction in the equilibrium potential of a galvanic cell. Any reader who is interested in learning more about potentiometry is directed to any of the books listed in the bibliography.

ELECTROLYSIS

In the discussion of electrode potentials, we have been dealing with spontaneous redox reactions. Potentiometry involves potential measurements of galvanic (i.e., electricity-producing) cells. In theory, these reactions are reversible, and many nonspontaneous reactions can be forced to proceed by the application of an external source of electrical energy. The process of causing reactions to occur in this manner is called *electrolysis*. Electrolysis is further defined to be those currents arising from Faradaic reactions.

The process of electrolysis is carried out in an electrolytic cell in which an external potential is applied to the working electrode of the electrochemical cell. Electrolysis cells are commonly used in chromatographic applications of electroanalytical chemistry. In an electrolytic cell, the potential applied to force the reaction to proceed is ideally equal to the magnitude of the calculated Nernstian potential, or galvanic potential, of the half-cell reaction required to decrease the reactant concentration to a minimal value. With the application of an electrode potential to the working electrode, a current develops as a result of the induced redox reaction at the electrode surface.

For a reversible electrode system, the equilibrium between oxidized and reduced species maintains a particular value for the potential difference across the interface. An ideal reversible electrode process is nonpolarizable, and the potential is expected to be independent of any current. Figure 2.7A shows that no matter how much current is present, the electrode potential does not change. If the cell has any internal resistance, current via Ohm's law will result in a potential loss or gain, which is reflected as a finite slope in the $i-E$ plot. Processes other than cell resistance that alter the cell potential are classified under the term of polarization. Figure 2.7B shows that the effect of polarization on the cell potential results in a plot with a finite slope. A purely polarizable interface is the opposite of an ideal reversible electrode process, and, as shown in Figure 2.7C, no current is observed over an extended potential range. Nonideal polarization effects result in departures from linearity, which are denoted by the dotted lines.

Figure 2.7B represents the typical electrochemical cell. Note that when polarization occurs in a galvanic cell, the observed potential is less than expected. This observation is a reflection of effects that impede the rate of the overall reaction or

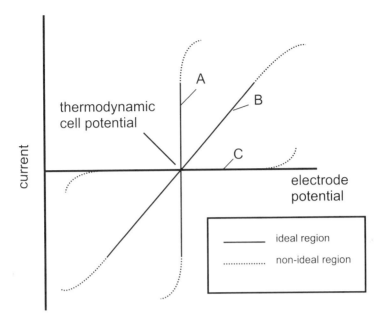

Figure 2.7. Diagram showing the effect of (*A*) a nonpolarized electrode, (*B*) a partially polarized electrode, and (*C*) a purely polarized electrode on current–potential response.

current, such as limits in mass transfer to the electrode, physical or chemical processes that interfere with the redox process, slow electron transfer, and ohmic losses. Therefore, in an electrolytic cell, a higher potential is required to achieve a given current. Since there are many electroanalytical techniques (e.g., amperometric detection and pulsed amperometric detection) in which currents occur in the electrochemical cell ($i > 0$), it is important that we discuss what current is and its effects on the applied cell potential.

Current

Unlike in electrical wires in which electricity is the movement of free electrons in metallic conductors, electricity in solutions is transported via the migration of ions. Hence, we can view current as the movement of ions, and this movement is induced by potential differences within the electrochemical cell. In addition, the resistance depends on the types and concentrations of all ions in the solution. Liquid junction potentials develop as a result of migration differences induced by concentration gradients when solutions of varying composition come in contact with each other in an electrochemical cell. Similarly, under constant applied potential, solution ions will migrate at differing rates toward the electrodes of opposite charge. Differences in the rates of ion migration H^+ versus Cl^- will result in a charge imbalance, which, in turn, results in an increase in solution resistance. This effect can be counteracted (overcome) by the addition of an excess of additional ions, which need not be involved in the half-cell reactions. These ions are typically salts, and they are added in enormous excess. The resulting solution is called the *supporting electrolyte*, because it supports the flow of charge (current) in the electrochemical cell. Thus, all the ions in a solution participate in conducting electrical charge, and the fraction carried by one particular ion may be markedly different from that carried by another. The fraction depends on the relative concentration of the ion and its inherent mobility. Therefore, the onus of being charge carriers does not fall completely on the half-cell reaction components.

Faradaic and Nonfaradaic Currents

Currents that are a result of the direct transfer of electrons via a redox reaction at the electrode–solution interface are called *faradaic currents*. Faradaic currents are a consequence of faradaic processes. Faradaic processes are governed by Faraday's law, which states that the amount of chemical change produced by application of an electric current is proportional to the quantity of electricity passed, or

$$w = \frac{it \, \text{MW}}{nF} \tag{2.41}$$

where w is the weight of a substance oxidized or reduced, i is the current in amperes, t is the time in seconds, MW is the molecular weight, n is the number of electrons transferred in the reaction per mole of compound, and F is the Faraday value. Faraday's law forms the basis of coulometry. *Coulometry* is a general term used to describe various techniques for measuring the *quantity* of electricity required to react, directly or indirectly, with an analyte of interest. Coulometry is typically conducted in a large electrolysis cell. For the purposes of this book, we will not discuss batch techniques in detail.

Coulometry is also used as a detection mode in HPLC. The analyte in the chromatographic band is completely reacted as it passes through the on-line electrochemical cell. Since 100% of the analyte reacts, coulometric systems can be very sensitive. Unfortunately, a large background signal, which is a result of the large electrode areas used to achieve 100% conversion, reduces the signal-to-noise (S/N) value. Limits of detection of <1 pmol injected are routine. In addition, the coulometric detector can provide an accurate mass assay independent of the use of a standard compound using Faraday's law.

If an electrode is immersed in a solution and a potential is applied, the solution ions adjacent to the electrode surface will rearrange to compensate for the charge developed at the electrode surface. As shown in Figure 2.8, the layer of solution adjacent to the electrode surface acquires an opposing charge. The charged solution layer forms a compact inner layer over which the potential decreases linearly and a diffuse layer, in which the decrease is exponential. The charged

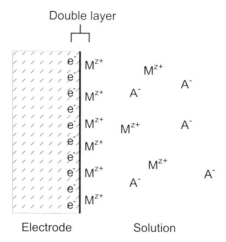

Figure 2.8. Illustration of the electrical double layer. M^{z+} and A^- represent ions of the supporting electrolyte.

electrode surface and the compact layer adjacent to the electrode surface are known collectively as the *electrical double layer*. The process described above is observed as a momentary surge of current, which rapidly decays to zero after the charge at the electrode surface is totally compensated for. This current is called a *charging current*, or *double-layer charging current*, and it is not the result of a redox reaction or faradaic processes. Hence, charging current is an example of a *nonfaradaic current*, which is a result of *nonfaradaic processes*. When the charging current drops to zero, the electrode is said to be *polarized*.

As discussed earlier, these departures from thermodynamic calculations as a result of current arise from several phenomena, which include solution resistance and polarization effects. The overall effect of these phenomena is to increase the potential needed to drive the desired reaction. The magnitude of the potential that must be applied to an electrolytic cell is composed of the opposing voltage of the galvanic cell and any additional voltage required to overcome solution resistance and polarization effects. Solution resistance and polarization effects act as impediments to current in electrochemical cells. In that respect, let us now look at the individual components of the applied potential in an electrolytic cell.

Potential

Galvanic Potential

If the electrode reactions along with the concentrations of reactants and products are known, the galvanic cell voltages can be calculated from the appropriate Nernst equations. Equation (2.42) shows that the galvanic cell potential

($\Delta E_{galvanic}$) is the equivalent to the equilibrium cell potential (ΔE_{cell}) discussed earlier:

$$\Delta E_{galvanic} = E_{cathode} - E_{anode} = \Delta E_{cell} \qquad (2.42)$$

The electrolysis cell is different from the galvanic cell in that the galvanic cell is typically under essentially equilibrium conditions (i.e., $i = 0$), and the electrolysis cell operates with current. The current will result in a departure from Nernstian behavior because the cell reactants are converted to products, and the overall effect is that the minimum applied potential needed to sustain the electrolysis must also increase.

Ohmic Potential Loss or iR *Drop*

The current in an electrochemical cell is sustained by ion migration within the supporting electrolyte. Resistance to ion migration is similar to resistance in metallic conduction, and it is measured in ohms. Hence, if a finite current is to be maintained in the electrolysis process, an ohmic potential (E_{ohmic}) will result from the product of the current (i) in amperes and the cell/solution resistance (R_{cell}) in ohms:

$$E_{ohmic} = iR_{cell} \qquad (2.43)$$

This is simply an application of Ohm's law. The ohmic potential is generally referred to as *iR drop*. The overall effect of *iR* drop is to increase the potential required to operate an electrolytic cell.

Overpotential and Polarization

An electrode whose actual potential is different from its calculated value is said to be polarized. The extent of polarization is reflected in the increased potential required to overcome these effects. The increased potential is known as the *overpotential* (η)

$$\eta = E - E_{eq} \qquad (2.44)$$

where E is the measured or applied potential and E_{eq} is the reversible or equilibrium potential of the electrode. The concept of overpotential is introduced as the measure of the difference between the observed potential and the reversible/equilibrium potential. Regardless of the source, overpotentials always operate to make the reaction at the electrode more difficult. The term *overvoltage* normally refers to the difference in the overpotentials between the two electrodes in an electrochemical cell.

Although the general half-cell reaction looks simple enough

$$oxd + ne^- \rightleftharpoons red \qquad (2.45)$$

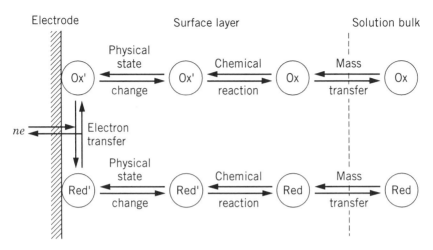

Figure 2.9. Illustration of the complexity of a simple electrochemical reaction. *Electrochemical Methods: Fundamentals and Applications*, A. J. Bard, L'. R. Faulkner, Copyright © John Wiley & Sons, 1980. Reprinted by permission of John Wiley & Sons, Inc.

the actual process by which the reaction occurs is more complex (see Fig. 2.9). Any one of the several intermediates steps shown in Figure 2.9 may limit the rate at which this overall reaction occurs and, thus, the magnitude of the current. To sustain the reaction, the oxd species must diffuse from the bulk of the solution to the electrode interface. This process, known as *mass transfer*, will be discussed in more detail later. If the rate of oxidation or reduction at an electrode surface is faster than the rate at which fresh reactants can be supplied to, or products removed from, the surface, then the process is impeded, and we have concentration polarization (η_{conc}). With concentration polarization, the concentration of the reactant and products at the electrode surface are different from those in the bulk of their solution. Any increase in current will exacerbate the process and will cause the overpotential to increase. The degree of concentration polarization increases with low concentrations of reactant concentration and high concentrations of total electrolyte. Mechanical agitation, such as stirring, will decrease η_{conc} because of increased mass transfer of analyte. A larger electrode will decrease concentration polarization, and this is one reason why auxiliary and counterelectrodes are typically quite large.

In the absence of concentration polarization, the possibility exists that the reaction proceeds via an intermediate chemical reaction such as oxd' or red', and this intermediate is then the actual participant in the electron-transfer process. Hence, when the current at the electrode surface is controlled by the rate of the electrode reaction rather than the rate of mass transfer as in concentration polarization, these processes come under the heading of kinetic, or reaction, polarization ($\eta_{kinetic}$). The physical process that is occurring in the reaction mechanism (e.g., adsorption, desorption, or crystallization) determines the type of polarization.

At the heart of every electrode reaction is the actual electron transfer between the reactant and the electrode. As with any of the other steps in an electrode reaction, the electron-transfer step can also be rate-limiting. Hence, when the rate of oxidation or reduction is insufficient as compared to theory, the electrode is experiencing charge-transfer polarization (η_e). Charge-transfer polarization generally increases with increasing current density (measured in amperes per square centimeter of electrode surface) and decreases with increases in temperature. In general, the magnitude of overpotential due to charge-transfer polarization cannot be predicted because it is a function of many variables (e.g., electrode material).

The Applied Potential

More often then not, a half-cell experiences a combination of polarization effects. Thus, as shown by the following equation

$$\eta_{total} = \eta_{conc} + \eta_{kinetic} + \eta_e \qquad (2.46)$$

the overpotential (η_{total}) is composed of concentration, kinetic, and charge-transfer polarizations. As shown in Figure 2.10, for an electrolytic cell the anode potential increases, the cathode potential decreases, and the overall cell voltage increases with increasing current or polarization effects.

In general, the presence of an overpotential of whatever origin will cause the electrode potential or cell voltage to be increased in order to maintain a specific current. Equation (2.47) illustrates that the minimum applied potential required to

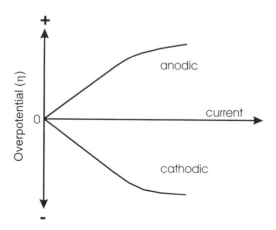

Figure 2.10.   Depiction of the effect of polarizations on the cell potential.

induce electrolysis is composed of the galvanic potential, sum of all polarization effects, and any ohmic potential losses at a given electrode:

$$\Delta E_{applied} = \Delta E_{galvanic} + \Delta E_{overpotential} + \Delta E_{ohmic} \qquad (2.47)$$

In more familiar terms, the applied cell potential takes into consideration the galvanic voltage of the cell, the sum of the overvoltages (i.e., η_{total}), and any ohmic potential losses (i.e., iR_{cell}):

$$E_{applied} = E_{galvanic} + \eta_{total} + iR_{cell} \qquad (2.48)$$

Overpotentials are significant typically for electrode processes that yield gaseous products and anodic oxygen-transfer reactions. The overpotentials associated with the evolution of H_2 and O_2 at inert electrodes have been extensively studied. Table 2.4 shows the magnitude of the overpotential for these gases at various current densities for several electrode materials. Note that polarization effects in the SHE are minimized by using a large surface-area electrode (platinized platinum), which greatly minimizes the current density. The large overpotential for H_2 reduction at Hg electrodes is why these electrodes are preferred for monitoring reduction reactions. In electroanalytical techniques, the usable potential range of an application is often limited by the onset of O_2 and H_2 evolution.

TABLE 2.4 Overpotentials for the Evolution of H_2 and O_2 at Inert Electrodes at 298 K

	Overpotential (V)					
	O_2			H_2		
Electrode	0.001	0.01	1	0.001	0.01	1
Material	(A cm^{-2})	(A cm^{-2})	(A cm^{-2})	(A cm^{-2})	(A cm^{-2})	(A cm^{-2})
Pt (bright)	0.721	0.85	1.49	0.024	0.068	0.676
Pt (platinized)	0.348	0.521	−0.7	0.015	0.030	0.048
Au	0.673	0.963	1.63	0.241	0.391	0.798
C (graphite)	—	—	—	0.600	0.799	1.220
Ag	0.580	0.729	1.131	0.475	0.762	1.089
Cu	0.422	0.580	0.793	0.479	0.584	1.269
Ni	0.353	0.519	0.853	0.563	0.747	1.241
Zn	—	—	—	0.716	0.746	1.229
Sn	—	—	—	0.856	1.077	1.231
Pb	—	—	—	0.52	1.090	1.262
Bi	—	—	—	0.78	1.05	1.23

Source: Data were taken from *Lange's Handbook of Chemistry*, 13th Edition, and D. A. Skoog and J. L. Leary, *Principles of Instrumental Analysis*, Saunders College Publishing, New York, 1992, p. 484.

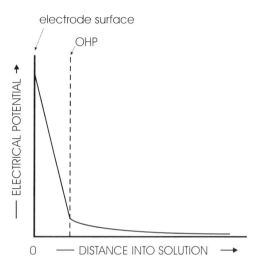

Figure 2.11. Plot of how the electric field drops off with distance from the electrode surface. OHP refers to the outer Helmholtz plane.

THE ELECTRODE INTERFACE AND MASS TRANSFER

Let us return to the electrode interface to which a positive potential is applied. The electric field stretches out into the solution from the electrode as shown in Figure 2.11. Initially the electrode field falls off linearly with distance through the Helmholtz–Perrin region. Figure 2.12 shows a detailed view of the elec-

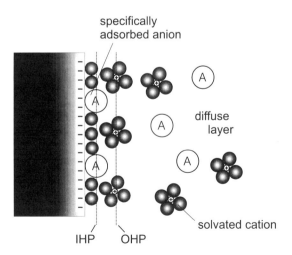

Figure 2.12. Detailed depiction of the electrode solution interface. Both the outer and inner Helmholtz planes are shown.

trode–solution interface. The electrode is covered by a sheath of oriented solvent molecules and ions and molecules absorbed directly on the electrode surface. These adsorbed species are said to be specifically adsorbed or contact absorbed, and they are only partially solvated. The imaginary plane that passes through the center of these molecules is called the *inner Helmholtz plane* (IHP). A second layer of molecules, which are fully solvated, is absorbed to the electrode surface but separated from the surface by a monolayer of water and solvent molecules. The imaginary line that passes through the center of these molecules is called the *outer Helmholtz plane* (OHP). This model considers the interface to resemble a capacitor; hence the reference to the electrode and Helmholtz–Perrin region as the double layer. The Helmholtz–Perrin region often contains a large percentage of the ionic charge necessary to balance the charge on the electrode surface. Beyond the compact layer defined by the OHP, the ions in the solution are under the influence of the electric field, which tries to order the molecules; and thermal forces, which tend to disorder the system. The thermal forces are based on a Boltzmann distribution of the ions. As a consequence of these opposing forces, a diffuse layer known as the *Gouy–Chapman diffuse layer*, is formed. It should be clear that as one proceeds farther from the electrode surface the effect of the applied potential on the electrode surface becomes negligible. At this point, the ions in the solution are unaware of the presence of the electrode, and the solution carries the connotation of being the *bulk solution*. Note from Figure 2.11 that the electric field drops off dramatically (exponentially) with distance from the electrode.

In an electrolysis cell, we are interested in developing a faradaic current, which is an instantaneous measure of the rate of an electrochemical reaction. The rate is essentially determined by the rate of the overall electron-transfer process at the electrode surface, and the rate of mass transport of the electroactive species from the bulk of the solution to the electrode surface.

Electron-Transfer Process

It is important to remember that the observed current (i_{obs}) at the working electrode in an electrolysis cell is composed of the summation of all the oxidations (anodic current) and reductions (cathodic current) occurring at a particular electrode potential.

$$i_{obs} = i_{anodic} + i_{cathodic} \tag{2.49}$$

In the simplest case of a singular half-reaction, we know from the Nernst equation that when the electrode potential is equivalent to the equilibrium potential, or

$$E_{app} = E_{eq} = E° - \frac{0.0592}{n} \log \frac{red}{oxd} \tag{2.50}$$

the concentration of the reduced species must be equal to the concentration of the oxidized species. Hence, the forward reaction rate (cathodic current) must be equal to the reverse reaction (anodic current) in order to be at equilibrium. By convention, the anodic current is negative and the cathodic current is positive, and as deduced from Equation (2.49), the overall observed current is zero ($i_{obs} = 0$).

If an external potential is applied to an electrolysis cell, the electrode potential changes from E_{eq} to E and a finite, or nonzero, current develops. The magnitude of the i_{obs} is also a function of E. To explore the effect of electrode potential on i_{obs}, we will first consider the situation where the rate of mass transport to the electrode surface is unlimited. Hence, we can isolate the electron-transfer process from mass-transport effects. Figure 2.13 shows a generalized redox reaction profile. For profile A, the activation energy barrier is low, and the electron-transfer reaction is expected to be very fast. Under these conditions, the current–potential plot of Figure 2.13A shows that only at E_{eq} does the $i_{obs} = 0$. As E moves away from E_{eq}, the current is either purely anodic or purely cathodic. In other words, the slightest difference in E from E_{eq} is enough to convert the reactant into product, and hence, only at E_{eq} do both anodic and cathodic currents occur. Reactions that fall into this category are considered to be reversible.

In contrast, if the activation energy barrier is very high (Fig. 2.13, profile B), the rate of electron transfer can be considered to be very slow. As a consequence, a significantly higher applied potential is required to overcome the activation energy barrier before a finite current is developed. The anodic and cathodic

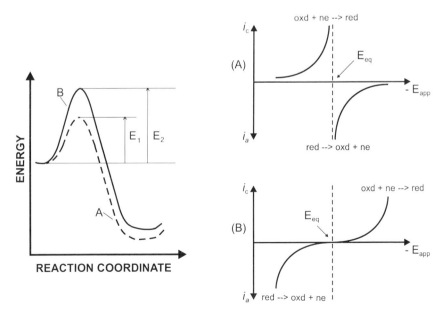

Figure 2.13.   Generalized redox reaction profiles and current–potential plots for a (A) fast and a (B) slow electron-transfer reaction. E_1 and E_2 represent the applied potentials required to overcome the activation energy barriers.

reactions will be well separated on the $i-E$ plot (Fig. 2.13B), and there is little or no region of mixed anodic–cathodic current. This effect is the same as the charge-transfer polarization discussed earlier, and the difference between $E_{applied}$ and E_{eq} is the overpotential. Reactions under these conditions are considered to be irreversible.

If an intermediate rate of electron transfer, or activation energy, is considered, some overpotential is required. It is important to keep in mind that the observed current is comprised of i_a and i_c, and the reactions of oxd $+ ne^- \rightarrow$ red and red $\rightarrow$ oxd $+ ne^-$ each occur individually. When E is more positive than E_{eq}, i_{obs} will be dominated by the anodic current (oxidation); if E is more negative than E_{eq}, then the i_{obs} will be dominated by the cathodic current (reduction). Thus, for a slow electron-transfer process, there is a wide range of potentials where the current is of mixed anodic–cathodic origin. Even though the electron-transfer rate is not infinite, these reactions are typically classified as reversible.

Mechanisms of Mass Transport

In the preceding discussion, we assumed that the rate of mass transfer was unlimited and the reactions were for pure electron-transfer reactions. Often when the electron-transfer rate is fast enough, the electrochemical reaction, or current, is limited by the rate at which ions or molecules are transported to or from the electrode surface.

There are three mechanisms of mass transport: diffusion, convection, and migration. These modes of mass transport account for the transference of electroactive material to and from the electrode surface. When any one of these mechanisms is insufficient to transport the reactant to, or the product from, the electrode surface at a rate demanded by the theoretical current, concentration polarization is observed.

Whenever a concentration difference develops between two regions of a solution, ions or molecules move from the more concentrated region to the more dilute region as a result of diffusion. During electrolysis and as a consequence of the ensuing electrode reaction, the concentration of electroactive species at the electrode surface will be decreased, and a concentration gradient will be developed. Hence, the formation of a concentration gradient is inherent to electrolysis. Diffusion is under the influence of a concentration gradient.

In the bulk of the solution, ion movement is caused primarily by the influence of the electric field. This process is known as migration. The influence of the electrostatic field on the reactant ions is diminished as the total concentration of electrolytes in the solution increases. At some concentration of total ions in the supporting electrolyte, the effect of the electric field on the reactant becomes negligible. For high-ionic-strength solutions, the effect of migration can be ignored. Migration is under the influence of a potential gradient.

The simplest and most obvious manner of transferring reactants to the electrode surface is by some mechanical means (e.g., stirring or agitation). Forced convection will decrease concentration polarization. When a solution is quiescent, only diffusion and migration are involved, and migration can be eliminated by the addition of supporting electrolyte. It is important to remember that natural convection (e.g., thermal or density gradients) is always present, and, as a consequence, voltammetric experiments in a quiescent solution can be performed only for a finite period of time (~5 min).

Current is the rate of electron flow, and can be expressed symbolically as dQ/dt. From Faraday's law, the number of electrons, or coulombs, is directly related to the amount of reactant:

$$i = nF \frac{dC}{dt} \tag{2.51}$$

where n is the number of electrons in the half-reaction, F is the Faraday constant, and dC/dt represents the rate of concentration change (in mol s^{-1}). Normalization of dC/dt for electrode area A (in cm^2) results in

$$\frac{dC}{dt} = AJ \tag{2.52}$$

the concentration flux J (in mol s^{-1} cm^{-2}). For a particular current to be maintained, the reactant must be conveyed from the bulk of the solution to the electrode interface at a rate determined by the expression

$$i = nFAJ \tag{2.53}$$

which is merely the combination of Equations (2.51) and (2.52). The overall concentration flux [Eq. (2.54)] results from a combination of the fluxes

$$J = J_{diff} + J_{mig} + J_{conv} \tag{2.54}$$

due to diffusion (J_{diff}), convection (J_{conv}), and migration (J_{mig}). Flux due to migration is generally negligible in the ionic-strength solutions used in typical electroanalytical experimentation.

The Diffusion Layer

Let us begin with a stationary electrode that is submersed in a solution supporting electrolyte and analyte in its reduced form. The electrode is held at a potential at which the analyte is not oxidized. Under these conditions, the analyte is distributed throughout the solution. If the electrode is stepped to a higher potential, the material at the electrode surface, or $C°$, is oxidized. The product begins to diffuse away. Over time, a concentration gradient of the analyte is produced, which is ultimately determined by the mass transfer of the species of interest. This zone near the electrode is known as the *diffusion layer.*

The flux of analyte due solely to diffusional mass transfer is described as follows:

$$J = -D \frac{dC}{dx} \qquad (2.55)$$

where dC/dx is the concentration gradient and D is the diffusion coefficient (in $um^2 \ s^{-1}$). This relationship is known as *Fick's first law*. This law simply states that the net transfer of solute mass per unit time across a plane intersecting a concentration profile will be proportional to the steepness of the profile. By combining Equation (2.53) with Fick's first law

$$i = nFAJ = nFAD \ \frac{dC}{dx} \qquad (2.56)$$

we can define the current due to mass transfer as a function of the concentration gradient adjacent to the electrode surface. Returning to our submersed electrode, we can determine the slope of the concentration gradient by simple Cartesian math (see Fig. 2.14). The concentration gradient, dC/dx, is assumed to be linear over the diffusion layer (δ), and, hence, the slope is derived from

$$\frac{dC}{dx} \cong \frac{(C^b - C^\circ)}{\delta} = \frac{C^b}{\delta} \qquad (2.57)$$

when the concentration at the electrode surface (C°) goes to zero.

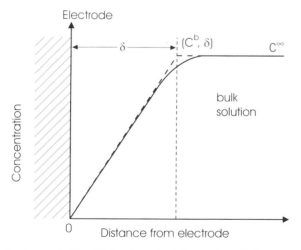

Figure 2.14. Development of the diffusion layer at a stationary electrode in contact with a solution of analyte. δ denotes the diffusion-layer thickness; C°, C^b, and C^∞ represent the solution concentration of analyte at the electrode surface (i.e., $C^\circ = 0$), bulk concentration at the diffusion-layer boundary, and at an infinite distance away from the electrode, respectively.

A characteristic feature of the voltammetric response is limiting current. The limiting current

$$i_{\text{lim}} = nFAD \frac{C^b}{\delta} \tag{2.58}$$

is inversely proportional to the thickness of the diffusion layer. For a planar electrode, which is uniformly accessible, the diffusion-limited current develops over time as the electrode potential is stepped at $t = 0$ from a potential at which there is no reaction to one where all the species that reach the electrode either are oxidized or reduced (see Fig. 2.15A). The diffusion-limited current is inversely dependent on the square root of time, which reflects the growing dif-

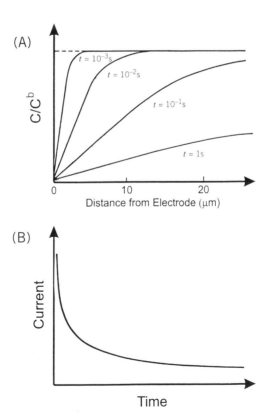

Figure 2.15. Schematic of the effect of (*A*) time on the diffusion-layer thickness at a stationary electrode and (*B*) a plot of the Cottrell equation.

fusion layer. The equation, which is known as the *Cottrell equation*, is given below:

$$i(t) = i_d(t) = nFAD^{1/2} \frac{C^b}{(\pi t)^{1/2}} \qquad (2.59)$$

The variation of current with time, which is called *chronoamperometry*, according to the Cottrell equation is shown in Figure 2.15*B*. At long times (i.e., seconds to minutes), the measured currents cannot be considered to be free of contributions from natural convection, which perturb the concentration gradient. In addition, the currents at small values of t are a combination of faradaic current and double-layer charging. Diffusion-limited current equations can be derived for a variety of electrode geometries (e.g., planar, spherical, tubular).

VOLTAMMETRY

Electroanalytical techniques that measure current as a function of potential are generally termed *voltammetry*. In contrast to coulometry, only a fraction of the analyte is consumed during the experiment. Voltammetry is typically performed using a three-electrode potentiostat, which accurately controls the applied potential. A generic three-electrode potentiostat schematic is shown in Figure 2.16. The redox reaction to be monitored ensues at the working electrode; the second electrode is a reference electrode, which maintains a constant potential through-

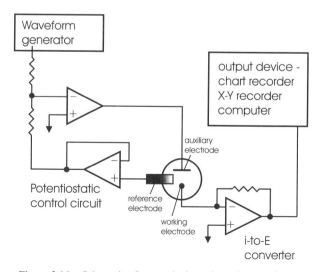

Figure 2.16. Schematic of a generic three-electrode potentiostat.

out the experiment; and the counterelectrode completes the electrical circuit. The counterelectrode, also known as the *auxiliary electrode*, is often much larger than the working electrode to minimize current density at the electrode surface.

The voltammetric experiment is typically carried out in an electrochemical cell containing analyte in an aqueous solution, or simply the *supporting electrolyte*. In spectroscopy, the range of wavelengths over which the spectrum of the analyte is collected is limited by the optical transparency of the diluent, or background solution. Similarly, the range of potentials over which the redox reaction of the analyte can be observed is limited by the electroactivity of the supporting electrolyte. In voltammetry, the potential window available to study electrochemical reaction is determined by both the supporting electrolyte and the electrode material used to perform the experiment. The potential ranges over which the common working electrode materials can be used will be discussed in Chapter 3. The positive and negative limits are usually determined by the potentials at which water is oxidized to form O_2 and reduced to produce H_2, respectively (see Table 2.4).

Voltammetry can be considered as the study of the relationship between current and voltage, or potential. The potential, or independent variable, is applied to the working electrode as a function of time in order to evoke a response in the form of current, or the dependent variable, which is unique to the analyte of interest. The pattern in which the potential changes with time is denoted as the *waveform*. Figure 2.17 shows several of the most common waveforms used in electroanalytical voltammetry today.

Historically, voltammetry was epitomized by polarography, and many of the potential–time waveforms were originally invented for use with polarography.

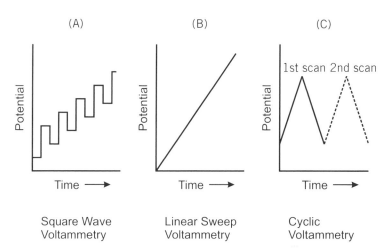

Figure 2.17. Potential–time waveforms used in various electroanalytical techniques. Waveforms based on (A) square, (B) linear, and (C) triangular potential-time patterns are used in Square Wave, Linear Sweep, and Cyclic Voltammetry, respectively.

Polarography is distinguished from other voltammetric techniques by its working electrode, which is in the form of a dropping-mercury electrode (DME). Many of the early applications of polarography were the determination of reducible metal ions in solution. Mercury electrodes afforded the analyst three very important advantages: (1) a large negative potential window over which the reduction of many metal ions occurs, (2) the metals are often reversibly reduced to amalgams at the electrode surface, and (3) a means to overcome fouling by simply forming a new drop; hence the development of dropping-mercury electrodes. The cyclic patterns of many of the waveforms used in polarography were timed to be in sync with the formation of a new drop. The entire electrochemical measurement (i.e., seconds to minutes) is often carried out at a clean electrode surface. Many of the techniques used in polarography derive their names from the type of potential–time waveform used to perform the experiment. For instance, *square-wave voltammetry* (SWV) is based on the application of a square potential–time waveform to the working electrode (see Fig. 2.17*A*). Other methods include *normal pulse voltammetry* (NPV), *differential pulse voltammetry* (DPV), *reverse pulse voltammetry* (RPV), *differential normal pulse voltammetry* (DNPV), and *double differential pulse voltammetry* (DDPV). Many of these applications have been oriented toward quantitative analysis, and as a consequence, the interest has been on the relationship of current versus potential versus concentration. Although polarography is still widely used, many of its early applications have been replaced by spectroscopic methods. Details concerning polarographic techniques can be found in the reference books listed at the end of this chapter. Only the techniques germane to the understanding pulsed electrochemical detection will be reviewed.

Stationary Electrodes

Several of the waveforms used in polarography are also used in voltammetric studies at solid electrodes. The simplest waveform is that of *linear sweep voltammetry* (LSV). As shown in Figure 2.17*B*, the waveform consists of a potential ramp, or linear potential scan. Important parameters include the initial potential (E_i), final potential (E_f), and the sweep rate (ϕ). Either a stationary, solid electrode or hanging mercury-drop electrode (HMDE) is often used in LSV. Since the Hg drop in DME is expanding in the supporting electrolyte, DME is classified as a hydrodynamic technique. The scan rate is typically between 20 and 400 mV s^{-1}. The current–potential plot is denoted as a *voltammogram*. Figure 2.18 shows the linear sweep voltammogram for a hypothetical analyte in the reduced form A_{red} at a stationary planar electrode that undergoes a reversible redox reaction. As potential is scanned positive, oxidation of A_{red} to A_{oxd} commences. The rate of oxidation, or current, continues to increase until it reaches a peak current (i_p) at the peak anodic potential ($E_{p,a}$). The peak current can be defined by the Randles–Sevcik equation:

$$i_p = 2.69 \times 10^5 n^{3/2} A D^{1/2} C^b \phi^{1/2} \tag{2.60}$$

where i_p is the peak current (amperes), n is the number of electrons transferred, A

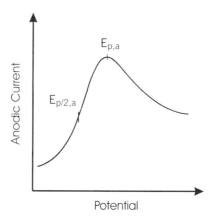

Figure 2.18. Linear sweep voltammogram of a hypothetical compound at a stationary electrode. $E_{p,a}$ is the potential for the peak anodic current.

is the area of the electrode (cm^2), D is the diffusion coefficient of the species being oxidized or reduced ($cm^2 \ s^{-1}$), C^b is the bulk solution concentration of the species being oxidized or reduced (mol cm^{-3}), and ϕ is the scan rate (V s^{-1}). At this point and beyond, the current is limited by the rate of mass transport to the electrode surface, and the current is said to be under diffusion control. Owing to depletion of A_{red}, the current begins to attenuate proportional to $t^{-1/2}$. The shape of the curve is typically sigmoidal, and is called a *voltammetric wave*. The midpoint of the curve, or $i_p/2$, is the half-wave potential ($E_{1/2}$). The peak potential for an oxidation, which is derived from the Nernst equation, is defined as

$$E_p = E'_a + \frac{0.0285}{n} \tag{2.61}$$

at 25°C, where E'_a is the formal potential corrected to the reference electrode being used. Hence, the $E_{1/2}$ is related, but not identical, to the thermodynamic standard potential of the redox reaction. The $E_{1/2}$ is very useful in characterizing compounds.

For a reversible reaction, i_p should be linear with the square root of scan rate, E_p is independent of scan rate, and $|E_p - E_p/2| = 56.6/n$ mV. An irreversible system behaves differently than a reversible system with respect to both i_p and E_p, and the resulting equations must be modified to reflect the overpotential required to drive the reaction. An irreversible system is easy to diagnose in that the voltammetric wave is shifted to more positive potentials for an oxidation, and E_p is no longer independent of scan rate. Hence, the voltammetric wave appears to be broader and lower.

Cyclic voltammetry (CV) is probably the most widely used technique of all the voltammetric methods by both electrochemists and nonelectrochemists alike.

Cyclic voltammetry allows the analyst to mechanistically study redox systems, especially the assignment and characterization of redox couples. The potential–time waveform for CV consists of a triangular waveform, which is cycled between two potential values (see Fig. 2.17C). The same waveform can be cycled indefinitely. Important parameters include the initial potential (E_i), initial sweep direction ($-/+$), the maximum potential (E_{max}), the minimum potential (E_{min}), the final potential (E_f), and the sweep rate (ϕ). The variation of any one of these parameters can be used to delineate electrochemical information.

The current–potential (i–E) plot is called a *cyclic voltammogram*. Figure 2.19 shows the cyclic voltammogram for a hypothetical compound (_____) in supporting electrolyte. If the CV is thought of as a forward scan and reverse scan of an LSV being coupled together, it is easy to extrapolate the findings in LSV to that of cyclic voltammetry. As potential is scanned positive, oxidation of A_{red} to A_{oxd} commences at ~200 mV. The rate of oxidation, or current, continues to increase until it reaches a peak current (i_p) at the peak anodic potential ($E_{p,a}$). At point E', the concentration of A_{red} equals the concentration of A_{oxd} present at the electrode surface, which satisfies the condition of the applied potential at this point being equal to the formal electrode potential of the redox couple. Note that this point is not at the halfway point. Beyond $E_{p,a}$, the concentration of A_{red} at the electrode surface is essentially zero, and the observed current is under diffusion control. Since the applied potential is still more positive than the formal potential of the redox couple, the current begins to attenuate at a rate proportional to $t^{-1/2}$, which is attributable to the diffusion layer extending farther into the bulk of the solution. If the scan is continued more positive, a potential is reached where the flux of A_{red} to the electrode becomes limited by its rate of diffusion to the

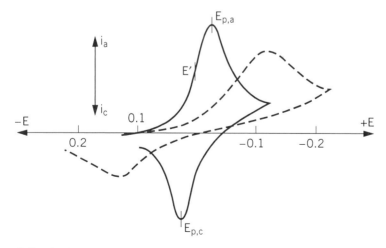

Figure 2.19. Cyclic voltammogram of a hypothetical compound (_____) at a stationary electrode showing the effect of increasing irreversibility (---------) on the shape of the i–E plot.

electrode surface. At E_{max}, the scan direction is reversed and the current remains anodic because the potentials are still sufficiently positive to induce oxidation. At some point oxidation of A_{red} ceases, and the reduction of A_{oxd} produced during the forward potential scan commences. The cathodic current profile follows the same course of events as that of the forward scan with the formation of a cathodic peak maximum ($E_{p,c}$) and a depletion of A_{oxd} near the electrode after $E_{p,c}$. The voltammetric waves of the forward and reverse curves in CV are sigmoidal.

Cyclic voltammetry can be used as a diagnostic tool to characterize electrochemical processes under various conditions. As with LSV for a reversible redox couple, i_p is proportional to $\phi^{1/2}$, and E_p is independent of scan rate. In CV, the following equations apply for a reversible redox couple with no kinetic complications:

$$\Delta E_p \cong \frac{0.057}{n} \tag{2.62}$$

$$E' = \frac{E_{p,a} + E_{p,c}}{2} \tag{2.63}$$

$$\frac{i_{p,a}}{i_{p,c}} = 1 \tag{2.64}$$

Since E_p is independent of scan rate for a reversible system, the ΔE_p separation of the peak maxima is also independent of scan rate.

Electrochemical irreversibility is caused by slow electron transfer of the redox species with the working electrode, and an irreversible system behaves differently than a reversible system in terms of both i_p and E_p. Hence, the preceding equations do not apply. Electrochemical irreversibility in CV is noted by a greater separation between the peak maxima than $0.058/n$ V, and the peak potentials are *dependent* on scan rate. Figure 2.19 also compares a reversible (_____) redox system with an irreversible (---------) redox system. Note the "smeared" appearance of the voltammogram. In general, as the heterogenous rate constant of electron transfer decreases, the ΔE_p increases. An interesting effect to note is that the more irreversible a system, the smaller the i_p of the reverse scan becomes. This is because the potential and time parameters in CV are coupled, and the greater separation in potential also reflects a greater separation in time. Hence, more of the analyte produced on the forward scan diffuses away before reaching a potential sufficiently negative to invoke reduction. It is also important to eliminate any uncompensated iR drop in the electrochemical cell, which causes a similar peak separation phenomenon that increases with ϕ. Looking ahead, most of the systems encountered in PED are irreversible redox systems.

Hydrodynamic Electrodes

Linear sweep voltammetry and cyclic voltammetry at stationary electrodes are important tools for obtaining qualitative information about electrochemical redox systems. At stationary electrodes, mass transfer occurs *only* via diffusional transport to the electrode surface. In electrochemical detection in chromatography, mass transfer to the electrode surface also occurs via forced convection. For example, the eluted analyte peak in a chromatographic system is delivered as a band to the electrochemical cell via a flowing stream. Hence, it is important to understand the effect of forced convection as a means of mass transport to the electrode surface. Transport of analyte to the electrode surface via forced convection can be accomplished by agitating the solution above the electrode or moving the electrode itself. Voltammetric techniques that involve transport via forced convection come under the heading of *hydrodynamic voltammetry*. Hydrodynamic systems have enhanced mass transfer, which translates to higher currents and, as a consequence, higher sensitivity.

In order to understand hydrodynamic systems better, let's take a closer look at the solution adjacent to the electrode surface. Previously we defined the diffusion layer as distance over which the concentration gradient of C^b to C^o exists adjacent to the electrode. This layer is more precisely denoted as the Nernst diffusion layer. One can develop a relationship between the mass-transfer coefficient and δ, when $C^o = 0$, as the following:

$$k_d = \frac{D}{\delta} \tag{2.65}$$

where k_d has the dimension of a rate constant (cm s^{-1}), D is the diffusion coefficient (cm^2 s^{-1}), and δ is the Nernst diffusion layer (cm). From the Cottrell equation, we know that the diffusion-layer thickness increases in proportion to $t^{1/2}$ on stepping from a potential of no reactivity to one where all the analyte reaching the electrode surface is oxidized. The smaller the δ value, the larger the concentration gradient, which leads to higher currents.

In a rapidly stirred solution, the solution–electrode interface is a bit more complex. Three regions exist at the electrode interface: (1) the turbulent flow region, in which the solution is in random motion; (2) a more ordered region where the transition to a laminar flow takes place; and (3) a stagnant solution region nearest the electrode, which is analogous to the Nernst diffusion layer with thickness δ cm (see Fig. 2.20). The layer nearest the electrode is stagnant as a result of friction between the electrode surface and the solution molecules, and has a typical thickness ranging from 10^{-2} to 10^{-3} cm. The stagnant layer and the transition layer are sometimes collectively called the *hydrodynamic layer*, or δ_H, which for a typical aqueous solution is approximately 10 times thicker than the stagnant layer.

For an unstirred solution at a stationary electrode, mass transport takes place by diffusion alone, and with time, depletion of the analyte extends farther and farther out into the bulk solution. As a consequence, the concentration gradient over the stagnant layer decreases. In a stirred solution, diffusion alone occurs over the stagnant layer, but beyond the stagnant layer analyte is continuously

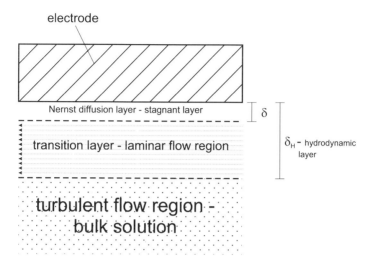

Figure 2.20. Electrode/solution interface in a stirred solution.

transported by forced convection (see Fig. 2.20). Hence, diffusion is limited to a narrow layer of liquid (δ_H), which, even with time, cannot be extended farther into the solution. Under stirred conditions, the concentration gradient over the stagnant layer is greater in magnitude, and the product from the reaction diffuses across the stagnant layer and is swept away. In addition, the hydrodynamically defined diffusion layer is quickly attained, and the overall system rapidly achieves steady-state, diffusion-controlled currents.

Figure 2.21 compares the (*A*) linear sweep and (*B*) cyclic voltammograms for

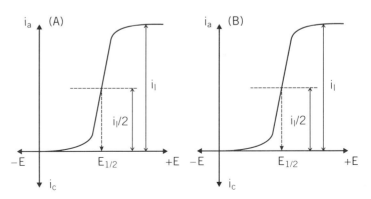

Figure 2.21. Voltammograms of a hypothetical compound under stirred conditions for (*A*) linear sweep voltammetry and (*B*) cyclic voltammetry.

our hypothetical compound under stirred conditions. Note that the voltammetric wave reaches a plateau and not a peak for both techniques. This is because the solution near the electrode surface does not continue to be depleted by the growing diffusion layer and the analyte is continuously being transported to the electrode surface by forced convection. A similar effect is noticed in the forward scan of our hypothetical compound for cyclic voltammetry, but note that the current response of the reverse scan is simply a retrace of the forward scan. This effect is due to the fact that the product formed in the forward scan is being continuously swept away. Hence, it is important to note that all forms of hydrodynamic voltammetry are steady-state methods (if the sweep rate is not too large), which means that the potential is independent of both scan direction and time.

A major drawback of stirred solution voltammetry is the lack of control of the hydrodynamic process afforded by external stirring. Precise control of the hydrodynamics can be accomplished with *rotating-disk voltammetry*. A typical rotating-disk electrode is shown in Figure 2.22. The hydrodynamic flow pattern is such that the liquid or supporting electrolyte is drawn axial to the electrode to replenish the fluid that moves horizontally out and away from the electrode's center. The hydrodynamic diffusion layer at the electrode has been determined to be

$$\delta_H = 1.61 D^{1/3} \omega^{-1/2} \nu^{1/6} \tag{2.66}$$

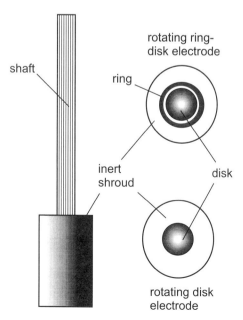

Figure 2.22. Diagram of a rotating-disk electrode and rotating-ring–disk electrode.

where D is the diffusion coefficient (cm^2 s^{-1}), ν is the kinematic viscosity (cm^2 s^{-1}), and ω is the rotation speed (radians s^{-1}). Note that the hydrodynamic diffusion layer becomes smaller as the rotation speed increases, which leads to higher currents. When combined with the limiting current equation [Eq. (2.58)], the resulting equation is known as the *Levich equation*, which is given below:

$$i_{\text{lim}} = 0.620nFD^{2/3}\omega^{1/2}\nu^{-1/6}C^b \qquad (2.67)$$

Figure 2.23 compares linear sweep voltammograms at (*A*) stationary and (*B*) rotating-disk electrodes. The current for the rotating electrode is larger than that for the stationary electrode because of enhanced forced convection. The current reaches a time-independent steady-state response as noted by the plateau for the RDE in contrast to the peak for the stationary electrode. In addition to the electroanalytical studies which can be done at stationary electrodes, the control of the rotation speed allows for a "time window" to be changed to monitor reaction kinetics.

All electrochemical reactions occur as a sequence of finite steps. In the simplest case, these steps include (1) the convective–diffusional transport of the analyte to the electrode surface, (2) adsorption of the analyte to the electrode surface, (3) dissociative or other chemical processes prior to electron transfer, and (4) the actual transfer of electronic charge across the electrode–solution interface. The electrode current is limited by any one particular step that is significantly slower than all the other steps. Hence, characterization of, and ultimately understanding, the electrode reaction depends on studying the nature of the rate-controlling processes.

The current (i) produced at the RDE under coupled transport–kinetic control is described by the Koutecky–Levich equation:

$$i = \frac{nFADC^b}{1.61D^{1/3}\nu^{1/6}\omega^{-1/2} + D/k_{\text{app}}} \qquad (2.68)$$

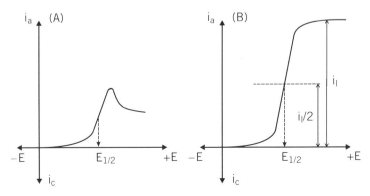

Figure 2.23. Comparison of linear sweep voltammograms at (*A*) stationary and (*B*) rotating-disk electrodes.

where n is the number of electrons (equiv mol^{-1}), F is the Faraday constant (96,496.6 coul equiv^{-1}), A is the geometric surface area of the electrode (cm^2), D is the diffusion coefficient of reactant (cm^2 s^{-1}), ω is the electrode rotational velocity (radians s^{-1}), ν is the kinematic viscosity of the supporting electrolyte (cm^2 s^{-1}), C^b is the bulk concentration of reactant (mol cm^{-3}), and k_{app} is the apparent rate constant of the heterogeneous reaction (cm s^{-1}). As shown in Figure 2.24A, when k_{app} is very large, the electrode response is under *transport* control, and the $i-\omega^{1/2}$ plot is linear with a zero intercept. Negative deviations, expected for large ω values, indicate a finite value of k_{app}. Under this circumstance, the response is stated as being under *mixed* control. For small values of k_{app}, the response becomes independent of ω. Hence, variations in rotation speed (ω), which allows for hydrodynamic control of the electrochemical system, lead to investigation of rate-limiting transport processes. By inverting Equation (2.68) as follows

$$\frac{1}{i} = \frac{1}{k_{app}nFAC^b} + \frac{1}{0.62nFAD^{2/3}\nu^{-1/6}C^b\omega^{1/2}} \tag{2.69}$$

we find that a plot of i^{-1} versus $\omega^{-1/2}$ yields a slope of $(0.6nFAD^{2/3}\nu^{-1/6}C^b)^{-1}$, and the intercept is proportional to k_{app}^{-1}. In Figure 2.24B, plots of a and b have different slopes but similar intercepts, which suggests that the number of electrons involved in the reaction are significantly different. On the other hand, plots of b and c have the same slope but different intercepts, which suggests that the

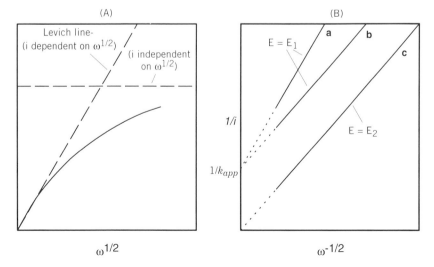

Figure 2.24. The effect of k_{app} on the plots of (A) i versus $\omega^{1/2}$ and (B) $1/i$ versus $1/\omega^{1/2}$.

rate constants of these two reactions are significantly different. In both cases, it is assumed that all the other parameters, including the individual diffusion coefficients, are the same.

The practical limits of RDE rotation speeds are $10 < \omega < 1500$ radians s^{-1}. At the lower ω, the boundary-layer thickness is on the order of the disk radius, and convections from room vibrations and thermal variations become significant. At higher ω, turbulence and air bubbles result, and the laminar flow assumption in the transition layer for the hydrodynamic theory breaks down.

Controlling the hydrodynamics of the redox system gives the analyst inside information to the mass-transport aspects of an electrochemical mechanism. These studies can be combined with scan rate studies to investigate surface-controlled processes. *Surface*-controlled processes correspond to a faradaic reaction of the electrode surface and/or reaction of species accumulated on the electrode surface at potential values when no reaction has occurred. These processes can be examined for the effect of variations in potential scan rate (ϕ). A linear response of $i-\phi$ usually indicates contribution of a surface-controlled process.

ELECTROANALYTICAL METHODS

This chapter has focused on the fundamental tenets of electrochemistry. Many electroanalytical methods exists that utilize the principles put forth in the chapter, and each technique has its own unique advantage. In principle, the wide array of electroanalytical techniques can be simplified to be a study of the effects of potential, current, and time on an electrochemical system. We have covered only those techniques that are relevant to the understanding of pulsed electrochemical detection. LSV and CV at stationary and hydrodynamic electrodes will form the foundation from which to understand how pulsed electrochemical detection works. These techniques focus on qualitatively characterizing electrochemical systems. In the following chapters, our interest will be focused on those electro-analytical techniques that offer the analyst the opportunity to acquire quantitative information.

BIBLIOGRAPHY

Bard, A. J., and L. R. Faulkner, *Electrochemical Methods: Fundamentals and Applications*, Wiley, New York, 1980.

Brett, C. M. A., and A. M. O. Brett, *Electrochemistry: Principles, Methods, and Applications*, Oxford University Press, Oxford, 1993.

Dahmen, E. A. M. F., *Electroanalysis: Theory and Applications in Aqueous and Non-aqueous Media and in Automated Chemical Control*, Elsevier Science Publishers B.V., Amsterdam, 1986.

Kissinger, P. T., and W. R. Heineman, eds., *Laboratory Techniques in Electroanalytical Chemistry*, Marcel Dekker, New York, 1984.

Koryta, J., and J. Dvorak, *Principles of Electrochemistry*, Wiley, Chichester, 1987.

Rieger, P. H., *Electrochemistry*, Prentice-Hall, Englewood Cliffs, NJ, 1987.

Riley, T., and C. Tomlinson, in *Principles of Electroanalytical Methods*, A. M. James, ed., Wiley, New York, 1987.

Sawyer, D. T., A. Sobkowiak, and J. L. Roberts, Jr., *Electrochemistry for Chemists*, Wiley, New York, 1995.

3 Amperometric Detection in HPLC

Let us separate and oxidize, for tomorrow we may pulse.

Inherent to all electroanalytical techniques is the selectivity afforded by the requirement of electroactivity of the compound. If the import of an experiment is to study a particular redox reaction, there is little need for additional selectivity, and many of the electroanalytical techniques, as described in Chapter 2, can be performed in a batch cell. On the other hand, if a mixture of electroactive compounds is to be assayed, some form of selectivity is required to extract quantitative information about the analyte(s) of interest from the sample matrix. Early researchers emphasized the chemistry in electro*chemistry*, and these efforts focused on manipulating the supporting electrolyte, masking agents, electrode material, and voltammetric selectivity. Polarographic methods used advanced waveforms (e.g., differential methods) to enhance analyte selectivity, and these methods worked beautifully for simple mixtures of metals ions and other electroactive compounds. The problem with batch-cell methods originates with the limited potential window in which to analyze an infinite number of compounds. For example, let us assume that a linear scan voltammetric experiment is performed over a 1000-mV range, which is a typical potential window available between dissolved O_2 reduction and the breakdown of water. Under ideal conditions, it would take ~60 mV to develop a voltammetric wave for an analyte with reversible redox chemistry. This means that if all the compounds had evenly spaced half-wave potentials, at most 16 compounds could be resolved. In addition, voltammetric resolution is practically nonexistent for electrocatalytic response; all analytes produce the same $E_{1/2}$. The disadvantage of limited voltammetric selectivity of batch-cell methods is clearly offset by the high sensitivity of electroanalytical techniques. Hence, we are left with a paradox of high sensitivity and poor ability to resolve a mixture of compounds.

Since *necessity* is said to be the *mother of invention*, the impetus to resolve and sensitively quantitate a complex mixture of electroanalytes presented itself in the form of research on the study of aromatic metabolism in the mammalian nervous system. A means to this end was to combine the high resolving power of high-performance liquid chromatography (HPLC) with the high sensitivity of electrochemical detection, and most of the papers published in the years to follow would involve neurochemical analyses. Today, electrochemical detection (ED) in flow-through cells following HPLC is one of the most powerful bioanalytical techniques available. Sometimes electrochemical detection is abbreviated EC, and the combined techniques are often termed simply LCEC.

Electrochemical detection exploits the property of a compound to undergo either oxidation or reduction at an electrode to which a potential has been applied. The rate of the electrochemical reaction is observed as current, and, hence, these techniques fall under the title of *amperometry*, which means literally the measurement of current. The output from an electrochemical detector may be measured in either amperes or coulombs, if the signal is integrated over time. The conversion efficiency, or the percent of analyte converted to product, is typically less than 5% [1]. If all the analyte is oxidized or reduced to product, then the technique is referred to as *coulometry*, and the output is measured in coulombs. Since 100% of the analyte is consumed, Faraday's law can be used to determine the quantity of analyte, which may eliminate the need for a standard to which to compare the analyte's response. In practical terms, any technique with a conversion efficiency of >95% is considered to be coulometric [1]. Techniques that operate at 5–95% conversion efficiency are considered to be quasi-amperometric.

When charged analytes pass through an electric field between two electrodes in an electrochemical cell, anions will migrate toward the positive electrode and cations will migrate toward the negative electrode. By implementing a rapidly oscillating electric field (i.e., >1 kHz) in an electrochemical cell, the mobility of the ions in solution, or conductivity, can be measured with great accuracy. *Conductivity detectors* (CDs) are based on these principles, and the units of conductivity are *siemens* (S). It is important to note that no oxidation or reduction reactions occur with conductivity detection. There are no electrons transferred to and from the analyte, but charging and discharging of the cell electrodes does occur and an alternating-current (ac) current is present. The focus of this book is current-based electrochemical detection techniques that involve the analyte, and CD will not be covered.

FUNDAMENTAL PRINCIPLES OF AMPEROMETRY

In HPLC–ED, the mobile phase of the chromatography system acts as the supporting electrolyte as it flows into the electrochemical cell. The electrochemical cell is the transducer between the potentiostat and the chromatography system. A three-electrode potentiostat is required for precise control of the applied potential. The transducer confines the working, auxiliary, and reference electrodes into a small volume, which is typically <20 µL [1]. Hence, the detection cell volume is insignificantly small in comparison to the peak volumes (~1 mL) in normal-bore chromatography. As an eluent band passes over the working electrode, it serves as an electron sink to either accept or donate electrons should the analyte be oxidized or reduced, respectively. The working electrode is part of an electrical circuit that amplifies and measures the flow of these electrons (i.e., current). The resulting current is proportional to the concentration of the analyte as it passes through the cell, which results in a chromatographic peak on a suitable output device. Figure 3.1 illustrates the concept of electrochemical detection following HPLC.

Since the supporting electrolyte, or the mobile phase, is in motion over the electrode surface, mass transfer is assisted by forced convection. Hence, all

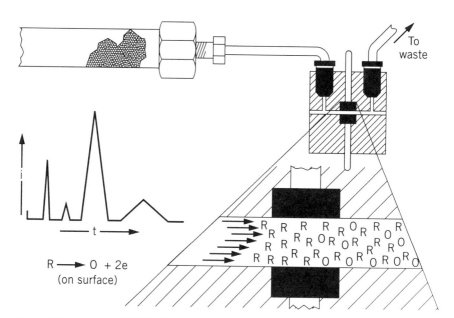

Figure 3.1. Concept of HPLC–ED. The eluant from an HPLC column passes into the electrochemical cell. Analyte in the reduced form is then oxidized, or vice versa, and the resulting current is monitored versus time. The separated compounds are observed as peaks where peak height is proportional to concentration.

amperometric techniques applied for HPLC detection can be classified as hydrodynamic voltammetric techniques. The simplest potential–time waveform is that of applying a constant potential as is shown in Figure 3.2. An amperometric technique that applies a constant potential is called *dc amperometry*, and it is the one most widely used in chromatographic applications. Since the electrode potential is held constant, the current is recorded with respect to time, which makes it a special case of *chronoamperometry*. The applied potential in amperometric detection is selected to operate the electrochemical cell in the steady-state limiting current region. Theoretically, for the LSV shown in Figure 3.3, any potential greater than E_1 could be used in the amperometric detection of the compound. As we shall discuss later, optimal detection will occur at E_2. Under these conditions, any analyte that passes through the cell is converted to product. As discussed in Chapter 2, in hydrodynamic voltammetry only the forward reaction of the redox reaction is observed, and i_{lim} attains a time-independent value that is proportional to the concentration. Since the chromatographic system provides the selectivity, voltammetric selectivity in electrochemical systems becomes irrelevant, although the limited voltammetric selectivity can be used to discriminate between analytes and characterize their responses.

ELECTROLYSIS CELL DESIGNS

At the heart of electrochemical detection in HPLC is the transducer. This transducer is typically an electrolysis cell in which the eluant stream passes either over, at, or through the working electrode. In the electrolysis cell, nonspontaneous electrochemical reactions are forced to take place by applying an external potential to the working electrode. Over the years many different cell designs have been described in the scientific literature, and four major configurations

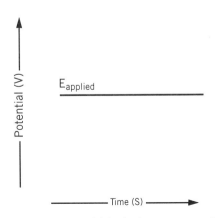

Figure 3.2. DC amperometry waveform which is simply a constant applied potential versus time.

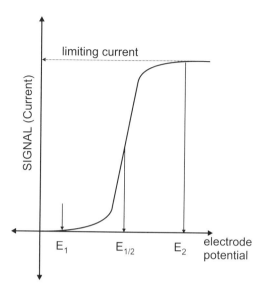

Figure 3.3. Linear sweep voltammogram of a hypothetical compound at a hydrodynamic electrode. E_1 and E_2 are possible applied potentials.

predominate flow-through electrochemical cell designs. These designs include the thin-layer, wall-jet, tubular, and porous electrode cells. Figure 3.4 shows a schematic representation of each of these cell designs. Note that all configurations utilize a three-electrode electrolysis cell. The major difference between each of the cells is centered about the working electrode. Of these cell designs, all can be used in the amperometric mode, whereas the porous electrode is most commonly used in the coulometric mode.

Thin-Layer Cells

Probably the most popular electrochemical detector cells are thin-layer cells, in which a working electrode is embedded in a wall of a channel defined by a gasket sandwiched between two blocks (see Fig. 3.4A). The working electrode is typically in the form of a disk. The thin-layer cell permits the electrochemistry to proceed while the supporting electrolyte and analyte pass over the working electrode in a flowing stream. Laminar flow of the supporting electrolyte is established as the eluant passes over the electrode. For the thin-layer cell design, the limiting current (i_{lim}) in the cell can be determined from the following equation [2]:

$$i_{lim} = 1.47nFC^bD^{2/3}\left(\frac{A}{h}\right)^{2/3}V_f^{1/3} \qquad (3.1)$$

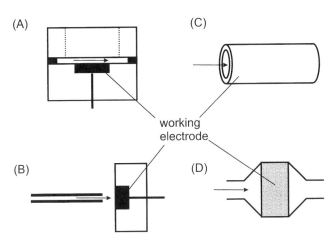

Figure 3.4. Most common flow-through cells are based on either (*A*) thin-layer, (*B*) wall-jet, (*C*) tubular, or (*D*) porous electrode configurations. The arrows denote direction of fluid flow.

where h is height of the channel (cm), A is the electrode area (cm)^2, V_f is the volume flow rate of solution (cm^3 s^{-1}), and each of the other parameters have the usual significance. Often the limiting current equation is given in terms of linear velocity (v) across the electrode surface, which is directly proportional to V_f. Cells are defined on the basis of their flow characteristics, and, mathematically, an amperometric determination is one where the current is dependent on the cube root of v. A coulometric response is linear with v, and everything in between is considered to be quasi-amperometric. Depending on how the thin-layer cell is configured and operated, it can also be operated in the coulometric mode.

The iR drop (ohmic resistance drop), or loss of voltage across the cell due to the cell's inherent resistance, which is commonly associated with thin-layer amperometry, is not a significant problem in HPLC–ED because the mobile phases are of high ionic strength. The iR drop is electronically compensated for by using positive feedback. The problem of iR drop cannot be completely overcome because there is always some distance between the auxiliary electrode and the working electrode. The current between the auxiliary and working electrodes passes along the thin-layer channel, including that portion adjacent to the electrode. Therefore the potential is not uniform across the surface of the electrode. The problem is usually minimized by placing the auxiliary electrode directly across from the working electrode. This geometry ensures that the uncompensated iR drop will be extremely small and that the potential is uniform across the electrode surface. Numerous other arrangements of the working, auxiliary, and reference electrodes have been investigated for the thin-layer cell, but there does not appear to be an optimum cell design or geometry that will hold up for all applications.

The thin-layer cell configuration is probably most common of all designs. It offers the advantages of being well characterized in the scientific literature, the thin layer over the electrode is easily established with a gasket, and the working electrode configuration is amenable to mechanical reactivation, or polishing.

Wall-Jet Cell

The wall-jet design (Fig. 3.4B) places the working electrode opposite the inlet of the fluid flow, and the working electrode is subjected to a high-velocity fluid jet. The layer of the fluid over the electrode is very turbulent, which results in a narrow hydrodynamic diffusion layer due to more effective mass transfer. The distance between the jet and the electrode, the size of the jet, and the radius of the electrode are all crucial to the characterization of the this type of electrode. In fact, the jet impinging on the electrode results in four distinct regions, one of which is the wall-jet region (Fig. 3.5). In a properly configured wall-jet cell, the limiting current in the cell is determined from the following equation [2]:

$$i_{lim} = 0.898nFC^bD^{2/3}v^{-5/12}a^{-1/2}A^{3/8}V_f^{3/4} \tag{3.2}$$

where v is the kinematic viscosity of the solution ($cm^2 \cdot s^{-1}$), a is the diameter of the impinging jet (cm), and A is the area of the electrode (cm^2).

As with the thin-layer configuration, the wall-jet configuration can utilize almost any electrode material, and the cell is easily dismantled for the purposes of mechanically reactivating the electrode. For several reasons the wall-jet configuration should be superior in performance to the thin-layer and tubular designs. These reasons include (1) superior rates of mass transfer, (2) smaller effective dead volume of the electrochemical cell, and (3) highest S/N when the electrode area is small. The last two reasons are critical when mating electrochemical detection with microchromatography systems. Even though the wall-jet configuration should be superior to other designs, the thin-layer cell routinely out performs the wall-jet in practice [1]. This may be a result of the critical nature of electrode placement.

The wall-jet configuration does appear to be less prone to electrode fouling, which manifests itself in less frequent polishing of the electrode. This observation may be because δ_H is smaller at wall-jet electrodes than at conventional hydrodynamic electrodes, and the rate of diffusion away from the electrode is faster than for conventional electrodes. Therefore, the surface residence time for intermediate and final detection products is decreased, and the rate of fouling is decreased. This ability to reduce fouling of the electrode surface may be especially advantageous in pulsed electrochemical detection, where electrode fouling is extensive. In fact, much of the work done by the Johnson group and the LaCourse group employ "homemade" modified wall-jet electrochemical cells. The one used by our group is shown in Figure 3.6. This cell has been used and tested with normal-bore, microbore, nanobore, and capillary liquid chromatography.

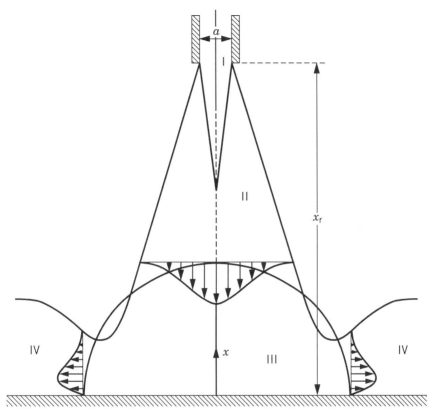

Figure 3.5. Jet region of a wall-jet cell design. Mass transfer at an impinging jet electrode: (I) central core potential region; (II) established flow region; (III) stagnation region; and (IV) wall-jet region. Reprinted with permission from D. T. Chin and C. H. Tsang, *J. Electrochem. Soc.* **125**, 1461–1470 (1978). Copyright 1978 Electrochemical Society.

Tubular Cells

Flow-through electrochemical cells based on a cylindrical geometry have been used for many years. Tubular cells can be configured so that the working electrode is an open tube or a wire electrode that is placed within the confines of a tube. Applications of the wire electrode configuration are on the rise because of its compatibility with the techniques of capillary liquid chromatography (CLC) and capillary electrophoresis (CE). In comparison to other cell designs, the mass-transfer equations for the cylindrical geometry are relatively easy to solve. In the tubular electrode cell (Fig. 3.4*C*), the fluid flow passes through a tubular electrode, in which laminar flow of the fluid is established. The limiting current in the cell is determined from the following equation [2]:

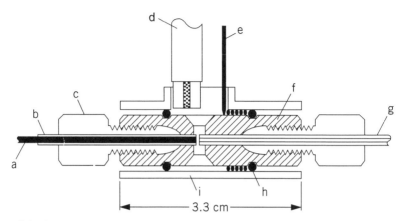

Figure 3.6. Schematic of a gold wire flow-through electrochemical cell: (*a*) Au wire, 1 mm diameter; (*b*) PTFE (polytetrafluoroethylene) tubing, $\frac{1}{16}$ in. outer diameter (o.d.); (*c*) tube-end fitting with ferrule; (*d*) reference electrode; (*e*) Pt wire auxiliary electrode, 0.25 mm; (*f*) Kel-f tubing coupler; (*g*) PEEK (polyetheretherketone) tubing, $\frac{1}{16}$ in. o.d., $\frac{5}{1000}$ in. inner diameter (i.d.); (*h*) Viton O-rings; (*i*) polyethylene casing. Reprinted from W. R. LaCourse and G. S. Owens, Pulsed electrochemical detection of thiocompounds following microchromatographic separations, *Anal. Chim. Acta.* **307**, 301–319 (1995) with kind permission of Elsevier Science–NL, Sara Burgerhartstraat 25, 1055 KV Amsterdam, the Netherlands.

$$i_{\lim} = 1.61 nFC^b D^{2/3} \left(\frac{A}{r}\right)^{2/3} V_f^{1/3} \tag{3.3}$$

where *r* is the radius of the tubular electrode (cm). Note that the current is proportional to the cube root of V_f and, consequently, υ, which classifies it as an amperometric detector.

The tubular design has several major drawbacks that have limited its overall analytical usefulness. First, the design of the electrode is discordant to the placement of the auxiliary and references electrodes to minimize uncompensated cell resistance, and large electrode surface area:volume ratio designs cannot be implemented without significant *iR* drop. Second, several of the commonly used electrode materials (e.g., Hg and carbon paste) cannot be easily used in the form of a tube or wire. Third, and possibly the most deleterious to their common use, is that tubular electrodes are not easy to polish in comparison to thin-layer or wall-jet cells. Since PED continuously cleans and reactivates the electrode surface electrochemically, the last problem may be relatively unimportant when using pulsed techniques.

Porous Electrodes and Other Cell Designs

Porous electrodes, in which the fluid flows through the electrode (Fig. 3.4*D*), are typically used for coulometric determinations. The porous electrode can be in the form of a porous solid (e.g., reticulated vitreous carbon), felts, wire meshes, or packed-bed electrodes. As discussed earlier, the analyte is almost totally reacted as it passes through the cell, and the coulometric efficiency is typically greater than 95%.

These types of electrodes have two major drawbacks. The electrodes are impossible to mechanically clean, and the background noise is typically high because of the large electrode surface area of the porous electrode. The large background signal typically reduces the overall S/N for porous electrode systems, and detection limits are seldom competitive with those achieved using amperometric detection. The near-100% conversion of compounds as they pass through coulometric cells makes them invaluable for electrochemical synthesis and preconditioning (e.g., removal of background interferences) of HPLC mobile phases.

A logical extension of all the polarographic techniques is to use a flow-through dropping-Hg electrode as a transducer for ED in HPLC. Since a new drop forms for each electrochemical experiment, mechanical cleaning of the electrode surface is irrelevant. The problem is not in the construction of the cells or the analytical cell volumes, but the high charging currents invoked from the constantly changing surface area of the Hg drop, and the ever-present concern of mercury toxicity and environmental contamination. Applications in which Hg electrodes offer a competitive advantage (e.g., need for a high hydrogen overpotential) are being replaced with Hg–Au amalgam electrodes.

Recently, Bioanalytical Systems (BAS, West Lafayette, IN) has introduced a new cell which they call the Uni-Jet electrode [3]. The Uni-Jet appears to be a cross between a wall-jet cell and a coulometric cell. The cell offers significant advantages for microchromatographic applications in that it provides superior response characteristics at flow rates of <200 μL/min. The thin layer radial flow pattern allows for increased reaction times and greater conversion efficiency of analytes.

The effective dead volumes of electrochemical cells are typically <10 μL. The volume of the solution in the active region of the electrode is ~1 μL. The commercial ED cell has a linear range of 200 ng–10 pg injected for typical small molecules (~200 MW). This is only a generality since the detector response is dependent on volume, flow rate, electron-transfer rate, chromatographic efficiency, ionic strength, applied potential, temperature, and background current. Under the proper HPLC conditions and electrode surface, the amount of analyte reacted on the order of picomoles has been achieved for a number of oxidizable compounds. The limits are about tenfold less favorable for reductive compounds because problems with dissolved oxygen [1].

OTHER CELL CONSIDERATIONS

The Working Electrode

The electrode surface is the most critical aspect to successful HPLC-ED operation. An ideal working electrode for HPLC has the following characteristics:

(1) a low potential for maximum analyte response, (2) a high potential before the commencement of electrolytic decomposition of the mobile phase, (3) electrochemical inertness of the electrode material, (4) inertness to mobile-phase constituents, (5) the analyte cannot diffuse into the electrode, and (6) it can be highly polished or mechanically reactivated. Electrodes that are in common use are glassy carbon, carbon paste, platinum, gold, gold/mercury amalgam, and silver.

Carbon-based electrodes are manufactured in a wide array of forms. One of the earliest carbon-based electrodes, and still a popular choice, is the carbon paste electrode, which is a mixture of graphite powder and a binder. The binder is typically a high-molecular-weight wax or a synthetic polymer. Unfortunately, these electrodes cannot be used under nonaqueous conditions. Carbon is available in other forms as well, including pyrolytic graphite, reticulated vitreous carbon, glassy carbon, and carbon fibers. The most versatile choice is glassy carbon. It is highly inert to nearly all solvents used in liquid chromatography, and it has a wide potential range. Many other carbon electrodes have been used for electrochemical detection, but they are not as common as those mentioned previously.

Mercury is better suited for reductive work in that it has an extended negative range and a limited positive range. The extended negative range is a result of the very high overpotential needed for H_2 evolution on such electrodes. In the presence of anions that form insoluble salts with Hg(I) (e.g., Cl^- and I^-), mercury oxidation will occur at potential less positive than approximately $+0.4$ v (vs. SCE), and this limits its positive range. Mercury is therefore better than glassy carbon when dealing with compounds that are difficult to reduce. In dc amperometry, amalgamated gold electrodes are popular in thin-layer cell applications.

Metals other than Hg are used in dc amperometry, but much less frequently. Platinum is relatively inert, and it can be used for the determination of chlorite, sulfite, hydrazine, H_2O_2, and other compounds. On the other hand, Ag electrodes are directly involved in the electrode reaction and can be used to detect complex or precipitate-forming ions. Hence, silver is used as a sacrificial electrode for cyanide, sulfide, and halide determinations. A sacrificial electrode is one where the electrode undergoes the redox reaction, and they are consumed in the process. For example, as Cl^- passes over the electrode, the redox reaction is as follows:

$$Ag^0 \text{ (the electrode)} + Cl^- \longrightarrow AgCl(s) + e^- \qquad (3.4)$$

Determination of analytes using sacrificial electrodes usually are more temperamental to the conditions of the experiments and exhibit poor reproducibility, not to mention that your electrode disappears with use. From that point of view, consider the fact that gold electrodes dissolve quite rapidly in the presence of chloride ions under applied potentials sufficient to generate surface oxide. Hence, gold electrodes, which are less inert than platinum, are almost never used in dc amperometry.

Electron-Transfer Kinetics

An electrode functions to accept or donate electrons to or from the analyte in solution. The rate of heterogeneous electron transfer at the electrode surface can be a limiting step in the overall detection process. For a kinetically fast electron-transfer reaction of an oxidation process, the electrode potential is usually set at least 120 mV more positive than the $E°$ of the reaction, where i_{lim} has been reached. At this potential, virtually all of the analyte that reaches the electrode surface is oxidized, and a quantitative determination can be done. Increasing the potential beyond this point results in no additional current from the analyte, but it may increase background current, which is detrimental to obtaining the highest S/N. For a reaction with slow electron-transfer kinetics, an even greater potential beyond the $E°$, or overpotential, must be applied to drive the reaction at a faster rate. These redox reactions are irreversible, and the more irreversible the reaction, the greater the overpotential that is required. The higher applied potential results in increased background noise and restricted voltammetric selectivity. Since organic electron-transfer reactions at surfaces are not well understood at a molecular level, reactions that may be fast and reversible at one electrode surface may be unfavorable at another. For instance, inorganic ions are typically more readily oxidized or reduced at a Pt electrode than at a glassy-carbon electrode.

In choosing an electrode in dc amperometry, it is desirable for the redox reaction to be as reversible as possible, and, additionally, it is desirable for the electrode material to remain active and resist fouling. Hence, the electrode should be inert. Glassy-carbon electrodes are more resistant to fouling than are other solid anode materials, and, as a consequence, glassy-carbon electrodes are the most popular electrode material in dc amperometry.

Mobile-Phase Limitations

Aside from its role in the separation process, the mobile phase is also the supporting electrolyte in the electrochemical detection system. The primary requirement of any mobile phase for use in ED is that it must have the ability to carry an ionic current. In other words, ions must be present. Since electrochemical detection involves complex surface reactions that are dependent on both the physical and chemical properties of the electrolyte, limitations are placed on the mobile-phase composition. Fortunately, ED is compatible with virtually all water-based chromatographic separations, including reversed-phase, ion-pair, and ion-exchange chromatography. The majority of HPLC separations are performed using these methods. Retention is determined by altering the organic modifier concentration, the pH, the ionic strength, or the presence of ion-pairing reagents.

In reversed-phase chromatography, the mobile phase is typically a mixture of highly purified water and an organic modifier. Alcohols (e.g., methanol, ethanol, isopropanol) and acetonitrile are the most commonly used solvents. The use of methanol can sometimes limit the positive potential limit of ED. Tetrahydrofuran

(THF) is used, but one must be aware of contamination by peroxides, which results from its prolonged exposure to air. Amine-based modifiers and salts should also be avoided. In general, normal or nonpolar phases are not compatible because of their inability to support an ionic strength sufficiently high to satisfy conductivity requirements of the electrochemical cell.

Control of pH is usually accomplished via the use of buffers and inorganic acids and bases. Buffered solutions are usually of high ionic strength (0.01–0.1 M) to provide adequate conductivity without adding unacceptable levels of background noise from reagent impurities. Some common buffer salts include phosphate, acetate, and citrate. Ion-pairing agents often are alkyl sulfonates as anionic agents and quaternary ammonium salts as cationic agents. Reagent impurities are especially a problem for reductive HPLC–ED. A major source of baseline noise is that of metal ions present as a result of the manufacturing process. The use of the highest-quality reagents and the purest solvents are the easy approach to overcome contaminant problems.

The mobile-phase composition, pH, and electrode material strongly influence the potential window available to perform electrochemical detection. Figure 3.7 shows an LSV for a hypothetical mobile phase under degassed (⸺⸺⸺) and

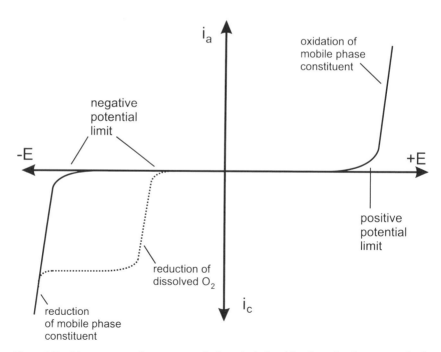

Figure 3.7. Linear sweep voltammogram of a hypothetical mobile phase showing commonly observed regions.

nondegassed (............) conditions. The negative and positive applied potential limits of the degassed mobile phase are where the mobile phase or supporting electrolyte is reduced or oxidized, respectively. Table 3.1 lists the potential limits for some common electrode materials under a variety of pH conditions. The overall potential window is shifted to more negative potentials under alkaline conditions and more positive potentials under acidic conditions. As the applied potential approaches either potential limit, background current and noise will increase, and, as a consequence, the S/N will be severely attenuated. If not removed, the background current produced from the reduction of dissolved oxygen may swamp the detector even at low potentials (see Fig. 3.7). The reduction of dissolved O_2 is as follows:

$$O_2 + 2H_2O + 4e^- \rightleftharpoons 4OH^- \quad \text{(alkaline conditions)} \quad (3.5)$$

$$O_2 + 4H^+ + 4e^- \rightleftharpoons 2H_2O \quad \text{(acidic conditions)} \quad (3.6)$$

Obviously, cathodic (reductive) detections are hindered by the deleterious effects of dissolved O_2 in samples and the mobile phase, which results in a cathodic signal. It is for this reason that anodic (oxidative) detection continues to predominate in HPLC–ED.

Electrochemical detection can be performed using totally nonaqueous mobile phases [e.g., acetonitrile, DMF (dimethylformamide), and DMSO (dimethylsulfoxide)] if the mobile phases contain ionic carriers. Typically, tetrabutylammonium hexafluorophosphate or tetrafluoroborate salts are used. As discussed earlier, the potential limits of a mobile phase are determined by the solvent breakdown, which is typically that of water in an aqueous system. Nonaqueous solvents are typically more resistant to oxidation and reduction, and, as a consequence, nonaqueous conditions offer the analyst a wider potential window to detect analytes that are very difficult to oxidize or reduce (e.g., alcohols and

TABLE 3.1 Useful Potential Limits for Common Electrodes under Various pH Conditions

Electrode	Supporting Electrolyte	Negative Limit (V)	Positive Limit (V)	ΔV
Pt	1 M H_2SO_4	−0.3	+1.2	1.5
	pH 7 buffer	−0.7	+1.0	1.7
	1 M NaOH	−1.0	+0.6	1.6
Au	1 M NaOH	−0.9	+0.8	1.7
Hg	1 M H_2SO_4	−1.1	+0.3	1.4
	1 M KCl	−1.9	+0.1	2.0
	1 M NaOH	−2.0	−0.1	2.1
C	1 M $HClO_4$	−0.2	+1.5	1.7
	0.1 M KCl	−1.3	+1.0	2.3

Source: Adapted from data contained within A. J. Bard and L. R. Faulkner, *Fundamentals and Applications*, Wiley, New York, 1980.

TABLE 3.2 Useful Potential Limits of Nonaqueous Systems at a Pt Electrode

Solvent	Electrolyte (0.1 M)[a]	Negative Limit (V)	Positive Limit (V)	ΔV
Acetonitrile	TBABF$_4$	-2.5	$+2.5$	5.0
Propylene carbonate	TEAP	-2.5	$+2.2$	4.7
Tetrahydrofuran	TBAP	-3.1	$+1.4$	4.5
Dimethylformamide	TBAP	-2.8	$+1.5$	4.3
Benzonitrile	TBABF$_4$	-1.8	$+2.5$	4.3
Methylene chloride	TBAP	-1.7	$+1.8$	3.5
Sulfur dioxide	TBAP	±0.0	$+3.5$	3.5
Ammonia	KI	-3.0	$+0.1$	3.1

[a]*Abbreviations:* TBABF$_4$—tetra-*n*-butylammonium fluoroborate; TBAP—tetra-*n*-butylammonium perchlorate; TEAP—tetraetylammonium percholrate; KI—potassium iodide.

Source: Adapted from data contained within A. J. Bard and L. R. Faulkner, *Fundamentals and Applications*, Wiley, New York, 1980.

aromatic hydrocarbons). Table 3.2 lists the potential windows available in various solvents. In addition to the difficulty associated with eliminating water from the mobile phase (<1 ppm), the drawbacks of nonaqueous electrochemical detection include higher cell resistance and background noise.

THE ORIGINS OF CURRENT

Current is an instantaneous quantification of the rate of an electrochemical process, which is dependent on how many electroactive species strike the electrode surface per unit time. Factors that influence the current signal include electrode surface area, reactant concentration, temperature, viscosity, velocity of solution, and the applied potential. The instantaneous current, which is measured as peak height in a chromatogram, is proportional to the instantaneous concentration of the detected species.

In HPLC–ED, the total current is composed of faradaic currents (i_F) for the sample and background and nonfaradaic currents as charging currents:

$$i_{total} = i_{F(analyte)} + i_{F(background)} + i_{NF(capacitive)} \qquad (3.7)$$

The faradaic background currents are derived from electrolysis of impurities, solvent, and the electrode material. All except the analyte current are undesirable, and their impact should be minimized. This is accomplished by subtracting the background response from the analyte response, which is essentially the determination of peak height in a chromatogram. The electrode–solution interface, or double layer, behaves like an electrical capacitor. In other words, it can store charge. The equation for the amount of charge (Q) stored is given below;

$$Q = CV \qquad (3.8)$$

$$i_{NF(capacitive)} = \frac{dQ}{dt} = C\frac{dV}{dt} \qquad (3.9)$$

where C is the capacitance in farads and V is the potential. The differential form of this equation shows that if the potential is changed over time, a current is observed. Fortunately, dc amperometry operates at a fixed potential, and $dV/dt = 0$. Hence, the charging current is negligible in HPLC–ED.

The limit of detection of a compound is ultimately determined by the ability of the detector to discriminate between the signal for the analyte of interest and the background signal, or noise. In dc amperometry, the analyte signal is mass-transport-limited, and the onus to increase S/N depends on the minimization of the noise component of the overall response. In addition to the optimal applied potential, the use of noise-free electronics, pulse-free pumping systems, high-purity reagents and solvents, inert electrodes and solvents, and temperature control is highly recommended for trace analysis.

OPTIMIZATION OF APPLIED POTENTIAL

The detection process is based on controlled potential amperometry. A predetermined potential is applied between the reference and working electrodes. The chosen potential is dependent on the redox behavior of the analyte, and it is the driving force of the reaction. The applied potential can be viewed as equivalent to electron pressure. The equilibrium of Reaction (3.10) can be disturbed by the stress placed on the system from the addition or removal of electrons; this effect is a simple application of LeChatelier's principle:

$$oxd + ne^- \rightleftharpoons red \qquad (3.10)$$

$$A^- + H^+ \rightleftharpoons HA \qquad (3.11)$$

This concept should be very familiar in that the equilibrium of a simple acid–base reaction is disturbed by the addition or removal of protons from the reaction. The effect of proton pressure in relation to a reference condition is given in the Henderson–Hasselbach equation [Eq. (3.12)]:

$$pH = pK_a - \log\frac{[HA]}{[A^-]} \qquad (3.12)$$

$$E = E° - \frac{0.059}{n}\log\frac{[red]}{[oxd]} \qquad (3.13)$$

Note the similarity of the Henderson–Hasselbach equation to the Nernst equation [Eq. (3.13)]. In both cases there is a unique situation in which the concentration of each form (oxd vs. red or A^- vs. HA) becomes equal. The electron pressure at which this occurs in one case is $E°$ and, in the other case, the pK_a. These equations provide a convenient index of the relative strengths of oxidant–reductant or acid–base. When the pH $<$ pK_a, there is a higher proton pressure, and the protonated form of the acid is favored. Along the same lines, if the electron pressure is higher than $E°$, the redox couple will favor the reduced form.

Hence, the electrode can be viewed as one of the half-reactions in a redox reaction, and the "half-reaction" denoted by the electrode is of controllable oxidizing–reducing strength. If the potential of the working electrode relative to the reference electrode becomes more negative, the surface of the electrode acts as a better reducing agent, and as the electrode becomes more positive relative to the reference, the electrode surface acts as a better oxidizing agent. In dc amperometry, a constant potential is applied to an electrode over time. Selection of the optimal applied potential can be accomplished by one of two methods.

The first method uses a rotating-disk electrode to simulate the hydrodynamics of the flowing eluant stream over a fixed electrode, and linear sweep voltammetry to scan a specified range of applied potentials. Scans are performed for the supporting electrolyte–mobile phase and the analyte. Figure 3.8A shows the hydrodynamic voltammogram (HDV), or $i–E$ plot, for a hypothetical compound. The large cathodic current at negative potentials is due to the reduction of solvent (e.g., water). Note that the background current begins to increase moderately at first and, as the applied potential increases, the response increases exponentially. The background current is attributable to oxidation of major mobile-phase constituents and solvent–reagent impurities. As discussed above, the potential limits of the experiment are determined by the breakdown of major mobile-phase constituents in the mobile phase. The dashed line indicates the response for a hypothetical analyte. If dc amperometry is performed at E_1, no response will be seen. At E_2, the applied potential is equivalent to the $E_{1/2}$ of the redox reaction. Although amperometric detection can be performed at this potential, any fluctuation in potential will dramatically affect the current, and, as a consequence, reproducibility will suffer. At E_3, the detector is operating at the minimum potential for mass-transport-limited current for the analyte. In contrast to E_2, the limiting current obtained at E_3 is insensitive to small variations in the applied potential. In addition, the background current is minimized by operating at the lowest possible potential to achieve I_{lim} for the analyte. Hence, E_3 exhibits the highest S/N, and it is therefore the optimum potential. The use of E_4 would not increase the signal of the analyte, but the background current and noise would be significantly higher. Thus S/N would be lower at E_4. Since the LSV method is run in a batch cell, this method is rapid and requires very little material. The effect of capacitance currents due to scanning of the applied potential is negligible since both the background and the analyte are scanned over the same range, and the associated charging currents are eliminated on subtraction of the two data files. Since the current is continuously sampled, no time parameters need be optimized. Hence, we are dealing with the optimization of a single detection variable.

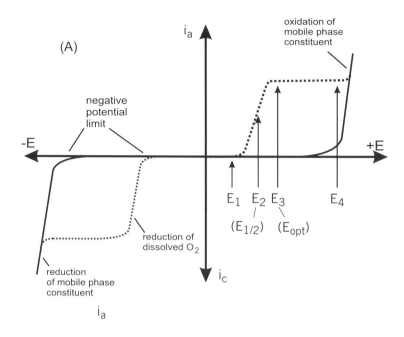

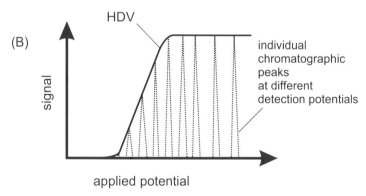

Figure 3.8. Linear sweep voltammogram of a (*A*) hypothetical compound in mobile phase (see Fig. 3.7) using a batch cell and (*B*) a hydrodymanic voltammogram derived via repetitive injections of analyte.

The second method uses the HPLC–ED system to make repetitive injections of the analyte while incrementally changing the detection potential. Hence, a hydrodynamic voltammogram is produced from which the optimal detection potential can be educed. A current to potential plot similar to that of the HDV is produced (Fig. 3.8*B*), and the optimal potential can be determined as explained above. Because of the repetitive nature of this method, it can be both time- and material-intensive.

A major advantage of the repetitive injection method is that several HDVs can be generated simultaneously by injecting a mixture of well-resolved standards. For a single analyte, the only goal for optimizing E_{app} is the sensitivity. For multiple analytes, selectivity can also be an important criterion.

The selectivity afforded by the chromatographic column is augmented by the electrochemical detector. Although a "universal detector" is useful for simple problems, a "selective detector" is more useful for complex samples in real-world matrices, and this is where the strength of ED lies. Other than the selectivity that can be gained from varying the electrode material, the ED is a tunable device that permits selectivity to be adjusted by varying the applied potential.

Figure 3.9 shows background-corrected HDVs for three analytes (i.e., *a*, *b*, and *c*), where compounds *a* and *c* are the analytes of interest and *b* is an interferant. The chromatograms derived from using the same applied potentials as denoted on the HDVs are also shown. The applied potential E_1 shows the greatest degree of selectivity, and for the most part, only compound *a* responds under these conditions. Since the response is not at the current-limited plateau, it is not operating at maximum S/N. Unfortunately, we do not see any response for analyte *c*. By increasing the applied potential to E_2, compound *a* is at its optimal for sensitivity, but compound *b* is now responsive, but it does not interfere with the quantitation of *a*, as noted the corresponding chromatogram for E_2. At E_3, all three peaks are observed in the chromatogram. The peak height of *a* is constant, peak *b* has become more significant, and peak *c* is electroactive at this potential. All the peaks are quantifiable, even though peak *c* is at less than optimal S/N. At E_4, all analytes are at their mass-transport-limited currents. Even though the analytes of interest are at their maximum sensitivity, the response of the interferant *b* overwhelms the chromatogram, and quantitation of *a* is deleteriously affected. Hence, for partial selectivity to discriminate against the interferant, it is best to use E_3 as the applied potential. Hence, for our theoretical compounds, a compromise has to be made between the lower potential, which gives more selectivity and less sensitivity; and the higher potential, which gives less selectivity and higher sensitivity.

In general, increasing the working electrode potential results in the following effects: (1) the response for the analyte of interest increases; (2) the background current/noise increases, resulting in decreased baseline stability; (3) at some point, the response for the analyte of interest reaches its mass-transport-limited value, which increases its stability of response; and (4) decreasing selectivity. Since the number of electroactive substances increases with increasing potential, the rule of thumb for optimal selectivity is to use the lowest working electrode potential possible for the sensitivity required.

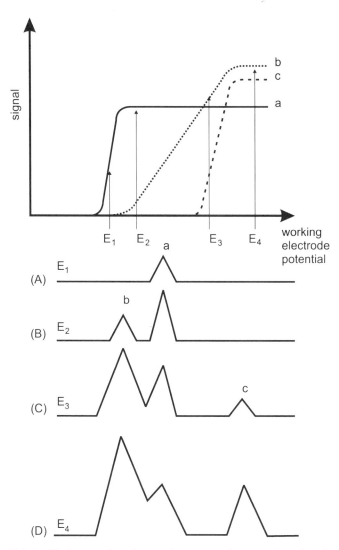

Figure 3.9. Relationship between the voltammetric response of compounds *a*, *b*, and *c* and the applied potential on selectively in HPLC.

SELECTIVITY: MULTIPLE ELECTRODES

Selectivity in HPLC-ED is accomplished via the chromatographic separation, the inherent voltammetric properties of electroactive species (discussed above), and the use of multiple electrodes in the electrochemical cell. This multiple-electrode approach has been exploited, and enjoyed much success, using thin-layer cell designs. Although many electrodes can be used, the majority of multiple-electrode applications have been performed with just two electrodes in either a parallel (Fig. 3.10A) or series (Fig. 3.10B) arrangement.

In the parallel arrangement, the chromatographic band flows over each of the electrodes independently. Thus, the response of one electrode is virtually unaffected by the response of the other. Hence, two potentials can be monitored simultaneously. One electrode can be set for high selectivity (low potential), and the other electrode can be set for high sensitivity (high potential). This arrangement functions in an analogous manner to that of a dual-wavelength uv absorption detector. In dual-wavelength uv absorption, the ratio of the response of the two wavelengths (Fig. 3.11A) can be outputted to give qualitative information about the analyte of interest. For instance, if a chromatographic peak is pure, the response ratio should be constant (see Fig. 3.11B). In addition, the ratio will be characteristic of that compound. In ED, if the potentials are selected between the onset and mass-transport-limited current of the reaction, a similar experiment can be performed to determine peak purity (see Fig. 3.11C), and give qualitative information about the analyte (Fig. 3.11D). The ratio can theoretically correspond to the slope of the voltammetric wave.

Figure 3.10B shows the series electrode arrangement. In this arrangement the products of the upstream electrode can be monitored at the downstream electrode. This can be used in a number of different ways to enhance selectivity and quantitation. The electrolysis products of the upstream electrode can be detected in a more favorable potential region at downstream electrode. For instance, a compound can be reduced at the upstream electrode at negative potential, which

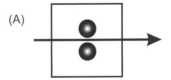

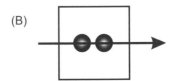

Figure 3.10. Dual electrodes are most commonly arranged either in (A) parallel or (B) series.

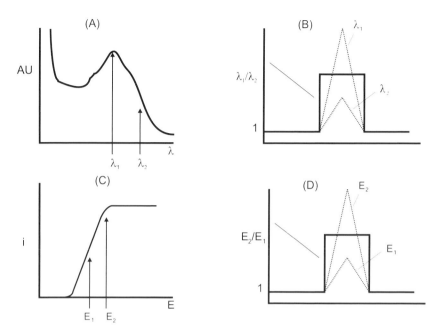

Figure 3.11. Both (*A*) dual-wavelength and (*C*) dual-potential monitoring can be used to determine peak purity (*B,D*) in HPLC.

is prone to interference from dissolved O_2, and the product can be oxidized at a positive potential, which is not affected by dissolved O_2. For example, disulfides are reduced at an upstream Hg/Au amalgam electrode, and then the thiols are determined via oxidation at positive potentials at a positive Hg/Au amalgam electrode. In addition, only redox reactions that are "reversible" can be detected at the downstream electrode and, as a consequence, the series arrangement can be used to discriminate between "reversible" and "irreversible" redox reactions.

By using a hybrid three-electrode cell design that takes advantage of series and parallel electrode arrangements simultaneously, only the thiols in the sample are detected at the parallel electrode, and all the disulfides and thiols are detected at the downstream electrode of the series arrangement. Thus, quantitation, selectivity, and compound identification are achieved with a single injection. In general, hybrid multiple electrode designs can incorporate any of number of electrodes, the electrode material can be varied, and the arrangement of the electrodes is also variable. These designs do not affect the chromatographic integrity of the peak, and they are amenable to multiple methods of applying the potential, such as differential pulse amperometry, offering additional advantages of unique methods of selectivity.

QUANTITATIVE ASPECTS

Figure 3.12 summarizes the response from a HPLC-ED system with respect to both time and potential. A chromatogram could be collected at each potential, and a voltammogram could be determined at each time point. Figure 3.13*A* shows the chromatogram at the optimum potential for peak *b*. The height of peak *b* corresponds to the current, which is proportional to concentration, and the area of peak *b* is the charge, which is also proportional to concentration. Note that the response of the analyte differs from that of the background current, which is inescapable. Often the background is electronically offset to zero, but it should be remembered that it is always present.

The sensitivity of a compound in HPLC–ED is defined in terms of "peak current per injected equivalents" [in nanoamperes per equivalent (nA equiv^{-1})]. The peak area per injected equivalent is another useful measure of sensitivity in that it easily permits the calculation of the conversion efficiency of the cell, which can be used to define the cell as operating amperometrically, quasi-amperometrically, or coulometrically. More familiar expressions include nanoamperes per nanogram or nanomole (nA ng^{-1} or nA nmol^{-1}).

The limit of detection (LOD) is dependent on the sensitivity and the baseline noise, and calculated limits should be supported by experimental values for confirmation. The noise in the baseline of Figure 3.13*A* is evaluated by measuring its peak-to-peak response. The LOD is typically given as that value which is three times the peak-to-peak response of the noise. The analytical aspects of detection are shown in Figure 3.13*B*. The slope of the line is equivalent to the sensitivity, and the LOD is determined typically at S/N = 3. One additional aspect of quantitation is the linear dynamic range (LDR) for the analyte of interest. Essentially, the LDR is defined by the upper limit of linearity and the LOD. Modern-day ED has a characteristically large linear dynamic range of (four to five orders of magnitude). In fact, the LDR is often limited by the dynamic range of the operational amplifiers used in the electronic circuits. LDR is dependent on the cell resistance, which, in turn, is dependent on the conductivity of the mobile phase and the electrode distance. Hence, in order to maintain a large LDR, highly conductive mobile phases should always be used, and the three electrodes should be kept in close proximity of one another.

APPLICATIONS: A BRIEF OVERVIEW

The electroactivity of a compound is dependent on several factors, including molecular structure, accessibility of filled and unfilled orbitals, and functional groups present. Among the electroactive organic compounds that are ideal candidates for ED are phenols, hydroquinones, catechols, aromatic amines, thiols, nitro compounds, and quinones. Some individual compounds that have been determined with good sensitivity are ascorbic acid, uric acid, and NADH (re-

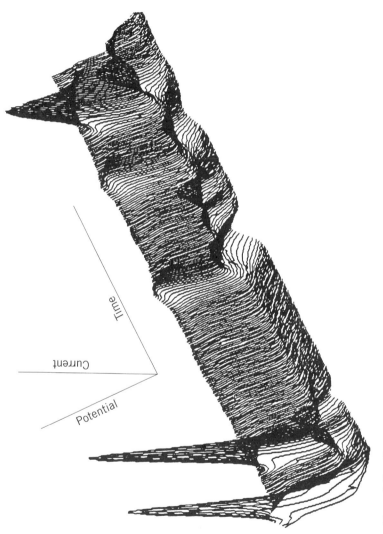

Figure 3.12. Chronovoltammogram showing the relationship between current versus potential versus time in HPLC. Reprinted from J. J. Scanlon, P. A. Flaquer, G. W. Robinson, G. E. O'Brien, and P. E. Sturrock. High-performance liquid chromatography of nitrophenols with a swept-potential electrochemical detector. *Anal. Chim. Acta.* **158**, 169 (1984) with kind permission of Elsevier Science–NL, Sara Burgerhartstraat 25, 1055 KV Amsterdam, the Netherlands.

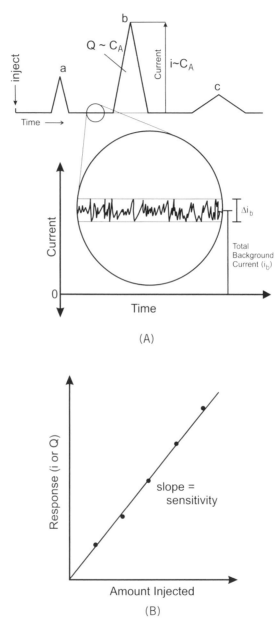

Figure 3.13. Quantitative aspects of a (A) separation of compounds a, b, and c using HPLC–ED. The response of compound b can be measured by either peak height (i.e., current i) or peak area (i.e., charge Q), either of which is proportional to the concentration of the analyte (C_A). A (B) calibration curve shows the relationship between response and the amount injected.

TABLE 3.3   Categories of Electrochemical Applications

Category	Compounds
	Oxidative Applications
Phenols	Hydroquinones, catechols, catecholamines, cresols
Aromatic amines	Phenylenediamines, benzidines, aminophenols
Thiols	Cysteine, glutathione, penicillamine, captopril, biotin
Miscellaneous	Ascorbic acid, uric acid, NADH, and phenothiazines
	Reductive Applications
Nitro Compounds	Nitrobenzene, nitrosamines, aliphatic/aromatic nitro compounds, nitrate esters, 2,4-dinitrophenylhydrazones
Miscellaneous	Azides, disulfides, quinones

duced nicotinamide adenine dinucleotide). Table 3.3 lists many of the compounds most commonly detected separated into various categories. Note the emphasis on neurotransmitters, which are predominantly bioactive aromatic alcohols and amines. As we shall see in Chapter 4, the electroactivity of these compounds is no coincidence. To take advantage of electrochemical detection, derivatization is now being used to make some compounds better candidates for ED. For example, nitrophenyl groups have been added to amino acids, carbonyl compounds, carbohydrates, and carboxylic acids. Up-to-date bibliographies of current LCEC applications and articles can be obtained from any of the instrumentation companies who manufacture electrochemical detectors.

The extent of HPLC–ED applications in the literature is much too broad to cover in any detail. Hence, the author has attempted to generalize these applications into categories of compounds.

REFERENCES

1. P. T. Kissinger, in *Laboratory Techniques in Electroanalytical Chemistry*, P. T. Kissinger and W. R. Heineman, eds., Marcel Dekker, New York, 1984, Chapter 22.
2. J. M. Elbicki, D. M. Morgan, and S. G. Weber, *Anal. Chem.* **56**, 978–985 (1984).
3. C. E. Bohs, M. C. Linhares, and P. T. Kissinger, *Curr. Seps.*, **12**(4), 181–186 (1994).

BIBLIOGRAPHY

Bard, A. J., and L. R. Faulkner, *Electrochemical Methods: Fundamentals and Applications*, Wiley, New York, 1980.

Brett, C. M. A., and A. M. O. Brett, *Electrochemistry: Principles, Methods, and Applications*, Oxford University Press, Oxford, 1993.

Kissinger, P. T., and W. R. Heineman, eds., *Laboratory Techniques in Electroanalytical Chemistry*, Marcel Dekker, New York, 1984.

4 Pulsed Amperometric Detection in HPLC

To Pulse or Not to Pulse? That is the question.

As discussed in Chapter 3, amperometric detection at solid electrodes (e.g., Au, Pt, and glassy carbon) under constant applied potentials can be used for the detection of a wide variety of compounds. A rudimentary examination of electroactive organic compounds quickly reveals that the majority exhibit some degree of aromaticity or conjugation (e.g., phenols, aminophenols, catechols, catecholamines) [1,2]. In contrast, the majority of polar aliphatic organic compounds (e.g., alcohols and amines) are not observed to produce a persistent signal under similar experimental conditions. This observation is more than just an interesting coincidence. Aromatic compounds have greater electroactivity than do polar aliphatic compounds with similar functional groups, due to a beneficial decrease in the activation energy barrier for the anodic mechanism because of stabilization by π-resonance of the free-radical products of one-electron oxidations [1]. For example, the voltammetric response for catechol (_____) at a stationary glassy-carbon electrode shows an anodic wave *a* in the region of approximately +300 to 1400 mV (vs. Ag/AgCl) (see Fig. 4.1). Note that the oxidation of catechol is reversible, and a cathodic wave *b* is observed at approximately +150 mV (vs. Ag/AgCl) for the reduction of the products formed during the anodic scan. Under the same experimental conditions, 1,2-cyclohexanediol (--------) doesn't show any significant electrochemical activity. Note that there is little or no difference

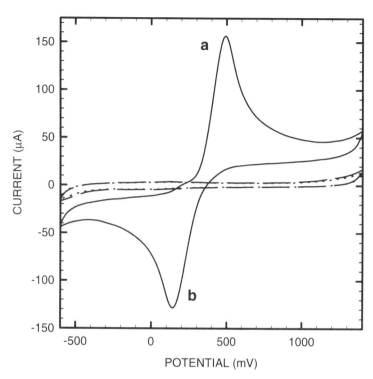

Figure 4.1. Cyclic voltammograms of catechol, 5 mM (——) and 1,2-cyclohexandiol, 5 mM (--------) in 100 mM acetate, pH 4.75 (.............), at a stationary glassy-carbon electrode. Scan rate, 250 mV s⁻¹.

between the response for 1,2-cyclohexandiol and the residual (.............). Further evidence illustrating the importance of conjugation to electroactivity is denoted in the cyclic voltammogram of caffeic acid (Fig. 4.2), which shows an anodic wave that peaks at approximately +300 mV (vs. Ag/AgCl). The anodic wave for caffeic acid is >100 mV less positive than the anodic wave of catechol. By extending the degree of conjugation, caffeic acid is more easily oxidized than catechol. Hence, aromatic compounds are self-stabilized via π-resonance, and the inert electrode need only act as an electron sink to accept (oxidation) or donate (reduction) electrons. This argument supports the heavy reliance of dc amperometry on glassy-carbon electrodes. In general, glassy-carbon electrodes do not strongly adsorb organic compounds or their redox products, and, therefore, they experience little fouling over relatively long periods of time. Glassy-carbon electrodes are for the most part noncatalytic, or inert.

An inherent electronic stabilization mechanism does not exist for polar aliphatic compounds, and the heterogeneous rate constants for their anodic mechanisms at inert anodes are much smaller than those for aromatic compounds. As a

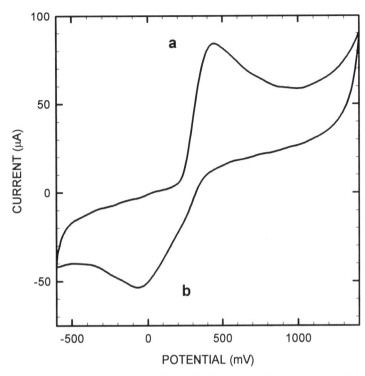

Figure 4.2. Cyclic voltammogram of caffeic acid (5 mM) in 100 mM acetate, pH 4.75, at a stationary glassy-carbon electrode. Scan rate, 250 mV s⁻¹.

consequence, researchers have considered aliphatic alcohols and amines not amenable to amperometric detection under constant applied potentials. Unfortunately for the analytical chemist, the photometric detection of these compounds is hindered by the lack of an inherent chromophore or fluorophore, which is also a result of the absence of π-bonding/conjugation. Along the lines of "the rich get richer and the poor get poorer," aromatic compounds, which may be electroactive, are also chromphoric, whereas polar aliphatic compounds appear to be neither.

Interestingly, virtually all organic compounds, and numerous inorganic species, are favored thermodynamically to be oxidized in aqueous solutions at potentials obtainable at commonly used solid electrode materials. The majority of these reactions require the transfer of oxygen from H_2O to the oxidation products and are generally kinetically inhibited because of their complexity. Oxidation of simple alcohols at noble-metal electrodes have been studied extensively, and none of these reactions has been found to produce a persistent reversible response, and the degree of irreversibility increases with extended exposure of the electrodes to the organic compound. It is important to understand that irreversibility of the reaction does not preclude analytical utility. Oxalic acid readily undergoes oxidization. Since CO_2 is formed as a reaction product, this reaction is understandably irreversible, yet the reaction is both rapid and reproducible. With a substantial overpotential, many irreversible reactions can be induced to proceed at a significant rate. However, if analytical precision is to be obtained, the reaction kinetics at the applied overpotential must either be fast enough to result in mass-transport-controlled signals or, if the heterogeneous rates are less than the rate of mass transport, be independent of time.

The point to be stressed is that the oxidation of polar aliphatic compounds at noble-metal electrodes does occur under constant applied potentials. In lieu of π-resonance stabilization, the activation barriers for oxidation of aliphatic compounds are decreased if the free-radical products of the anodic reactions are stabilized via adsorption to noble-metal electrode surfaces. The tendency for Pt to adsorb free radicals and other species is easily understood from the electronic structure of the metal, designated as $[Xe]6s^{1}4f^{14}5d^{9}$. Hence a partially filled d-orbital exists at the metal surface, thus facilitating the adsorption of free radicals and compounds with nonbonded electron pairs. At Pt, molecular H_2 is adsorbed and dissociated. The resultant species are presumed to be adsorbed radicals. As a consequence, Pt is well known as an effective hydrogenation–dehydrogenation catalyst [3]. Metals with saturated surface d-orbitals such as Au, or $[Xe]6s^{1}4f^{14}5d^{10}$, are relatively ineffective as hydrogenation–dehydrogenation catalysts [3].

As in life and the Second Law of Thermodynamics, you don't get something for nothing. The beneficial effects (i.e., electrocatalytic activity) of these interactions are accompanied by the undesirable tendency for the electrode surface to become fouled by the accumulation of adsorbed carbonaceous materials. The most common observation reported for anodic detection at solid electrodes, especially Pt, is a rapid loss of electrode response. This loss in activity is usually attributed to the strong adsorption of reactants and/or free-radical reaction products [4–7]. In some instances, free radicals at high concentration in the diffusion layer undergo polymerization reactions with subsequent coverage of the electrode by a polymeric film. Even in purified solvents, the presence of organic compounds at trace levels is responsible for observed losses in electrode activity. Interestingly, historical conclusions of nonreactivity for many polar aliphatic compounds at Au and Pt electrodes under dc conditions are flawed. The nonreactivity of aliphatic alcohols and amines is not attributable to low electroactivity, but instead, it is a direct consequence of *high electrocatalytic activity* at clean noble-metal electrodes, which results in fouling of electrodes within a few seconds or minutes. If only the noble-metal electrodes would remain clean, then a sustained, reproducible current would occur for the oxidation of polar aliphatic compounds.

As mentioned earlier, a desirable attribute of electrodes used in amperometric detection under constant applied potential is inertness. In other words, the electrode acts only to provide and receive electrons with no involvement in the reaction mechanism. In contrast, for the detection of polar aliphatic compounds, interaction between analyte and electrode surface is a requirement for oxidation. Faradaic processes that benefit from electrode surface interactions within the reaction mechanism are described as being *electrocatalytic*. Carbon electrodes do not enable the oxidation of polar aliphatic compounds. This is attributable to the absence of appropriate catalytic properties of carbon surfaces that must exist to support the anodic reaction mechanisms of aliphatic compounds.

In order to understand PED, which is based on electrocatalytic detections at noble-metal electrodes, we must first examine their electrochemical response characteristics.

THE NOBLE-METAL ELECTRODE

When we think of the precious metals Pt and Au, images of jewelry might come to mind. Part of the desirability of items made from precious metals is durability and inertness. Imagine buying an expensive Pt or Au ring and having it corrode or pit within a couple of months. I dare say that would be the last item made from a precious metal one would ever buy. Although noble metals are quite inert under typical atmospheric conditions, they are not inert at attainable potentials under aqueous conditions. In fact, the electrochemical profiles of noble-metal electrodes are quite active, and, from an electrochemist's point of view, the term *noble* (inert) *metal* is an oxymoron.

Figure 4.3 shows the current–potential (i–E) plot for a Pt RDE in 0.1 M NaOH with (............) and without (_____) dissolved O_2. The anodic signal for the positive scan corresponds to formation of surface oxide (wave a, ca. -400 to $+600$ mV). The formation of surface oxide at Pt electrodes has been studied extensively [8]. Beginning at a potential more negative than is required for oxide formation, the electrode surface is considered to be "bare" or "clean" Pt. In reality, the surface is covered with adsorbed water. Ions from the supporting electrolyte are also present, but our focus will be on the oxide formation process. As we begin to make the potential more positive, the surface becomes populated with adsorbed hydroxy radicals [3] by the reaction

$$Pt + H_2O \dashrightarrow PtOH + H^+ + e^- \tag{4.1}$$

The generation of adsorbed hydroxy radicals is a reversible process. Next, a place exchange between the adsorbed ·OH and Pt atoms takes place. This process, or

$$PtOH \dashrightarrow OHPt \tag{4.2}$$

is fast, and it is conjectured to be the basis of voltammetric hysteresis [3], which is observed in voltammograms of Pt. A further oxidation step generates the oxide phase:

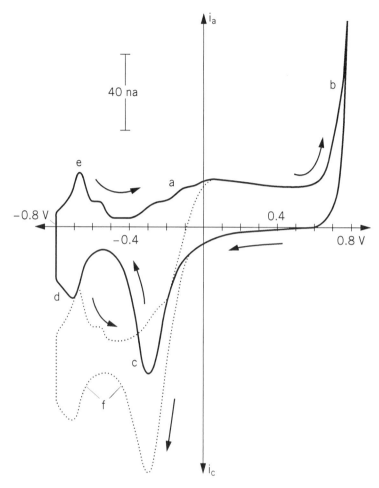

Figure 4.3. Residual cyclic voltammetric response (*i–E*) for a Pt RDE in 0.1 M NaOH. Conditions: rotation speed, 900 rev min⁻¹; scan rate, 200 mV s⁻¹; Ag/AgCl reference electrode. Solutions: (_____) 0.1 M NaOH, deaerated and (............) 0.1 M NaOH. Reprinted from W. R. LaCourse, Pulsed electrochemical detection at noble metal electrodes in high performance liquid chromatography, *Analusis* **21**, 181–195 (1993) with kind permission of Elsevier Science.

$$PtOH + OHPt \longrightarrow PtO + H^+ + e^- \qquad (4.3)$$

The positive-potential limit is determined by the anodic discharge of H_2O to produce O_2 (wave *b*, > ca. +650 mV). The cathodic peak signal for the subsequent negative scan corresponds to the reduction of the surface oxide (wave *c*, ca. −100 to −500 mV). Peak *d* (ca. −600 to −800 mV) corresponds to the formation of adsorbed hydrogen on the electrode surface. The negative-potential limit is denoted by the onset of bulk hydrogen evolution ($<$ ca. -800 mV). Peaks e (ca. -800 to -500 mV) correspond to the oxidation of adsorbed hydrogen. If the supporting electrolyte is not degassed, a cathodic wave is observed for dissolved O_2 during the positive and negative scans (wave f, $<$ ca. -50 mV).

Figure 4.4 shows the current–potential (i–E) for a Au RDE in 0.1 M NaOH with (............) and without (_____) dissolved O_2. The anodic signal for the positive scan corresponds to the formation of surface oxide (wave a, ca. $+150$ to $+700$ mV), and the anodic discharge of H_2O to produce O_2 (wave b, $>$ ca. $+700$ mV). The cathodic peak signal for the subsequent negative scan corresponds to

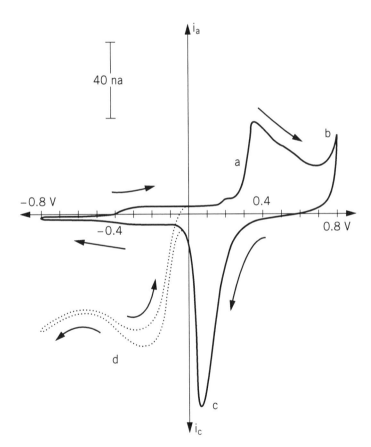

Figure 4.4. Residual cyclic voltammetric response (i–E) for a Au RDE in 0.1 M NaOH. Conditions: rotation speed, 900 rev min^{-1}; scan rate, 200 mV s^{-1}; Ag/AgCl reference electrode. Solutions: (_____) 0.1 M NaOH, deaerated and (............) 0.1 M NaOH. Reprinted from W. R. LaCourse, Pulsed electrochemical detection at noble metal electrodes in high performance liquid chromatography, *Analusis* **21**, 181–195 (1993) with kind permission of Elsevier Science.

the reduction of the surface oxide (wave c, ca. $+300$ to -100 mV). A cathodic wave is observed for the reduction of dissolved O_2 during the positive and negative scans (wave d, $<$ ca. -50 mV). Wave d for dissolved O_2 reduction is well resolved from waves a and c. This is in sharp contrast to the residual CV response for Pt electrodes (Fig. 4.3), where there is significant overlap of the wave for O_2 reduction with those for oxide formation and dissolution. As we shall see, this difference is a significant factor in the preference of Au over Pt electrodes in pulsed amperometric detection (PAD). Except for hydrogen adsorption, evolution, and oxidation, all other features of Pt electrodes are similar to those of Au electrodes in 0.1 M NaOH. Other metals, such as Pd (Fig. 4.5), show features similar to those of Au and Pt.

Another similar feature shared by Pt and Au electrodes is how each responds to changes in pH. Figure 4.6 shows cyclic voltammograms for AuO formation at several pH levels. Note that these voltammograms are shown using an alternate format where positive potentials are to the left and anodic current is down. The potential for onset of oxide formation is observed to decrease as pH increases at

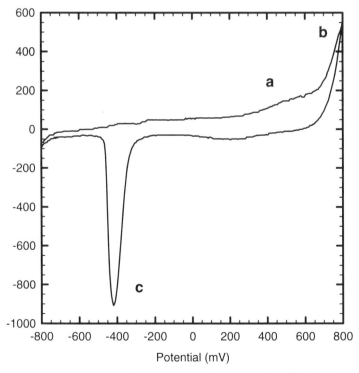

Figure 4.5. Cyclic voltammogram of Pd in 0.1 M NaOH at a rotating-disk electrode. Rotation speed, 900 rev min^{-1}; scan rate, 250 mV s^{-1}. Signals a, b, and c correspond to surface oxide formation, O_2 evolution from solvent breakdown, and surface oxide reduction, respectively.

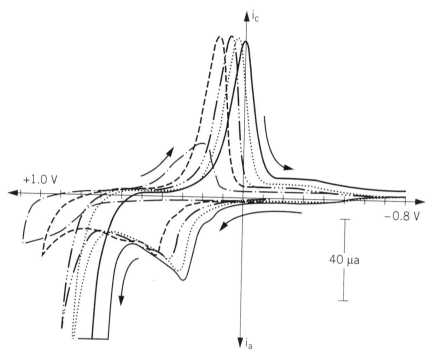

Figure 4.6. Effect of pH on voltammetric *i–E* response of the Au RDE in 0.1 M NaNO₃. Conditions: rotation speed, 1000 rev min⁻¹; scan rate, 200 mV s⁻¹; and Ag/AgCl reference electrode. Solution pH: (_____), pH 14; (............), pH 13; (_._._._), pH 12; (--------), pH 11; and (___·___), pH 7. Note that the axes have been reversed where positive potentials are to the left and anodic current is down. Reprinted from W. R. LaCourse and D. C. Johnson, Optimization of waveforms for pulsed amperometric detection (p.a.d.) of carbohydrates following separation by liquid chromatography, *Carbohydr. Res.* **215**, 159–178 (1991) with kind permission of Elsevier Science–NL, Sara Burgerhartstraat 25, 1055 KV Amsterdam, the Netherlands.

approximately −60 mV per pH unit. The lower potentials of oxide formation signify that it is easier to form surface oxide as the pH is increased. Hence, this is the expected trend because H_2O is consumed in the oxidation reactions with production of H^+, and, as a consequence the oxide reactions proceed more readily:

$$Au + H_2O \longrightarrow AuO + 2H^+ + 2e \tag{4.4}$$

Up to this point, we have focused on the electrochemical response of the supporting electrolyte. At the heart of PED is the ability of noble-metal electrodes to effect electrocatalytic oxidations of polar aliphatic functional groups. A sound understanding of how the various functional groups respond will be of great assistance in the understanding of PED as it is applied to liquid chromatography.

Monoalcohols and Glycols

In addition to the formation and dissolution of surface oxide being pH-dependent, the current of anodic reactions for polar aliphatic compounds is also observed to increase as pH increases, and the minimum required pH values are highly dependent on the choice of electrode material. For example, pH above ~12 is required for electroactivity of polyalcohols and carbohydrates at Au electrodes, whereas these same compounds respond at Pt electrodes in highly acidic media (pH below ~3). This difference in response is attributed to the difference in electronic occupancy of the surface d-orbitals of the two electrode materials. Pt atoms have a d^9 electronic configuration and, as a result, the metal surface has a high affinity for adsorption of free radicals [3]. Platinum will even cause aliphatic compounds to undergo dehydrogenation to produce adsorbed free radicals. In contrast, Au atoms have a d^{10}-electronic configuration, and radical oxidation products are only weakly adsorbed at a Au surface and, hence, alkaline media are required to produce appreciable rates for anodic reactions at this electrode.

These effects are dramatically illustrated for solutions of ethanol by $i–E$ curves (---------) shown in Figure 4.7A obtained at a Au RDE in 0.15 M NaOH (9% ACN) and Figure 4.7B obtained at a Pt RDE in 0.05 M HClO₄. Note that the oxide formation and dissolution signals are displaced to more positive potentials at Pt in 0.05 M HClO₄ in comparison to the same experiment performed in 0.1 M NaOH (Fig. 4.3). This shift illustrates the effect of pH on the oxide formation–dissolution process discussed earlier. Shown also are the residual $i–E$ curves (_____) obtained in the absence of ethanol, and residual curves (...........) obtained with dissolved O₂ present. At both electrodes the oxidation of ethanol (i.e., wave a at Au and waves a and b at Pt) occurs in the oxide-free region of the electrode, at approximately −500 to +400 mV for the Au electrode and ~100–800 mV for the Pt electrode. Since the —CH₂OH group is substantially more hydrophilic than the alkyl chain, it has been conjectured that the alcohol molecule at the electrode surface is preferentially oriented with the —CH₂OH group pointing toward the bulk of the solution [9]. This effect, in conjunction with the weaker radical adsorption characteristics of the Au electrode, may result in the observed diminished response. The oxidation mechanism of simple alcohols at Pt electrodes has been concluded to involve surface-catalyzed dehydrogenation, which is driven by the stabilization of the intermediate radicals formed in the redox process by adsorption on the electrode surface. As denoted by the large anodic peak at approximately +800 to +1200 mV, the strongly adsorbed species are simultaneously removed with the formation of surface oxide (wave c) [10,11]. The absence of signal on the negative scan in the region of approximately +1400 to −400 mV indicates the absence of activity of the oxide-

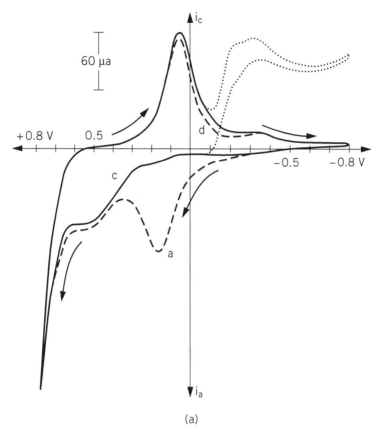

(a)

Figure 4.7.   Voltammetric response for ethanol at (*A*) a Au RDE in 9% ACN/150 mM NaOH and at (*B*) a Pt RDE in 50 mM HClO$_4$. Conditions: rotation speed (*A*) 200 rev min^{-1}, (*B*) 400 rev min^{-1}; scan rate 200 mV s^{-1}. Solutions: (———) deaerated supporting electrolyte; (...........) with dissolved O$_2$; and (--------) 27 mM ethanol. Note that the axes have been reversed where positive potentials are to the left and anodic current is down. Reprinted with permission from W. R. LaCourse, D. C. Johnson, M. A. Rey, and R. W. Slingsby, *Anal Chem* **63**, 134–139 (1991). Copyright 1991 American Chemical Society.

covered surface for ethanol oxidation. When the oxide is cathodically dissolved on the negative scan, the surface activity for ethanol is immediately returned and an anodic peak *d* is observed for the oxidation of ethanol. All the anodic waves are observed to increase with increases in the ethanol concentration [9].

Aldehydes and Polyalcohols

Figure 4.8 shows the voltammetric response of glutaric dialdehyde (...........), sorbitol (--------), and glucose (-.-.-.-.-.) at a Au RDE in 0.1 M NaOH. The residual response (———) is also shown for comparison. Oxidation of the al-

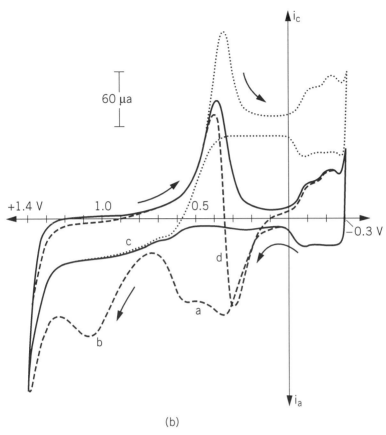

(b)

Figure 4.7. (*Continued*)

dehyde groups of glutaric dialdehyde and glucose occurs during the positive scan for $E >$ approximately -600 mV. Oxidation of one or more of the alcohol groups in each of the compounds occurs in the approximate potential region -300 to $+300$ mV. Oxidation of all three compounds is dramatically attenuated for $E >$ $+300$ mV, indicating that formation of the surface oxide strongly interferes with their electrochemical detection mechanisms. For the subsequent negative scan, oxidation of all three compounds commences when a significant portion of surface oxide has been removed from the electrode surface at approximately $+100$ mV.

The plateau current for aldehydes (i.e., < 1 mM, which is typical of concentrations found in HPLC) is observed to vary linearly with the square root of rotation speed ($\omega^{1/2}$) of the Au RDE [12]. This behavior is characteristic of fast heterogenous kinetics such that the amperometric response is under the control of convective–diffusional mass transport. The peak current for glycerol and glucose at approximately $+200$ mV exhibits a strong dependence on rotation speed;

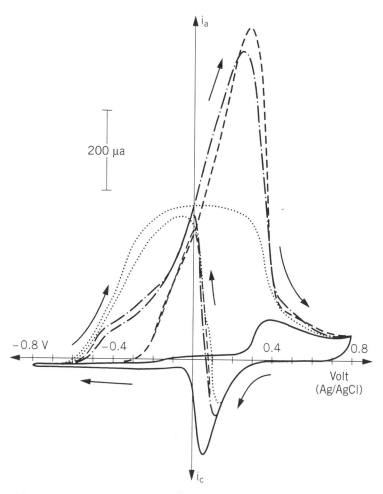

Figure 4.8. Voltammetric response of (............) glutaric dialdehyde, (-.-.-.-.-.) glucose, and (---------) sorbitol at a Au RDE in (_____) 0.1 M NaOH. Conditions: rotation speed, 1600 rev min⁻¹; scan rate, 250 mV s⁻¹, and Ag/AgCl reference electrode. Solutions: 1 mM. Reprinted with permission from D. C. Johnson and W. R. LaCourse, *Electroanal.* **4**, 367–380 (1992). Copyright 1992 VCH Publishers.

however, if the voltammetric scan is interrupted and the electrode potential is held constant at approximately +200 mV, the current begins to decay almost immediately. This is an indication of electrode fouling by adsorbed intermediate products in the oxidation mechanisms.

The anodic response of glucose at a Au electrode in 0.2 M NaOH for dilute solutions (≤1 mM) is conjectured to occur via a sequential process (see Fig. 4.9) [13]. The first step involves the oxidation of the C_1-aldehyde group to produce

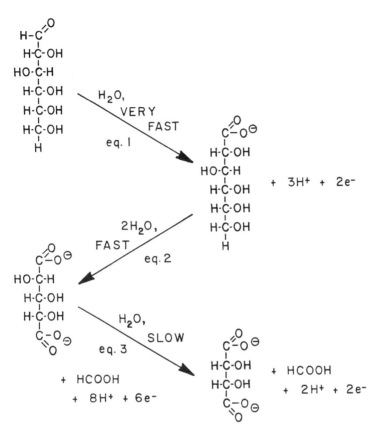

Figure 4.9. Conjectured reaction mechanism of glucose detection at a Au electrode under alkaline conditions. Reprinted with permission from D. C. Johnson and W. R. LaCourse, *Anal. Chem.* **62**, 589A–597A (1990). Copyright 1990 American Chemical Society.

the gluconate anion. This reaction is extremely fast. Concertedly, the C_6-alcohol group of the gluconate anion is oxidized to produce the glucarate dianion. This reaction is slower than the first. Subsequently, decarboxylation can occur to produce electroinactive formate ions and the corresponding dicarboxylate dianions. The extent of decarboxylation is dependent on the amount of time glucose remains at the electrode surface, which is a function of the supporting electrolyte, concentration, and convective–diffusional transport.

The i–E curves of other carbohydrates are sufficiently similar to suggest that analogous electrode processes are involved in the oxidations of virtually all carbohydrates. Hence, a single PAD waveform can be designed to detect all carbohydrates under alkaline conditions. On closer study, rotation speed and potential scan studies indicate significant differences in reaction dynamics [14]. The —OH response for glucose and glucitol is linear with the square root of the rotation speed with no dependence on scan rate. These observations support the conclusion stated earlier that their mechanisms are primarily under mass-transport control. In contrast, the anodic responses for sucrose and maltose increase with increases in scan rate, yet show negligible change as a result of variations in rotation speed. These observations indicate that the anodic response is under surface control. Fructose is under mixed control. One ramification of the differences in controlling mechanisms for carbohydrate response is reflected in the shape of current–concentration ($i–C^b$) plots. Carbohydrates that undergo transport-controlled reactions produce linear plots over larger ranges of C^b than for those reactions under surface control. Calibration data are shown in Figure 4.10 for five carbohydrates studied at a Au RDE in 0.1 M NaOH [14]. As expected, the plots of glucitol (▲) and glucose (●) are more linear than those for sucrose (○) and maltose (△). The response for fructose (■) is intermediate to that for glucose and sucrose. Two tentative explanations are offered for the nonlinear $i–C^b$ plots for reactions characterized as being under surface control: (1) the reactant molecules can be strongly adsorbed, and as a consequence, response is controlled by the adsorption isotherm for that reactant; and (2) if detection

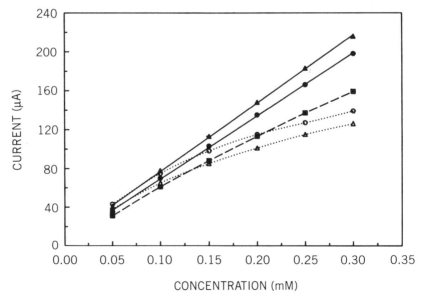

Figure 4.10. Maximum current response (*i*) as a function of concentration for (▲) glucitol, (●) glucose, (■) fructose, (○) sucrose, and (△) maltose. Conditions: rotation speed, 1000 rev^{-1}; scan rate, 200 mV s^{-1}, and Ag/AgCl reference electrode. Reprinted from W. R. LaCourse and D. C. Johnson, Optimization of waveforms for pulsed amperometric detection (p.a.d.) of carbohydrates following separation by liquid chromatography, *Carbohydr. Res.* **215**, 159–178 (1991) with kind permission of Elsevier Science–NL, Sara Burgerhartstraat 25, 1055 KV Amsterdam, the Netherlands.

products are strongly adsorbed with the result of surface fouling, the current response will be attenuated more abruptly during the positive scan for large values of C^b for which full coverage of the surface is achieved more quickly.

For all alcohols and aldehydes, the anodic response occurs in the oxide-free region of the electrode. The oxide-free region at Pt is obscured with the reduction wave for dissolved O_2. In contrast, a window of potentials at which there is little or no contribution from the reduction of dissolved O_2 is available for the Au electrode. It is for this reason that Au electrodes have become the most significant of the noble-metal electrodes for pulsed electrochemical detection techniques. Unfortunately, this window does not extend sufficiently negative (i.e., ca. -0.3 V in 0.1 M NaOH) for the selective detection of the aldehyde group over the alcohol group.

Amine-Based Compounds

Voltammetric studies have shown that primary, secondary, and tertiary amines are electroactive at both Au and Pt electrodes, whereas quaternary ammonium compounds are electroinactive [15–17]. From this observation and other studies, it has been determined that the adsorption of amines is required prior to their electrocatalyzed oxidations. Quaternary ammonium compounds do not have a free pair of electrons available for adsorption to the electrode surface. In fact, all electrocatalytically oxidized compounds require adsorption to the electrode surface prior to their oxidation. This adsorption requirement should serve to remind us that we are dealing with a heterogeneous system.

Figure 4.11 shows the $i–E$ curve of n-propylamine (---------) obtained at a Au RDE in 0.1 M NaOH. Anodic detection of the amine group occurs simultaneously with oxide formation. Amine oxidation is catalyzed by the formation of surface oxide. This conclusion is supported by the observation that no anodic current occurs during the negative scan in the region of approximately $+600$ to $+400$ mV. At this point, oxide formation on the surface has ceased, and the resultant oxide is quite inert. The net anodic current of amine response increases generally with increases in potential scan rate, but shows very little change as a result of variations in electrode rotation speed. This observation is consistent with a mechanism that is under the control of an electrode surface process; in other words, the reaction is surface oxide–catalyzed.

The mechanism for detection of amines involves preadsorption of the amine group through the nonbonded electron pair. It is possible to exploit the strength of this adsorption to promote oxidation of other more weakly adsorbed functional groups attached to the same molecule [18,19]. Figure 4.11 also shows the response for 20 mM of n-propanol (............) and 0.5 mM of 3-amino-1-propanol (-.-.-.-.-.) at a Au electrode in 0.1 M NaOH. Even though the concentration for n-propanol is 40 times greater than that for 3-amino-1-propanol, the anodic signal for oxidation of the alcohol group in the range of approximately -400 to $+300$ mV is significantly larger for the alkanolamine. This is as expected; since,

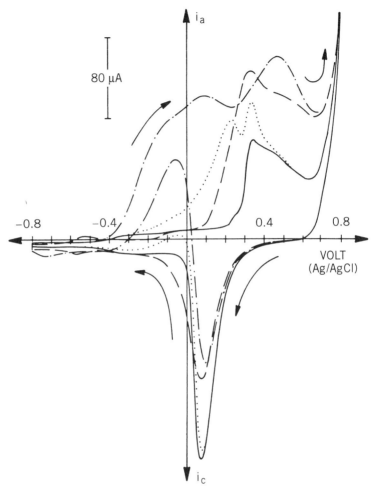

Figure 4.11.   Voltammetric response of an alcohol, an amine, and an alkanolamine at a Au RDE in 0.1 M NaOH. Conditions: rotation speed, 1600 rev min⁻¹; scan rate, 250 mV s⁻¹; and Ag/AgCl reference electrode. Solutions: (_____) residual in deaerated 0.1 M NaOH; (............) 20 mM propanol; (---------) 0.5 mM propylamine, and (—·—·—) 0.5 mM 3-amino-1-propanol. Reprinted with permission from D. C. Johnson and W. R. LaCourse, *Electroanal.* **4**, 367–380 (1992). Copyright 1992 VCH Publishers.

simple alcohols are only weakly electroactive at Au electrodes. Hence, the strongly adsorbed alkanolamine significantly increases the residence time of the alcohol group at the electrode surface in comparison to the residence time of the weakly adsorbed amine-free alcohol. The second wave of 3-amino-1-pro-panol corresponds well with oxidation of the amine group as noted by the similarity of the response of *n*-propylamine (---------) (Fig. 4.11). Interestingly, the product of the oxidation of ethanolamine in the region of approximately -200 to $+200$ mV at Au in 0.1 M NaOH is the corresponding aminoacid [19]. The carboxylate anion is significantly more hydrophilic than the precursor and, therefore, is less strongly adsorbed than the unreacted alkanolamine. Hence, the anionic oxidation product is desorbed with adsorption of fresh akanolamine, and the anodic process is perpetuated. This conclusion is supported by the observation that the anodic response attributed to the alcohol of an alkanolamine increases markedly with increases in electrode rotation speed, yet exhibits negligible change with variations in the potential scan rate. These observations are consistent with a mechanism producing anodic response that is under mass-transport control.

Thiocompounds

The electrochemical response of sulfur-containing compounds at Au electrodes in alkaline media is also catalyzed by formation of surface oxide. All thiocompounds that do not have a fully oxidized sulfur group are conceivably detectable under these conditions. In addition, thiocompounds can be detected at Au electrodes over the entire range of pH (0–14), whereas amine compounds can be detected at Au electrodes only in alkaline media (pH > ca. 9). This observation is attributable to the fact that the lone pair of electrons on the N atom in amines is protonated for pH below ~9 and, therefore, adsorption to the Au surface is blocked. The analogous protonation does not occur for thiocompounds and, therefore, these compounds will adsorb and be detected over the entire range of pH values. Figure 4.12 shows $i–E$ plots of thiourea (---------) and urea (............) at a Au RDE in 0.1 M NaOH [12]. The anodic prewave from approximately $+100$ to $+300$ mV is conjectured to correspond to the one-electron oxidation of the sulfur atom followed by dimerization to produce the disulfide [20]. The second anodic wave is attributed to the direct oxidation of the sulfur atom to sulfate [20]. As with amines, scan rate and hydrodynamic studies of the response of sulfur compounds are consistent with a mechanism that is under the control of an electrode surface process. In addition, note that the anodic response for thiourea is dramatically different from the response of urea. Oxidation of amines is suspected to produce hydroxylamines, for which $n_{net} = 2$ equiv mol^{-1} for urea, whereas thiourea oxidation is suspected to produce sulfate ion. Production of the sulfate ion would correspond to an $n_{net} \geq 8$ equiv mol^{-1} [21]. Thus, the increased response can be attributed to the larger number of electrons transferred in the oxidation process of thiocompounds as compared to nitrogen containing compounds. In general, mechanisms of detection of thiocompounds are quite complex, and they are under continued study.

All of these voltammograms represent steady-state, reproducible responses of these organic compounds. The same traces would be obtained for the fifth or the fiftieth scan. How can this be if high electrocatalytic activity of polar aliphatic compounds is typically accompanied by fouling of the electrode surface? Let's

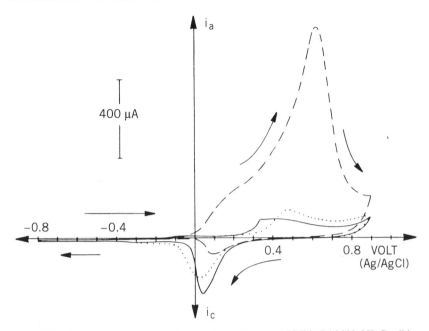

Figure 4.12. Voltammetric response of urea and thiourea at a Au RDE in 0.1 M NaOH. Conditions: rotation speed, 1600 rev min^{-1}; scan rate, 250 mV s^{-1}; and Ag/AgCl reference electrode. Solutions: (————) deaerated 0.1 M NaOH; (............) 0.5 mM urea; and (---------) 0.5 mM thiourea. Reprinted with permission from D. C. Johnson and W. R. LaCourse, *Electroanal.* **4**, 367–380 (1992). Copyright 1992 VCH Publishers.

think about what happens during a cyclic voltammetric experiment. The potential is first scanned in a positive direction well into oxide formation just until O_2 generation (i.e., the breakdown of water), and then the scan is reversed past the point where the formed oxide layer is cathodically dissolved from the electrode surface. This process is inherently self-cleaning because the formation of surface oxide is very effective in achieving oxidative desorption of adsorbed organic and inorganic species. The following reactions illustrate that oxide formation and electrocatalytic desorption are similar processes;

$$S + H_2O \rightarrow S[O] + 2H^+ + 2e^- \qquad (4.5)$$

$$S[R] + H_2O \rightarrow S[O] + RO + 2H^+ + 2e^- \qquad (4.6)$$

$$S[O] + S[O] \rightarrow S + O_2 \qquad (4.7)$$

where S is the electrode surface, [O] is semistable surface oxygen, [R] is adsorbed reactant, and RO is the oxidation product. This electrocatalytic link between activated surface oxygen species and O-transfer reactions has been implicated by numerous studies [3]. The strongest evidence perhaps comes from studies of the oxidation of Br^- and I^- at rotated ring–disk electrodes in acidic media. Adsorbed Br^- is oxidatively desorbed as $HOBr$ [22], and adsorbed I^- (existing as I^o) [23] is converted into IO_3^- [24,25] simultaneously with oxide formation. It is concluded that oxygen from adsorbed $\cdot OH$ is transferred to Br^- and I^o simultaneously with anodic electron transfer to produce $HOBr$ and IO_3^-.

Hence, when the electrode surface has a *preadsorbed* species present, the "activated-oxygen" species promotes oxygen transfer to the adsorbed compound. This process can represent detection of the compound of interest, or removal and/or cleaning of the electrode surface. The fully formed surface oxide layer is thermodynamically stable and inert. The unreactivity of the oxide-covered electrode protects the electrode from further fouling until the oxide layer is dissolved, and the detection process can begin again. The inertness of an oxide covered metal is also evident in the world around us. Take, for instance, copper-clad roofs or the Statue of Liberty. The Lady of Liberty is covered in copper plate, and copper is considered to be a very active metal. With acid rain, the Statue of Liberty should end up in New York harbor, but something protects it. That protection is the copper oxide/sulfate outer layer, or patina, which is inert and extremely stable. This is also why the Statue of Liberty and copper-clad roofs are green and not copper-colored.

CONCEPTS OF PULSED ELECTROCHEMICAL DETECTION

The current–potential (i–E) response is shown in Figure 4.13 for a Au RDE in 0.1 M NaOH with and without glucose in the absence of dissolved O_2. With the presence of glucose (_____), a reducing sugar, an anodic wave is observed on the postive scan beginning at approximately -600 mV (wave E). This wave corresponds to oxidation of the aldehyde group to the carboxylate anion in this alkaline medium. A much larger anodic signal is obtained for the combined oxidations of the alcohol and aldehyde groups in the region of approximately -200 to $+400$ mV (wave F). The anodic signal is attenuated abruptly during the positive scan with the onset of oxide formation (wave A). The signal for approximately $+400$ to $+600$ mV (wave G) in addition to that for oxide formation results from the anodic desorption of adsorbed glucose and/or intermediate products simultaneously with the formation of surface oxide on the Au electrode. The absence of the negative scan in the region of approximately $+800$ to $+200$ mV indicates the absence of activity for the oxide-covered electrode surface. Following cathodic dissolution of the oxide on the negative scan to produce wave C, the surface reactivity for glucose oxidation is immediately returned and an anodic peak (H) is observed for oxidation of alcohol and aldehyde groups on glucose. Anodic waves E, F, G, and H are all observed to increase in signal intensity with increases in glucose concentration.

If the goal is to detect glucose after its separation from sucrose and lysine, we

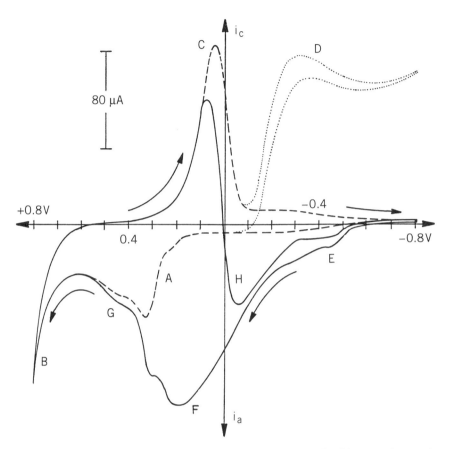

Figure 4.13. Voltammetric response (*i–E*) for glucose at a Au RDE. Conditions: rotation speed, 1000 rev min⁻¹; scan rate, 200 mV s⁻¹; and Ag/AgCl reference solution. Solutions: (--------) deaerated 0.1 M NaOH; (...........) with dissolved O_2; and (_____) 0.2 mM glucose. Note the axes have been reversed where positive potentials are to the left and anodic current is down. Reprinted from W. R. LaCourse and D. C. Johnson, Optimization of waveforms for pulsed amperometric detection (p.a.d.) of carbohydrates following separation by liquid chromatography, *Carbohydr. Res.* **215**, 159–178 (1991) with kind permission of Elsevier Science–NL, Sara Burgerhartstraat 25, 1055 KV Amsterdam, the Netherlands.

set up a chromatographic system followed by an electrochemical detector with a Au working electrode. The first order of business is to determine the appropriate detection potential. An examination of the cyclic voltammogram of glucose (Fig. 4.13) shows that the anodic response reaches a maximum at approximately +200 mV. If an attempt is made to perform amperometric detection at a constant applied potential of +200 mV at a Au working electrode, the result is a series of chromatograms with diminishing peak responses (see Figure 4.14*B*). By the fifth

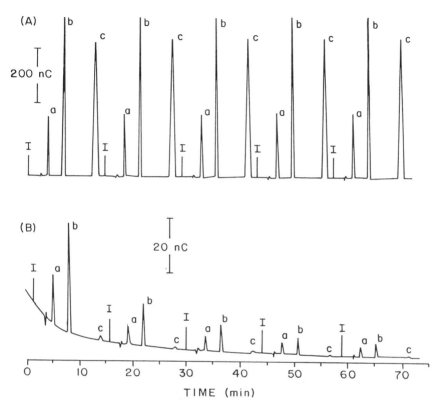

Figure 4.14. Comparison of (*A*) PAD and (*B*) dc amperometry in HPLC. Solutions: (*a*) lysine, 30 ppm; (*b*) glucose, 10 ppm; (*c*) sucrose, 40 ppm. Reprinted from W. R. LaCourse in *Electrochemical Detection in Liquid Chromatography and Capillary Electrophoresis* (P. Kissinger, ed.), in press, by courtesy of Marcel Dekker.

injection the response is of no analytical utility. This loss of response is attributable to fouling of the electrode surface. What is needed is some form of on-line cleaning. This is where pulsed electrochemical detection (PED) techniques come into play

These techniques exploit the high electrocatalytic activity of noble-metal electrodes by combining amperometric detection with pulsed potential cleaning of the electrode surface. Fouling species, which are typically adsorbed carbonaceous material produced during a short detection potential step, can be oxidatively desorbed quite efficiently from noble-metal electrodes by the application of a large positive-potential pulse to generate surface oxide (i.e., AuO and PtO). This can be related to the positive-potential scan in cyclic voltammetry. After the anodic potential pulse, the oxide-covered electrode is inert and must be cathodically reduced by a negative-potential pulse (the negative scan in cyclic voltammetry) to restore the native activity of the "cleaned" noble-metal electrode

TABLE 4.1   PAD Waveform and Table of Waveform Parameters for Carbohydrates in 0.1 M NaOH at a Au Electrode

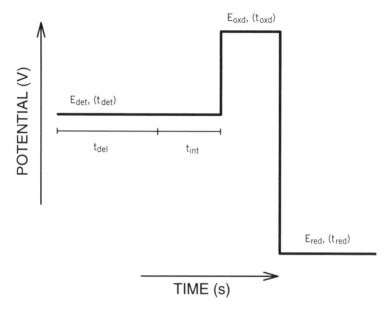

	Potential (mV vs. Ag/AgCl)			Time (ms)	
Parameter	General	Optimized	Parameter	General	Optimized
E_{det}	−200 to +400	+200	t_{det}	>40	440
			t_{del}	>20	240
			t_{int}	>20	200
E_{oxd}	+300 to +800	+800	t_{oxd}	>60	180
E_{red}	−800 to +100	−300/−800[a]	t_{red}	>60	360

[a]For complex samples which may contain strongly fouling species, an E_{red} value of −800 mV is recommended to remote cathodic cleaning of the electrode surface.

surface. By utilizing a simple three-step potential waveform (see Table 4.1) reproducible detection of polar aliphatic compounds can be achieved (see Fig. 4.14*A*). This technique is popularly known as *pulsed amperometric detection* (PAD).

Pulsed electrochemical detection exploits faradaic processes that benefit from participation of the electrode surface within the reaction mechanism. These processes are described as being *electrocatalytic*. In conjunction with oxide formation and dissolution by alternated anodic and cathodic polarizations, three modes of anodic electrocatalytic detection are obvious at noble-metal electrodes.

Mode I: Direct Detection at Oxide-Free Surfaces.   Oxide-free surfaces have an affinity for the adsorption of organic aliphatic compounds. As shown in Figure 4.15*A*, the analyte is brought to the electrode surface via convec-

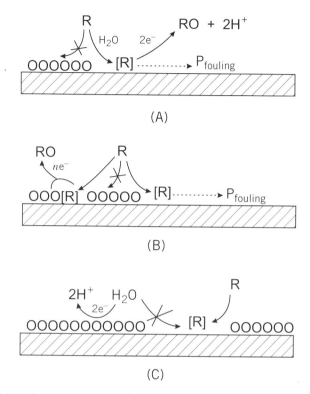

Figure 4.15. Schematic representations of (*A*) mode I, (*B*) mode II, and (*C*) mode III detections. In mode I detections, the reactant (*R*) is adsorbed on the oxide-free electrode surface from which it can either be oxidized to (i.e., RO) or foul ($P_{fouling}$) the electrode. In mode II, the reactant is again adsorbed to the electrode surface, but it is only oxidized concomitantly with surface-oxide formation. As with mode I, the adsorbed species can also foul the electrode surface. In mode III, the signal from oxide formation is suppressed by the adsorption of the reactant, which results in a negative response. Species: O, active oxygen intermediate; R, reactant/analyte; [R], adsorbed intermediate; and P, product.

tive–diffusional mass transport. While at the electrode surface, surface-stabilized oxidation results in a product that may leave the diffusion layer, readsorb for further oxidation, or foul the electrode surface. In mode I detections, oxidation of the compound can occur with little or no concurrent formation of surface oxide. Hence, background currents originate primarily from double-layer charging, which quickly decays to an insignificant level. Any surface oxide that does form inhibits the detection process. Mode I detections are typically performed at Au electrodes under alkaline conditions and Pt electrodes under alkaline and acidic conditions for compounds containing alcohol and aldehyde groups (see Appendix B).

Mode II: Direct Oxide-Catalyzed Detection. Mode II exploits the electro-catalytic activity of the activated oxygen intermediates in the oxide formation process (i.e., PtOH and AuOH) to enhance the rate of anodic oxygen transfer from H_2O to the compound of interest. The analyte is brought to the electrode surface via convective–diffusional transport, and it is adsorbed to the electrode surface at applied potentials prior to their detection. At the detection potential, surface-stabilized oxidation with anodic-oxygen transfer occurs concomitantly with the formation of surface oxide (see Fig. 4.15B). Oxidation of preadsorbed analyte is the primary contributor to the analytical signal; however, simultaneous catalytic oxidation of analyte in the diffusion layer is not excluded. The oxidation products may leave the diffusion layer or foul the electrode surface. Readsorption of analyte and its detection products is attenuated by the constantly forming surface oxide. Background signals are composed primarily of oxide formation currents. Since the electrode surface is undergoing continuous change, double-layer charging currents can never be totally dissipated, and, as a consequence, double-layer charging currents are minor contributors to background signals of mode II detections. In general, the background signals are significantly larger for mode II detections as compared to mode I detections. This phenomenon has a deleterious effect on the limits of detection. Compounds containing amine groups and thiocompounds are detected by mode II at Au and Pt electrodes (see Appendix B).

Mode III: Indirect Detections at Oxide Surfaces. Essential to mode I and mode II detections is the preadsorption of the analyte at oxide-free surfaces prior to electrocatalytic oxidation of the analyte itself. Species that adsorb strongly to the electrode surface and are electroinactive interfere with the oxide formation process. Preadsorbed species reduce the effective surface area of the electrode surface, and the analyte signal originates from a suppression of oxide formation (see Fig. 4.15C). The baseline signal results from anodic currents from surface oxide formation at a "clean electrode" surface, and the suppression of the oxide by the analyte results in a negative peak. Sulfur-containing and inorganic compounds have been detected by mode III.

In addition, indirect detection of strongly adsorbing compounds can be achieved via suppression of the response of weakly adsorbing PAD-active compounds or dissolved-O_2 reduction. A negative signal results similarly to the mode III detection, except the mode I detection of the weakly adsorbing compound is used in lieu of the oxide formation process. As a consequence, indirect detection can be performed at oxide-free regions of the electrode surface. Along the same line, suppression of dissolved-O_2 reduction signal can be used for indirect detection of strongly adsorbing compounds. Since the indirect signal is a result of the suppression of a reduction process, the overall signal mathematically results in a positive value. Neither of these indirect detection processes is considered to be a mode III detection, which must originate from the nature of the noble-metal electrode itself.

Voltammetric resolution of complex mixtures is almost futile, since electrocatalysis-based detection of various members within a class of compounds is controlled primarily by the dependence of the catalytic surface states on the electrode potential rather than by the redox potentials (E°) of the reactants. Most frequently, PED is applied at Au electrodes under alkaline conditions (pH > 12). Superimposed on the cyclic voltammogram of 0.1 M NaOH (degassed) at a Au RDE in Figure 4.16 are the various potential regions of response by modes I and II. All aldehydes, including the so-called reducing sugars, are anodically detected during the positive-potential excursion at the oxide-free surface in the region of approximately −600 to +200 mV (mode I). Large anodic signals are obtained

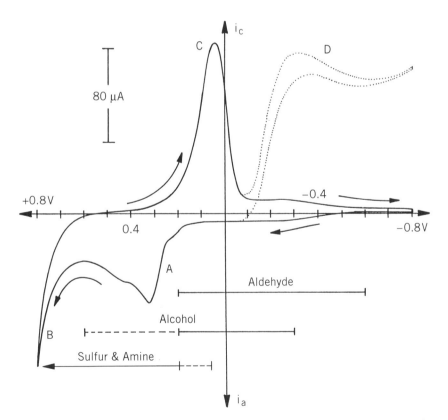

Figure 4.16. Residual voltammetric response (*i–E*) for a Au RDE showing regions of response for various functional groups. Note that the axes have been reversed where positive potentials are to the left and anodic current is down. Reprinted with permission from D. C. Johnson and W. R. LaCourse, *Anal. Chem.* **62**, 589A–597A (1990). Copyright 1990 American Chemical Society.

for alcohols, polyalcohols, and nonreducing sugars in the region of approximately -300 to $+200$ mV (mode I). Amines, aminoacids, and sulfur-containing compounds, for which a nonbonded electron pair resides on the N and S atoms, are adsorbed at oxide-free Au surfaces for E below approximately $+100$ mV and can be anodically detected by oxide-catalyzed reactions during the positive scan for E above approximately $+100$ mV (mode II). The region of approximately $+100$ to approximately $+200$ mV only "mimics" mode I detection because of the low density of surface oxide formed at these potentials. Detections at E above approximately $+600$ mV are not recommended because of the deleterious effects of coevolution of O_2. Hence, general selectivity is achieved via chromatographic separation prior to electrocatalytic detection. This conclusion does not preclude limited selectivity from control of detection parameters.

The simplest way to exploit electrocatalytic detection at noble-metal electrodes in a flow-through cell is with the application of a triple-step potential–time waveform at a frequency of ~0.5–2 Hz., which is generally appropriate for detection in conventional high-performance liquid chromatography. Amperometric detection under the control of a triple-step waveform is known as *pulsed amperometric detection* (PAD). Although PAD is most effective for oxide-free detection (mode I), it is also applicable to oxide-catalyzed detections (mode II).

PULSED AMPEROMETRIC DETECTION

PAD's triple-step potential waveform is illustrated in Table 4.1. The detection potential in the potential–time waveform is chosen to be appropriate for the desired surface-catalyzed reaction (Fig. 4.16), and the electrode current is sampled during a short time period (t_{int}) after a delay of t_{del}. The delay time is necessary to overcome double-layer charging currents, which would dramatically affect S/N. The combination of t_{del} and t_{int} constitutes the detection period (t_{det}). Following the detection process, adsorbed carbonaceous species are oxidatively desorbed simultaneously with anodic formation of surface oxide following a positive-potential step to the value E_{oxd} for a duration of t_{oxd}. Unlike a CV scan, the oxidation step can be held as long as necessary to fully "clean" and develop the oxide layer. The activity of the "clean," but inert, electrode surface is then regenerated by a subsequent negative-potential step to E_{red} for a duration of t_{red} to achieve cathodic dissolution of the oxide film prior to the next cycle of the waveform. See Table 4.1 for general and accepted values of PAD waveform parameters for carbohydrate detection in 0.1 M NaOH at a Au electrode. Note that the general requirements allow the analyst a great deal of latitude in the selection of waveform parameters. Thus, one can be up and running experiments in no time. Optimization of these parameters will allow PAD to perform with the most sensitivity, reproducibility, and insensitivity to fouling of the electrode surface. Chapter 6 is devoted to a discussion of waveform optimization.

The use of on-line, pulsed potential cleaning and reactivation is sufficient to maintain reproducibly high electrode activity. The significance of PAD within chemical and biochemical analysis can best be appreciated in view of the commonly held impression, based on attempted detections at constant (dc) applied potential, that polar aliphatic compounds are generally not electroactive. A comparison of the responses for PAD (Fig. 4.14*A*) and constant (dc) applied potential (Fig. 4.14*B*) for a chromatographic separation of lysine, glucose, and sucrose illustrates that the benefit of on-line, pulsed cleaning and reactivation of noble-metal electrodes is sufficient to maintain a reproducibly high electrode activity. Note the difference in scale between the PAD and dc amperometric chromatograms. The sensitivity is enhanced by orders of magnitude.

The origin of detection peaks in HPLC–PAD based on mode I is illustrated in Figure 4.17 by generic chronoamperometric (*i–t*) response curves following the step from E_{red} to E_{det} at an oxide-free electrode. The residual response from double-layer charging (curve *a*) decays very quickly, and the baseline signal in HPLC–PAD is very small for t_{del} above ~100 ms. Curve *b* represents the *i–t* response for the presence of analyte. An important feature of mode I detection is that, at the proper detection potential, the response with the analyte present is always greater than the background response. In a flow-through detector, the background response is continuously monitored. This signal may be electronically "bucked" or zeroed out. This signal is recorded as the baseline. When an analyte passes through the detector, the difference between the analyte response and the background is measured. The arrow represents the corresponding signal expected for HPLC–PAD with the indicated value of t_{del} in the waveform.

Anodic detection of amine- and sulfur-containing compounds occurs in a potential region where there is a significant signal from the concurrent formation of surface oxide. Hence, this difference from mode I (oxide-free) detections is clearly evident in the *i–t* response curves for oxide-catalyzed detections (mode II) (see Fig. 4.18). Here, the *i–t* response (curve *a*) corresponds to double-layer charging and to the formation of surface oxide. The current from surface oxide formation decays much more slowly than the current from double-layer charging. As a reminder, the current from double-layer charging is also always changing, since the electrode surface is constantly changing in response to the forming oxide layer. Although this is only a minor contributor, it is still present. One could suggest waiting until the current from oxide formation is minimal, but this would mean that the surface would then be covered with an oxide film and, hence, inert. The oxide-catalyzed detection requires the activated surface oxygen species available only during the transient formation of surface oxide. Hence, baseline signals for mode II typically have a nonzero value over a large range of t_{del} values (see Fig. 4.18, curve *a*). The *i–t* response corresponding to the presence of adsorbed analyte is represented as curve *b*. For small delay times, an almost unexpected "negative" peak is obtained. This effect is attributable to the fact that the analyte is preadsorbed to the electrode surface, and, as a consequence, the oxide formation process is initially inhibited and the active electrode is effectively smaller, which lowers the double-layer charging currents. At longer

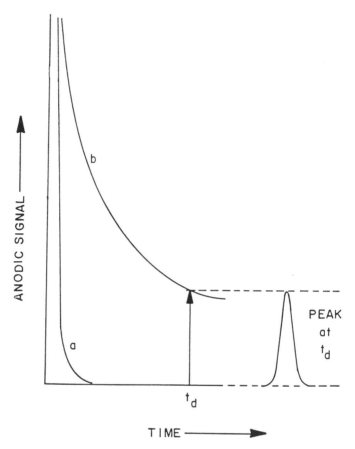

Figure 4.17. Chronoamperometric response (i–t) following a potential step from E_{red} to E_{det} in the PAD waveform (a) without and (b) with analyte present to illustrate the origins of chromatographic baseline and peak signals in HPLC–PAD (mode I). The delay prior to current sampling is denoted by t which is the same as t_{del}. Reprinted from W. R. LaCourse and D. C. Johnson, Optimization of waveforms for pulsed amperometric detection (p.a.d.) of carbohydrates following separation by liquid chromatography, *Carbohydr. Res.* **215**, 159–178 (1991) with kind permission of Elsevier Science–NL, Sara Burgerhartstraat 25, 1055 KV Amsterdam, the Netherlands.

delay times, a "positive" chromatographic peak is obtained. This effect occurs because charging currents are less, oxide formation currents are less, and sufficient oxide has been produced to catalyze the anodic reaction of the adsorbate. In the case of an unfortunate choice of an intermediate value of t_{del}, a detection peak might not be observed at all. A choice of t_{del} over ~150 ms is usually sufficient to assure "positive" HPLC–PAD peaks based on mode II detections.

The use of PAD for oxide-catalyzed detections has a number of distinct disadvantages, which fall into two categories. First are baseline, or background,

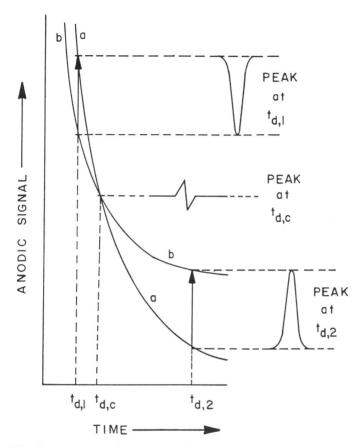

Figure 4.18. Chronoamperometric response (i–t) following a potential step from E_{red} to E_{det} in the PAD waveform (a) without and (b) with analyte present to illustrate the origins of chromatographic baseline and peak signals in HPLC–PAD (mode II). The delay prior to current sampling is denoted by t which is the same as t_{del}. $t_{d,1}$, $t_{d,2}$, and $t_{d,c}$ correspond to delays times (t_{del}) considered to be short, long, and at the crossover point. Reprinted from W. R. LaCourse in *Electrochemical Detection in Liquid Chromatography and Capillary Electrophoresis* (P. Kissinger, ed.), in press, by courtesy of Marcel Dekker.

sensitivities due to the presence of the forming oxide film. Any changes or gradients in pH, ionic strength, organic modifier, or temperature may lead to baseline drift. The baseline drift is attributable to variations in the extent and/or rate of surface oxide formation. The second category are those that occur only in the presence of analyte. Mode II detections with PAD suffer from poor S/N ratios. The sample current is only a fraction of the total signal, and the background signal is often 10–100 times greater than that of the analyte signal. Additionally, the oxide formation signal tends to be quite noisy. Another phenomenon that occurs only in the presence of the analyte is the effect known as

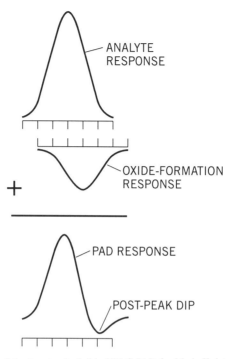

Figure 4.19. Origin of the "postpeak dip" in HPLC-PAD for Mode II detections. Reprinted from W. R. LaCourse in *Electrochemical Detection in Liquid Chromatography and Capillary Electrophoresis* (P. Kissinger, ed.), in press, by courtesy of Marcel Dekker.

"postpeak dipping." As shown in Figure 4.19, the presence of the analyte at the electrode surface interferes with surface oxide formation. The rate of oxide formation is reduced by the presence of the analyte and, as a consequence, an attenuated anodic current is observed as a "negative" peak. The shape of the "negative" background peak is a reflection of the oxide-catalyzed "positive" peak of the analyte, except that the "negative" peak is temporally offset from the analyte peak. The total effect is a dip after the chromatographic peak. Since the desired detection cannot by carried on in the absence of the forming oxide, one must find a better way. In PED, practically all advanced waveforms have been pursued with the goal of overcoming problems with mode II detections. Advanced waveforms are discussed in detail in the next chapter.

Even with all these complications, the judicious selection of waveform parameters allows for reasonable detections of amine and sulfur compounds using PAD. Figure 4.20 shows the chromatogram of several aminoacids from an over-the-counter health pill [26]. Although this chromatogram shows well-defined peaks, it still suffers from poor limits of detection.

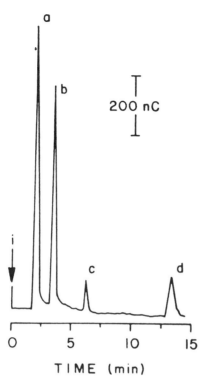

Figure 4.20. Chromatogram of aminoacids separated on an anion-exchange column with PAD at a Au electrode. Peaks (100 ppm): (*a*) arginine; (*b*) lysine; (*c*) leucine; (*d*) phenylalanine. Reprinted with permission from D. C. Johnson and W. R. LaCourse, *Electroanal.* **4**, 367–380 (1992). Copyright 1992 VCH Publishers.

Sampling of the Electrode Response in PAD

As with any detection system, an important consideration is the minimum detectable signal that can be recovered. The problem of differentiating analytical signal from noise becomes increasingly difficult as the signal source becomes weaker (i.e., trace-level concentrations). The ability of an instrument to discriminate between signal and noise is expressed in the S/N ratio of the detection system. Any reduction in noise represents an increase in the S/N ratio and an enhancement of the detection system.

The S/N ratio for measurements of transient amperometric signals is influenced by the instrumental strategy used for sampling the electrode current. A major noise component of the chronoamperometric signal is sinusoidal and correlated with 60-Hz line frequency. Hence, a common strategy for current sampling in PAD involves some form of signal averaging over the period of one 60-Hz oscillation (i.e, 16.7 ms). Since there is no contribution to signal strength from the 60-Hz noise, the time integral is zero. Extension of this strategy to the integration of an integral number (m) of 16.7-ms periods results in a significant increase in analytical signal strength while maintaining a 60-Hz noise signal of zero [27]. Figure 4.21 shows the separation and detection of five sugars using t_{int} values of 20 ms (A) and 200 ms (B). The S/N ratios for fructose in chromatograms A and B are 16 and 87, respectively. Typically, the integration period (t_{int}) is 200 ms, which is $m = 12$. In addition, since 200 ms is an integral multiple of one period of 50-Hz oscillation (i.e., 20 ms), 50-Hz line noise is also minimized. (In other countries, 50-Hz power systems are standard.) Because the signal output for an integrated amperometric response has units of coulombs, the corresponding technique was originally called *pulsed coulometric detection* (PCD) [27,28]. Presently, integrations of any time period fall under the umbrella of PAD. Although longer integration intervals are possible, there is a limit to the benefits of extending the integration interval for two reasons: (1) the time con-

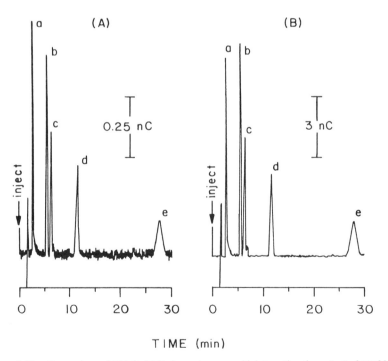

TIME (min)

Figure 4.21. Comparison of HPLC–PAD chromatograms with integration times (t_{int}) of (A) 20 ms and (B) 200 ms. Peaks (100 pmol each): (a) sorbitol; (b) glucose; (c) fructose; (d) sucrose; (e) maltose. Reprinted from W. R. LaCourse and D. C. Johnson, Optimization of waveforms for pulsed amperometric detection (p.a.d.) of carbohydrates following separation by liquid chromatography, *Carbohydr. Res.* **215**, 159–178 (1991) with kind permission of Elsevier Science–NL, Sara Burgerhartstraat 25, 1055 KV Amsterdam, the Netherlands.

straints of the overall cycle of the PAD waveform will eventually become large enough to compromise chromatographic peak resolution and (2) the longer the analyte is detected, the more electrode fouling occurs, and, as a consequence, S/N can actually become worse.

It is important to note that the signal output in PAD, which is self-described as an "amperometric" technique, can correspond simply to the time integral of the current, with the units of coulombs; or, alternately, to the average current response (i.e., the time integral divided by the integration period) in ampere units.

Effect of pH, Ionic Strength, and Organic Modifiers

PAD is dependent on chromatography for the resolution of complex mixtures, and fortunately, PAD is compatible with all water-based chromatographic systems. In general, these separations will be controllable through changes in pH, ionic strength, and organic modifiers.

The consideration of pH-gradient elution must recognize the effect of pH change on the background signal as well as the choice of E_{det} for maximum sensitivity in HPLC–PAD. The potential for onset of oxide formation at noble-metal electrodes shifts to more negative values with increases in pH at a rate of approximately -60 mV pH^{-1}. The effect of pH on the oxide formation process is attributable to the pH-dependent nature of Au oxide formation, specifically

$$Au(H_2O)_{ads} \longrightarrow Au-OH + H^+ + e$$

Because the optimal choice of E_{det} corresponds approximately to the value for onset of oxide formation, values of E_{det} and all other waveform potential parameters should be adjusted by the amount of approximately -60 mV pH^{-1} from the optimized value of detection recommended, for instance, at 0.1 M NaOH.

The negative shift in oxide formation with increasing pH can be reflected by a large baseline shift in HPLC–PAD under pH-gradient elution when E_{det} remains constant throughout the gradient. This effect can be alleviated to a great extent by substitution of a pH-sensitive glass-membrane electrode for the more popular Ag/AgCl reference electrode in the electrochemical cell. Because the response of the glass-membrane electrode is approximately -60 mV pH^{-1}, the value of E_{det} is automatically adjusted during execution of pH gradients [29,30]. If one stands at a fixed reference point, which is analogous to a Ag/AgCl reference electrode in an electrochemical cell, for observing a train, one would observe the train go by over time. On the other hand, if one observes the train from a moving car matching the speed of the train, which is analogous to the glass-membrane reference electrode, the train would appear to be standing still. Glass-membrane reference electrodes are high-impedance electrodes, and they tend to be a source of increased system noise.

Under ionic strength conditions suitable for electrochemical detection (i.e., $\mu > 50$–100 mM), the effect of changing ionic strength is reflected as only minor perturbations in the background signal from oxide formation. This effect is not noticed under isocratic HPLC conditions. Under gradient conditions (e.g., increasing acetate concentration), both positive and negative baseline drifts have been observed.

In comparison to ionic strength effects, changes in the concentration of organic modifiers can have a much greater effect on the baseline signal in HPLC–PAD. Electroactive organic modifiers can be used only in limited quantities under isocratic conditions. Even for electroinactive organic additives the baseline can be affected, because the modifiers are frequently adsorbed to the electrode surface with a resulting suppression of the oxide formation process. In addition to alteration of the HPLC–PAD baseline, adsorbed organic modifiers can severely attenuate the analytical signal for carbohydrates by interfering with access to specific adsorption sites on the electrode surface.

REFERENCES

1. R. N. Adams, in *Electrochemistry at Solid Electrodes*, Marcel Dekker, New York, 1969.

2. P. T. Kissinger, in *Laboratory Techniques in Electroanalytical Chemistry*, P. T. Kissinger and W. R. Heineman, eds., Marcel Dekker, New York, 1984.

3. D. C. Johnson, J. A. Polta, T. Z. Polta, G. G. Neuburger, J. Johnson, A. P-C. Tang, I-H. Yeo, and J. Baur, *J. Chem. Soc., Faraday Trans. 1* **82**, 1081–1088 (1986) and the references cited within.

4. S. Gilman, *J. Phys. Chem.* **67**, 78 (1963).

5. M. W. Breiter, *Electrochim. Acta* **8**, 973 (1963).

6. J. Giner, Electrochim. *Acta* **9**, 63 (1964).

7. S. Gilman, in *Electroanalytical Chemistry*, Vol. 2, A. J. Bard, ed., Marcel Dekker, New York, 1967.

8. R. Woods, in *Electroanalytical Chemistry*, Vol. 9, A. J. Bard, ed., Marcel Dekker: New York, 1976.

9. W. R. LaCourse, D. C. Johnson, M. A. Rey, and R. W. Slingsby, *Anal. Chem.* **63**, 134–139 (1991).

10. B. Beden, I. Cetin, D. Kahyaoglu, D. Takky, and C. Lamy, *J. Catalysis* **104**, 37 (1987).

11. P. Ocon, C. Alonso, R. Celdran, and J. Gonzalez-Velasco, *J. Electroanal. Chem.* **206**, 179 (1986).

12. D. C. Johnson and W. R. LaCourse, *Electroanalysis* **4**, 367–380 (1992).

13. L. A. Larew and D. C. Johnson, *J. Electroanal. Chem.* **262**, 167 (1989).

14. W. R. LaCourse and D. C. Johnson, *Carbohydr. Res.* **215**, 159–178 (1991).

15. J. Polta and D. C. Johnson, *J. Liq. Chromatogr.* **6**, 1727 (1983).

16. J. A. Polta, D. C. Johnson, and K. E. Merkel, *J. Chromatogr.* **324**, 407 (1985).

17. L. E. Welch, W. R. LaCourse, D. A. Mead, Jr., D. C. Johnson, and T. Hu, *Anal. Chem.* **61**, 555–559 (1989).

18. W. R. LaCourse, W. A. Jackson, and D. C. Johnson, *Anal. Chem.* **61**, 2466 (1989).

19. W. A. Jackson, W. R. LaCourse, D. A. Dobberpuhl, and D. C. Johnson, *Electroanalysis* **3**, 1 (1991).

20. P. J. Vandeberg, J. L. Kawagoe, and D. C. Johnson, *Anal. Chim. Acta* **260**, 1–11 (1992).

21. T. Z. Polta and D. C. Johnson, *J. Electroanal. Chem.* **209**, 159 (1986).

22. D. C. Johnson and S. Bruckenstein, *J. Electrochem. Soc.* **117**, 460 (1970).

23. R. F. Lane and A. T. Hubbard, *J. Phys. Chem.* **79**, 808 (1975).

24. D. C. Johnson, *J. Electrochem. Soc.* **119**, 331 (1972).

25. D. S. Austin, D. C. Johnson, T. G. Hines, and E. T. Berti, *Anal. Chem.* **55**, 2222 (1983).

26. L. E. Welch, W. R. LaCourse, D. A. Mead, Jr., and D. C. Johnson, *Talanta* **37**, 377–380 (1990).

27. G. G. Neuberger and D. C. Johnson, *Anal. Chim. Acta* **192**, 205–213 (1987).

28. G. G. Neuberger and D. C. Johnson, *Anal. Chem.* **60**, 2288–2293 (1988).

29. D. A. Mead, Jr., L. A. Larew, W. R. LaCourse, and D. C. Johnson, in *Advances in Ion Chromatography* Vol. 1, P. Jandik and R. M. Cassidy, eds.), Century International, Franklin, MA, 1989, pp. 13–34.

30. W. R. LaCourse, D. A. Mead, Jr., and D. C. Johnson, *Anal. Chem.* **62**, 220–224 (1990).

5 Integrated Pulsed Amperometric Detection and Other Advanced Potential–Time Waveforms

Out, damned oxide! Out, I say!

Without a doubt, the majority of applications using PED in HPLC involve detections at oxide-free surfaces, or mode I detections. The background signal in mode I detections originates primarily from double-layer charging, which quickly decays to an insignificant level, and it does not contribute to the final "sampled" detection signal. PAD is the technique of choice for mode I detections. In contrast, the background signal for oxide-catalyzed detections, or mode II detections, is composed primarily of current from the process of surface oxide formation. The rate of surface oxide formation (i.e., current) is sensitive to changes in pH, organic modifier concentration, ionic strength, and temperature. As a consequence, the baseline in PAD for mode II is very intolerant of any changes and/or gradients of the mobile-phase constituents, and this intolerance is reflected in drifting baselines and system artifacts. In addition, PAD of mode II detections suffers from "analyte-induced" artifacts and poor S/N.

As noted in Chapter 4, PAD can be used for the mode II detections, which encompass amine- and sulfur-based compounds, with the judicious selection of waveform parameters. The waveform parameters are highly specific for the classes of compounds targeted and the conditions used.

The best scenario would be to perform the detections in the absence of the forming surface oxide, but this is impossible. Labile oxide intermediates are needed to effect anodic oxygen transfer to the analyte of interest. Hence, we are forced to contend with the forming surface oxide, which is reflected in high background signals and poor chromatographic baselines. Battles may be lost with PAD for oxide-catalyzed detections, but victory can be ours with the use of advanced waveforms. In general, two strategies have been pursued to reduce the deleterious aspects of oxide formation currents. The first approach involves forming surface oxide and freezing its formation, which results in low background currents. The second method relies on subtracting the electronic charge due to the formation of surface oxide from the overall signal.

Although the majority of advanced waveforms have been developed to overcome the deleterious effects of surface oxide formation currents, these waveforms have also been designed to offer enhanced sensitivity, selectivity, and specificity. An examination of oxide formation and dissolution kinetics is needed for a full understanding of the advanced waveforms. In addition, the frequency at which a waveform can be applied in PED is fundamentally limited by the rate of oxide formation and dissolution. The frequency of the PED waveform is crucial to microchromatographic applications, where peak bandwidths are narrow. In order to maintain the integrity of the separation, the data must be collected at a faster rate than in normal-bore chromatography. Therefore, understanding the rate of oxide formation and dissolution will also assist in promoting the application of PED to microchromatographic separations.

OXIDE FORMATION AND DISSOLUTION KINETICS

Figure 5.1 shows the residual response for a (*A*) Pt RDE in 0.1 M H_2SO_4 and a (*B*) Au RDE in 0.1 M NaOH with (..........) and without (_____) dissolved O_2 present obtained at a relatively slow rate of potential scan (200 mV s^{-1}). The assignment of each of the waves to a particular electrode process was discussed in Chapter 4. Of interest to our discussion is the rate at which the oxide is formed and dissolved from the electrode surface.

The earliest work in this area goes back to studies by Gilroy [1,2], which focused on the oxide formation process at Pt electrodes in 1 M H_2SO_4. Gilroy's experiments involved pulsing the applied potential from a reduced (i.e., no oxide present) electrode surface to a voltammetric region at which oxide formation would commence. It was found that the amount of oxide formed on the electrode surface was logarithmically related to the length of time the potential pulse was

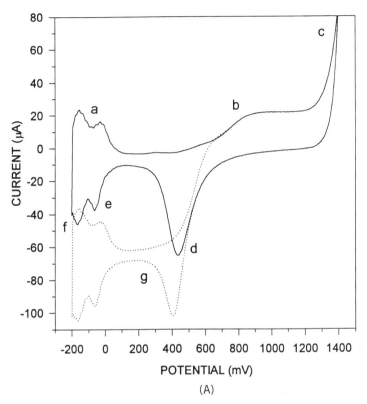

Figure 5.1. Residual response for a (*A*) Pt RDE in 0.1 M H_2SO_4 and (*B*) Au RDE in 0.1 M NaOH with (............) and without (_____) dissolved O_2 present. Wave: (A) a, e, f, hydrogen adsorption, oxidation, and reduction; b, surface oxide formation; c, O_2 evolution; d, surface oxide reduction; and g, dissolved O_2 reduction. (B) a, surface oxide formation; b, O_2 evolution; c, surface oxide reduction; and d, dissolved O_2 reduction. Reprinted from W. R. LaCourse, Pulsed electrochemical detection at noble metal electrodes in high performance liquid chromatography, *Analusis* **21**, 181–195 (1993) with kind permission of Elsevier Science.

applied. Hence, the amount of oxide formed (Q_a) is linearized with respect to time (t) by the following equation:

$$Q_a = A_a + (B_a \eta)\log \left[\frac{t}{t_0} \right] \qquad (5.1)$$

where t_0 is a normalization factor for unit time and η is the applied overpotential, or $E-E_{o,a}$, for oxide formation. The dependence of oxide formation rate on applied potential was further investigated to show that if the potential is initially pulsed to a particular value (E_2) and concertedly stepped to a lower potential (E_1), where oxide formation is still favored, the growth of oxide on the electrode surface could be halted for a finite length of time (see Fig. 5.2). Over this period

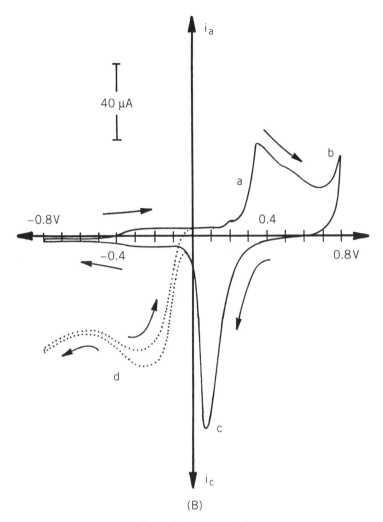

Figure 5.1. (*Continued*)

of time the rate of oxide formation is zero, and, hence, the current is also zero. Oxide formation will commence again at the rate corresponding to that observed for the application of the lower potential. This work has been repeated and corroborated in numerous publications [1–6]. More recently and of more importance to PED, the Johnson group has studied the kinetics of oxide formation and dissolution at Au microdisk electrodes in 0.1 M NaOH [7]. This work is briefly reviewed here, as it forms the foundation for many of the advanced waveforms.

The cyclic voltammogram in Figure 5.1*B* shows that the formation of surface oxide on Au commences at approximately −550 mV, which is denoted by the

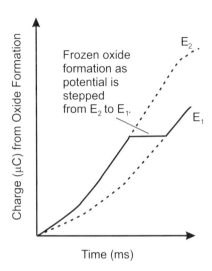

Figure 5.2. Plot showing that oxide formation can be stopped by the application of potential pulse (E_2) and concertedly stepped to a lower potential (E_1).

small anodic wave. The observation that it quickly reaches a plateau suggests that only a finite amount of oxide is formed. Even at these low levels, the catalytic activity of the electrode is noted by the onset of alcohol and aldehyde oxidation. Vitt et al. [8] showed that numerous catalytic processes are initiated at this potential. At potentials less than approximately -550 mV, the electrode surface is considered to be oxide-free, and electrocatalytic activity is typically not observed. The growth of surface oxides (AuOH and AuO) proceeds rapidly at potentials greater than approximately $+150$ mV. At approximately $+650$ mV, it is conjectured that a monolayer of oxide is formed on the electrode surface. The sharp increase in current at approximately $+650$ mV is attributable to the anodic breakdown of water to form O_2. Evidence that surface oxide formation has ceased is reflected in the current, which is the rate of an electrochemical process, being 0 nA on the reverse scan at approximately $+650$ mV. The cathodic peak at approximately $+100$ mV denotes dissolution of the surface oxide from the electrode surface. All oxide processes respond linearly with potential scan rate up to a point, which is indicative of surface-controlled process. In addition, variations in rotation speed have little or no effect on oxide formation or dissolution processes. At higher potentials (ca. $+800$ mV), the formation of high gold oxides (Au_2O_3) may result in surface reconstruction, which is reflected in an increase in surface roughness and response irreproducibility. For this reason, and the problem of bubble formation in flow-through electrochemical cells, potentials greater than approximately $+800$ mV are avoided in PED.

The PAD waveform can be thought of as a series of three potential steps each of increasing potential (i.e., $E_{red} < E_{det} < E_{oxd}$) (see Fig. 5.3*A*). Hence, the study of oxide formation on stepping from a lower potential is of the utmost importance to understanding detection at and cleaning of the electrode surface. Figure 5.4 shows the effect of stepping the potential from a clean electrode surface (i.e., ca -800 mV) to a potential of oxide formation (E_1) on the amount of surface oxide formed. Note that surface oxide is formed in three different phases. Phase I_a ($t <$ ca. 3 ms) reflects the initiation of oxide growth on the electrode surface. The most active sites on the electrode surface will "nucleate" first, and this region does not show a strong dependence on the applied potential. Phase II_a (3 ms $< t$ $<$ ca. 20 ms) indicates a rapid growth in the amount of oxide on the electrode surface. This phase is attributed to the growth of AuOH on the electrode surface via Reaction (5.2):

$$Au + OH^- ---\!\!> AuOH + e^- \qquad (5.2)$$

The increase in Q_a at a fixed potential is approximately linear with time. The phase II_a region also shows a linear dependence on the applied potential. Roberts and Johnson [7] found the slope, or ($B_a \eta$), of Equation (5.1) to give a B_a value of

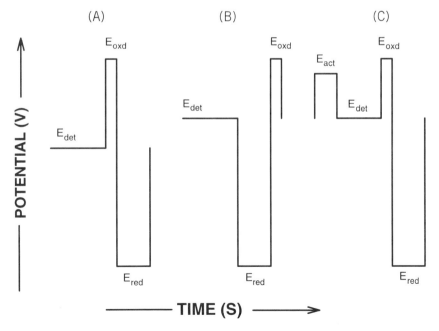

Figure 5.3. Generic (*A*) PAD, (*B*) RPAD, and (*C*) APAD waveforms. E_{det}, E_{oxd}, E_{red}, and E_{act} correspond to the detection, oxidation, reduction, and activation potentials of PED waveforms.

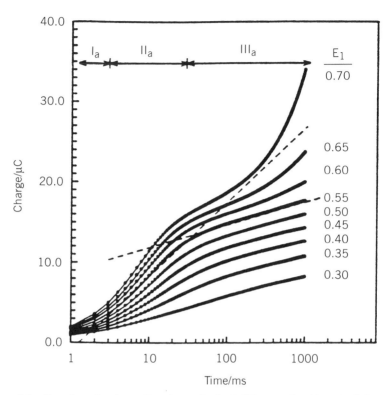

Figure 5.4. Plot of anodic change [in microcoulombs (μC)] versus log {time (ms)} for oxide formation at the Au minidisk electrode in 0.1 M NaOH. Values of E_1 are shown as V versus SCE. Linear approximations (---------) are indicated for regions II_a and III_a at E_1 = +0.55 V. Reprinted with permission from R. Roberts and D. C. Johnson, *Electroanal.* **4**, 741–749 (1992). Copyright 1992 VCH Publishers.

~26 μC V^{-1} and the potential for oxide formation ($E_{o,a}$) to be about +0.20 V over the applied potential range of +0.30 to +0.55 V. The $E_{o,a}$ value agrees well with the onset of oxide formation for the voltammogram in Figure 5.1*B*. Although the strong dependence of the rate of oxide growth on the applied potential agrees well with the work of Gilroy, the phase II_a rate of oxide formation on Au under alkaline conditions was found to be almost an order of magnitude greater than the rate of oxide formation on Pt under acidic conditions found by Gilroy [1]. A portion of this enhancement must be attributed to effects of pH on the oxide formation process, which is given by Reactions (5.2) and (5.3):

$$AuOH + OH^- ---\!\!\gg AuO + H_2O + e^- \tag{5.3}$$

In contrast to Gilroy's work at Pt electrodes under acidic conditions, the rate of oxide growth at t over approximately 20 ms is linear and proceeds much more slowly. This region is called *phase III$_a$*, and the rate of oxide growth appears to be independent of the applied potential. Phase III$_a$ represents the transformation of AuOH to the fully developed AuO, which proceeds at a slower rate than the formation of AuOH. The accumulated charge in the phase III$_a$ region at $E \geq$ 0.60V and longer times is attributable to O$_2$ evolution, which agrees well with the cyclic voltammogram in Figure 5.1B.

In PED, oxide formation is only half the process. Once the oxide layer is formed, the electrode is inert, and the oxide layer must be removed. Experiments similar to those described in the formation of the oxide layer can be performed for the removal of the oxide layer by stepping the applied potential to the electrode from a region of a fully developed oxide layer to one where the oxide layer is cathodically dissolved. As noted in Figure 5.5, the shapes of the oxide dissolution plots are sigmoidal. As with the oxide formation curves, the kinetics

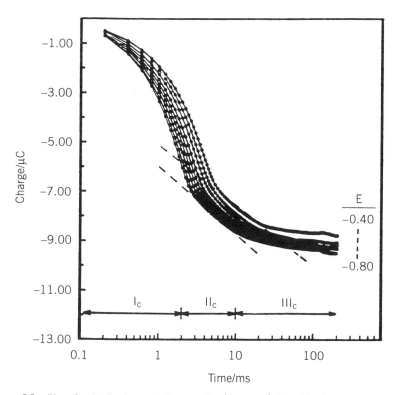

Figure 5.5. Plot of cathodic charge (μC) versus log {time (ms)} for oxide dissolution at the Au minidisk electrode in 0.1 M NaOH using a reverse potential-pulse waveform. The potential pulse was varied by 50-mV intervals in the range -0.40 to -0.80V. Linear approximations (---------) are indicated in region II$_c$ for $E = -0.40$ and -0.80 V. Reprinted with permission from R. Roberts and D. C. Johnson, *Electroanal.* **4**, 741–749 (1992). Copyright 1992 VCH Publishers.

are divided into three different regions. Phase I_c corresponds to the initiation of oxide removal ($t <$ ca. 2 ms). The initial removal of the oxide layer as reflected in the charge (Q_c) is linearly dependent on t. In addition, the slope of the line was found to be linear with applied potential. The more negative the applied potential the bigger the slope, or, in other words, the faster the oxide is reduced. Phase I_c accounts for the dissolution of 80–90% the surface oxide. The remaining 10–20% of the surface oxide is dissolved during phase II_c (ca. 2 ms $< t <$ ca. 10 ms) and phase III_c ($t >$ ca. 10 ms). Dissolution of the surface oxide during phase II_c is rapid, and Q_c is a linear function of log t. Surface oxide dissolution during phases II_c and III_c is independent of the applied potential.

In summary, oxide formation is initially rapid, resulting in the formation of a monolayer of AuOH. This process is strongly dependent on the applied potential. The conversion of AuOH to AuO proceeds at a much slower, potential-independent rate. Oxide formation results in a monolayer of oxide. It has also been conjectured that O_2 evolution can only occur after the formation of AuO on the electrode surface. At a typical potential of approximately $+700$ mV, the formation of a monolayer of AuOH takes about 30 ms to form while the conversion to AuO requires another 150–200 ms. The resulting oxide is then quickly removed in less than 30 ms, with the initial rate of dissolution being dependent on the applied potential.

On the surface (pun intended), the oxide formation/dissolution rates bode well for high-frequency PED waveforms, but it is important to remember that these studies were all performed in the absence of analyte. Adsorption of the analyte to the electrode surface requires a finite amount of time, which may drastically extend the overall waveform time. In addition, the adsorbed analyte will also interfere with the "neat" oxide formation and dissolution processes. High-frequency PED waveforms will be discussed in relation to PED following micro-chromatographic separations in Chapter 9.

ROLE OF THE OXIDE IN PED MECHANISMS

As discussed in Chapter 4, oxide formation is essential in "cleaning" the electrode surface in PED. In addition, oxide intermediates and the fully developed oxide can both catalyze or inhibit the oxidation of analyte. From our discussion of oxide formation, we can summarize the surface oxide formation process as follows:

$$S + H_2O \dashrightarrow S[OH] + H^+ + e^- \quad \text{(phases I and II)} \quad (5.4a)$$

$$S[OH] \dashrightarrow SO + H^+ + e^- \quad \text{(phase III)} \quad (5.4b)$$

where S is the electrode surface, S[OH] is a semistable surface oxide intermediate, and SO is surface oxide. It is important to remember that the exact chemical nature of the species involved is not known, and the mechanisms presented here are generalizations.

The first step in the detection process is the adsorption of the analyte to the "clean" electrode surface. This step occurs for both mode I and mode II detections. In the case of mode I, the adsorbed analyte (S[OR]), which is typically an alcohol-containing compound, is catalyzed by S[OH] as shown by the following reactions:

$$S + ROH \dashrightarrow S[OR] + H^+ + e^- \qquad (5.5a)$$

$$S + H_2O \dashrightarrow S[OH] + H^+ + e^- \qquad (5.4a)$$

$$\underline{S[OH] + S[OR] \dashrightarrow 2S + ROOH} \qquad (5.5b)$$

$$ROH + H_2O \dashrightarrow ROOH + 2H^+ + 2e^- \qquad (5.5c)$$

Note that Reaction (5.5a) is analogous with the phase I_a, S[OH] formation reaction. In other words, analyte replaces water in the phase I_a reaction. Anodic oxygen transfer takes place between the surface adsorbed species, or ROH, and S[OH]. Product analysis of ethanolamine oxidation showed that the corresponding carboxylic acid was produced [9]. After product formation, the surface is again "clean" and can be repopulated with either S[OR] or S[OH]. Since the rate of SOH formation at mode I potentials is rather slow, the rate of adsorption of the analyte is competitively favored, and catalytic oxidation of carbohydrates can proceed at nearly diffusion-transport-limited rates. Formation of SO blocks the further adsorption and oxidation of ROH. For oxide-catalyzed detection (mode II), both S[OH] and SO are conjectured to contribute to the overall signal for oxide-catalyzed detections.

Johnson and co-workers have conjectured that O_2 evolution from the oxidation of H_2O proceeds only from the SO-covered electrode surface. This process can be envisioned by the reactions shown below:

$$2S + 2H_2O \dashrightarrow 2S[OH] + 2H^+ + 2e^- \qquad (5.4a)$$

$$2S[OH] \dashrightarrow 2SO + 2H^+ + 2e^- \qquad (5.6a)$$

$$\underline{SO + SO \dashrightarrow 2S + O_2} \qquad (5.6b)$$

$$2H_2O \dashrightarrow O_2 + 4H^+ + 4e^- \qquad (5.6c)$$

Note the similarity between the mechanisms of analyte oxidation [Reaction (5.5c)] and O_2 evolution [Reaction (5.6c)]. In fact, there has been extensive published research that suggests that many anodic-oxygen-transfer reactions are merely competitive pathways of the oxide formation and oxygen evolution reactions [10].

A significant surface coverage of reactive intermediates (i.e., S[OH] and/or SO) is required for the analyte to be oxidized at a significant rate in mode II detections. Obtaining these active oxide intermediates is further hampered by the strong adsorption characteristics of amine- and sulfur-containing compounds, which block the actual formation of surface oxide. Hence, voltammetric studies of these compounds often show the suppression of the onset of surface oxide. Furthermore, higher applied potentials are used to activate the detection process and obtain the higher-surface-oxide intermediate coverages. The enhanced rate of labile oxide intermediates at these higher potentials is the source of the high-background currents observed in oxide-catalyzed detections.

It is important to keep in mind that the analyte must be preadsorbed to the electrode for detection to occur. At the higher potentials required for mode II detections, the surface is quickly repopulated with either S[OH] or SO, and readsorption of the analyte is less likely. Hence, the majority of signal for the analyte is obtained early in the detection process and at potentials where the rate of adsorption of the analyte is competitive with the repopulation rate of the surface oxide species. Figure 5.6 illustrates that the oxide-catalyzed detection process is essentially based on a set of competitive kinetic pathways for the surface. If the potential is too low, adsorption of the analyte will be optimal, but reactive surface species are minimal or not present. This effect is especially evident with strongly adsorbed analytes, which actually suppress or delay the onset of surface oxide formation as in the case of thiocompounds. At high potentials, the oxide forms very rapidly, which inhibits the readsorption and oxidation of analyte. Hence, the signal originates from the analyte preadsorbed during the "reduction" step, and it quickly is attenuated. At intermediate potentials, the probability of either analyte being readsorbed or an oxide intermediate forming are equivalent, and maximal mass-transport dependent response is expected. Figure 5.7 shows that the response under mass-transport control is at intermediate potential values for alanine at a Au RDE using *modulated hydrodynamic voltammetry* (MHDV), a technique that outputs the current as derived from the difference between voltammograms, which have been obtained at two different rotation speeds. Hence, only signals from mass-transport-dependent phenomena will be observed.

All in all, mode II detections occur in the presence of forming surface oxide. The rate of formation of the oxide is reflected as the background current, and the background current is high. The presence of the high-background currents results in poor S/N and system artifacts. The focus of most advanced waveforms is to overcome the effect of oxide interference. We cannot perform the assay in the absence of the oxide, but we can mitigate its effect.

REVERSE PULSED AMPEROMETRIC DETECTION

According to the work of Gilroy [1,2], growth of oxide on the electrode surface can be frozen if the potential is initially pulsed to a particular value and concertedly stepped to a lower potential, where oxide formation is still favored. As

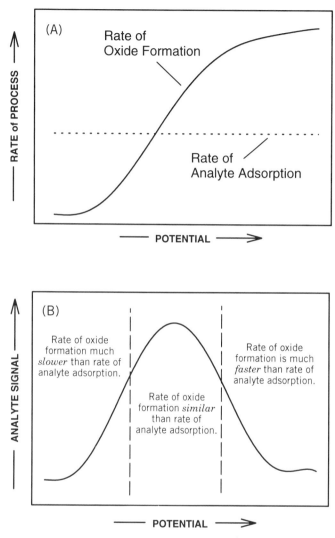

Figure 5.6. Effect of the competition between (*A*) the rates of oxide formation and analyte adsorption on (*B*) analyte response for mode II detections as a function of applied potential.

shown in Figure 5.2, the current is less as the potential-dependent rate of oxide growth changes from the higher to lower plot. It is during this region that high-sensitivity detections by mode II can be performed with significantly lower backgrounds. Polta and Johnson [11] first exploited this effect by reversing the E_{oxd} and E_{red} pulses of a PAD waveform (see Fig. 5.3*B*); hence the term *reverse pulsed amperometric detection* (RPAD). Note that E_{det} now takes place after a higher potential pulse (i.e., E_{oxd}) that satisfies the criteria established by Gilroy. It

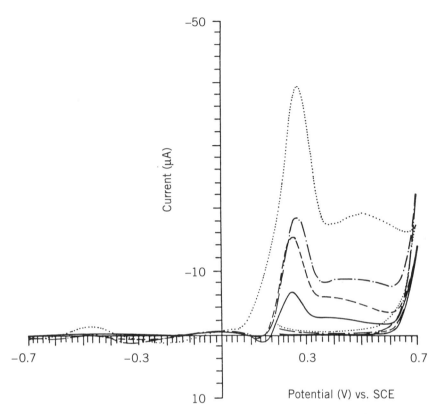

Figure 5.7. Modulated hydrodynamic voltammograms (MHDVs) of alanine at various concentrations showing the maximum response coincident with the onset of oxide formation. Concentrations (mM): (———) 0.02, (--------) 0.05, (-.-.-.-.-.) 0.10, and (............) 0.20.

is the preformed surface intermediates generated at E_{oxd} that function to effectively catalyze the mode II detections. The use of this waveform resulted in significantly lower baselines for the detection of sulfur compounds at a Au electrode [11,12]. These results were corroborated by Williams and Johnson for the oxidation of As(OH)$_3$ at a Pt electrode under acidic conditions [13]. One guideline for the effective use of RPAD waveforms is to minimize t_{oxd} so that surface oxide intermediates are not substantially converted to the inactive fully developed surface oxide (i.e., SO). If t_{oxd} is too long, most of the current from oxidation of preadsorbed analyte will be lost before the application of E_{det}. The use of E_{oxd} to partially form surface oxide intermediates is at odds with its original purpose to produce a fully developed oxide layer to "clean" the electrode surface. Hence, the RPAD waveform inherently provides insufficient oxidative cleaning of the electrode surface, which often results in distortion of chromatographic peaks in HPLC–RPAD.

ACTIVATED PULSED AMPEROMETRIC DETECTION

The problem of RPAD can be easily overcome by adding a fourth potential pulse (E_{act}) prior to E_{det} (see Fig. 5.3C). This more complex waveform combines the effect of activating the oxide, as in RPAD, and of maintaining the cleaning and adsorption characteristics of the PAD waveform. The waveform is known as either "activated" PAD (APAD) [13] or four-step PAD [12]. The positive potential pulse initiates the formation of surface oxide, and the rate of oxide formation is arrested on stepping back to a less positive detection potential. As discussed above, lower-background currents are achieved for mode II detection. Of equal importance are the oxidative cleaning and the reductive restoration of the oxide-free surface by the subsequent applications of $E_{oxd} > E_{act}$ and $E_{red} << E_{det}$, respectively. Hence, Williams and Johnson [13] were able to achieve a transport-limited signal for the oxidation of $As(OH)_3$ at a Pt electrode only by using an APAD waveform.

Both RPAD and APAD suffer from three drawbacks intrinsic to the basic premise that surface oxide formation stops completely if the applied potential is pulsed to a particular value and concertedly stepped to a lower potential, where oxide formation is still favored. First, Roberts and Johnson [7] determined that oxide formation does not completely stop when a Au electrode is used; and, unfortunately for PED, the majority of applications are performed at a Au electrode. Second, the length of time that oxide formation can be stopped or lessened before it begins to grow significantly again is usually less than needed for optimal detection conditions. Third, as oxidation of an analyte ensues, surface oxide is consumed, which results in either a changing background in the presence of the analyte or a less-than-maximal signal. The Welch group has been the most prominent in testing advanced waveforms for mode II detections [12]. Welch and co-workers found PAD performance for the detection of penicillins and related compounds was as good or superior to that of either RPAD and APAD.

INTEGRATED PULSED AMPEROMETRIC DETECTION (IPAD)

In addition to the limitations described above, suspending the formation of surface oxide formation is ideally applicable only under system conditions that are constant. In HPLC–PAD for mode II detections, baseline drifting is often attributable to changes or gradients in pH, ionic strength, organic modifier, and/or temperature. In addition, a large baseline current is frequently observed to drift to more anodic values, especially for new or freshly polished electrodes. This drift occurs because of slow growth in the true electrode surface area that is ascribable to surface reconstruction caused by the oxide on/off cycles induced by the applied multistep waveform. These baseline sensitivities would also be apparent with the use of either RPAD or APAD.

Baseline offset and drift can be significantly diminished by use of the waveform shown in Table 5.1. Here, the signal current is integrated throughout a rapid

Table 5.1   IPAD waveform and summary of parameter functions.

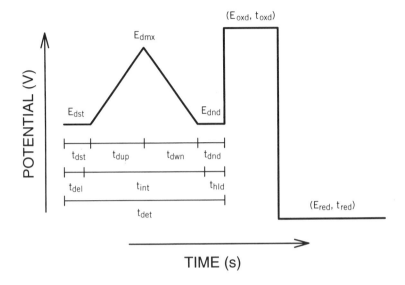

Parameter	Description
E_{dst}	Starting potential of scan— prior to onset of oxide formation
E_{dmx}	Maximum potential of scan for optimal analyte oxidation
E_{dnd}	Ending potential of scan—more negative than oxide dissolution
E_{oxd}	Oxidation potential to initiate formation of "cleaning" oxide
E_{red}	Reduction potential to initiate dissolution of inert oxide
t_{dst}	Time at E_{dst}
t_{dup}	Time for scan up from E_{dst} to E_{dmx}
t_{dwn}	Time for scan down from E_{dmx} to E_{dnd}
t_{dnd}	Time at E_{dnd}
t_{det}	Total time of detection step
t_{del}	Delay time required to overcome double-layer charging
t_{int}	Integration time for signal collection and background rejection
t_{hld}	Hold time to complete oxide dissolution at E_{red}
t_{oxd}	Time at E_{oxd} to achieve complete formation of "cleaning" oxide
t_{red}	Time at E_{red} to achieve complete dissolution of inert surface oxide

cyclic scan during the detection potential step within a pulsed potential–time waveform. The potential scan proceeds into (positive scan) and back out of (negative scan) the region of the oxide-catalyzed reaction for detection by mode II. The anodic charge for oxide formed on the positive sweep during the detection period tends to be compensated by the corresponding cathodic charge (opposite polarity) for dissolution of the oxide on the negative sweep. Hence the "background signal" on the electronic integrator at the end of the detection period can be virtually zero and is relatively unaffected by the gradual change of electrode area. More importantly, the majority of the analyte, which is preadsorbed to the electrode surface during E_{red}, is oxidized within the first 100 ms. If a typical delay time of 200–300 ms is used as in PAD, the majority of analytical signal is lost. In integrated IPAD, the sweep starts prior to oxidation of the analyte, sweeps into oxide formation with the concomitant mode II detection of the analyte, and proceeds back to the original starting potential. Hence, the majority of analyte signal is conserved while electronically rejecting the oxide background. The detection procedure based on the waveform in Table 5.1 combines cyclic voltammetry with pulsed potential cleaning to maintain uniform electrode activity. The function of each of the waveform parameters is also listed. The waveform was originally called potential sweep–PCD [14]; however, it has become known as *integrated pulsed amperometric detection* (IPAD) [15]. With hindsight being 20/20, a better name for this technique would have been *integrated voltammetric detection* (IVD).

At the heart of the IPAD waveform is the potential scan in the detection step, which takes place in a relatively short period of time. Scan rates are typically high (i.e., 1000 mV s^{-1}). High scan rates can lead to problems due to the effect of double-layer charging (i.e., capacitive) currents, which are induced with potential sweeps. Also, the larger the electrode size, the more deleterious the effect of capacitive current on the overall response. Figure 5.8 shows the cyclic voltammograms of the oxide background for electrodes with 5 mm (............), 3 mm (-.-.-.-.-.), 1 mm (_.._.._), and 50 μm (_____) diameters. Note the "smearing" of the typical oxide formation–dissolution signal for the fast scan rates for the larger electrodes due to increased capacitance. This effect would force the use of wider potential ranges, which would lead to the need for even faster scan rates. The smaller-diameter electrodes (i.e., 1 mm and 50 μm) minimize the effect of scan rate on the cyclic voltammograms. Hence, electrode diameters of <1 mm are best suited for the application of IPAD waveforms.

An additional consideration in IPAD relates to the electrochemical reversibility of the detection reaction. Clearly, the anodic signal is expected to be at a maximum when there is no cathodic contribution to the net current integral from reduction of the oxidation product during the negative portion of the cyclic scan of E_{det}. It is fortunate that all detection processes pertinent to this book are irreversible, that is, the oxidation products cannot be detected cathodically. However, even for a reversible redox system, there is sufficient loss of soluble oxidation products from the diffusion layer at the electrode by convective–diffusional mass transport that the cathodic charge from reduction of detection products will not be equivalent to the anodic charge from the detection process.

Since IPAD was designed to apply a waveform that coulometrically rejects the oxide background by summing the charges due to oxide formation and subsequent oxide dissolution, IPAD can virtually eliminate drift and changes associated with small variations in pH and composition of the mobile phase and

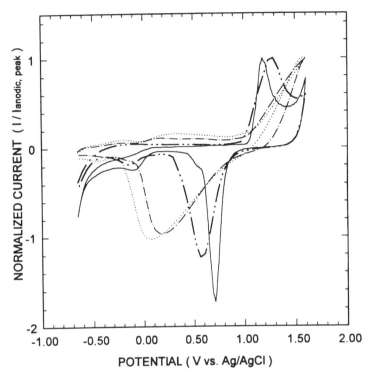

POTENTIAL (V vs. Ag/AgCl)

Figure 5.8. Comparison of voltammograms of 95% 100 mM phosphate buffer (pH 3), 5% CH₃CN at (_____) 50-μm wire and (_··_··_) 1-mm-, (-·-·-·-·-·) 3-mm-, and (............) 5-mm-diameter Au RDE. Rotation speed, 900 rev min⁻¹; scan rate, 10,000 mV s⁻¹. Reprinted from W. R. LaCourse and G. S. Owens, Pulsed electrochemical detection of thiocompounds following microchromatographic separations, *Anal. Chim. Acta.* **307**, 301–319 (1995) with kind permission of Elsevier Science–NL, Sara Burgerhartstraat 25, 1055 KV Amsterdam, the Netherlands.

changes in the total surface area of the noble-metal electrode surface, as well as analyte-induced effects. Figures 5.9*A* and 5.9*B* compare chromatograms using PAD and IPAD, respectively, for the determination of lysine. Note that the postpeak "dip" when using PAD, discussed in Chapter 4, is completely eliminated with the use of IPAD. In addition, the negative solvent front, due mostly to the suppression of oxide formation by electroinactive compounds, becomes a small positive peak when using IPAD.

Figures 5.10*A* and 5.10*B* compare PAD and IPAD, respectively, at a Au electrode for the isocratic separation of three aminoacids on an anion-exchange column [15]. The baseline drift in the PAD chromatogram is attributable to room-temperature fluctuations to which the oxide is sensitive. Clearly, IPAD is preferred over PAD to minimize baseline offset and drift. HPLC–IPAD results for the same three compounds offer detection limits of in the range of 1–10 pmol injected.

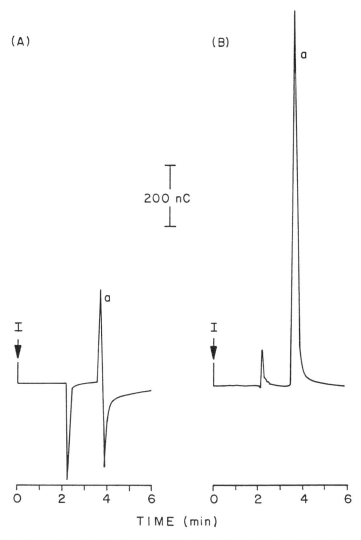

Figure 5.9. Comparison of (*A*) PAD and (*B*) IPAD for the detection of (*a*) lysine, 30 ppm. Reprinted from W. R. LaCourse in *Electrochemical Detection in Liquid Chromatography and Capillary Electrophoresis* (P. Kissinger, ed.), in press, by courtesy of Marcel Dekker.

A pH change causes a shift in the onset of the anodic wave for oxide formation (Fig. 4.6) by the amount approximately -60 mV pH^{-1}. Consequently, the baseline obtained for PAD increases (anodically) for an increase in pH. The use of IPAD results in a significant decrease in the baseline shift for small changes in pH (i.e., ΔpH < 2). However, optimum potential values for the IPAD waveform also shift with change in pH (ca. -60 mV pH^{-1}). The negative consequences of

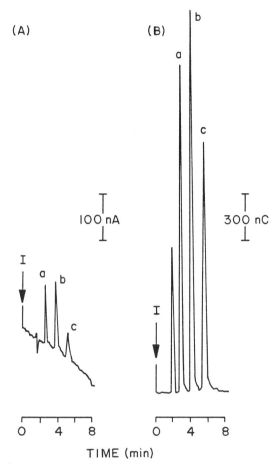

Figure 5.10. Effect of (*A*) PAD versus (*B*) IPAD on aminoacid detections. Peaks: (*a*) lysine, 280 pmol; (*b*) asparagine, 320 pmol; (*c*) 4-hydroxyproline, 351 pmol. Reprinted from W. R. LaCourse, Pulsed electrochemical detection at noble metal electrodes in high performance liquid chromatography, *Analusis* **21**, 181–195, 1993 with kind permission of Elsevier Science.

this fact can be decreased significantly by use of a glass-membrane, H$^+$-selective electrode as the reference in place of the conventional pH-independent electrodes, (e.g., mercury–mercurous chloride (SCE) or silver–silver chloride reference electrodes) [16,17]. Because the pH dependence of the "glass electrode" is the same as for the processes at the Au electrode, the voltammetric response at the Au electrode appears to be pH-independent. The use of a pH-sensitive reference electrode as compared to a SCE reference electrode results in a ~90% decrease in baseline drift for a pH gradient from 11 to 13 in HPLC–PAD (from

1500 to 150 nA). For the same comparison in HPLC–IPAD, a 100% decrease in baseline drift is observed for the same pH gradient (from 1200 to 0 nC).

The extent of oxide alteration varies with concentration and identity of the organic modifier in an unpredictable manner. A similar effect is observed for changes in ionic strength. Even more effectively than for pH gradients, IPAD virtually eliminates the baseline drift due to changes in organic modifier concentration and ionic strength. For example, it was not possible to find isocratic conditions for resolution of all aminoacids, and a gradient procedure was developed that incorporated a change in pH, organic modifier, and ionic strength. Attempts to use LC–PAD with a pH reference for the separation of a 17-component protein hydrolyzate resulted in severe baseline distortion, and no useful chromatogram was obtained. Chromatographic results are shown in Figure 5.11

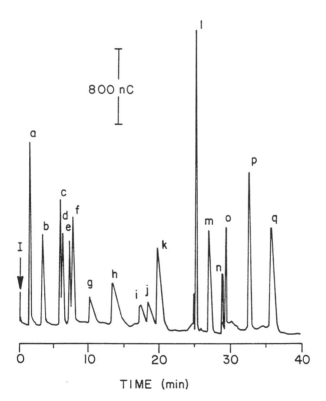

Figure 5.11. Aminoacid separation with IPAD. Peaks (25 nmol each): (*a*) arginine, (*b*) lysine, (*c*) threonine, (*d*) alanine, (*e*) glycine, (*f*) serine, (*g*) valine, (*h*) proline, (*i*) isoleucine, (*j*) leucine, (*k*) methionine, (*l*) histidine, (*m*) phenylalanine, (*n*) glutamic acid, (*o*) aspartic acid, (*p*) cystine, (*q*) tyrosine. Reprinted with permission from W. R. LaCourse and D. C. Johnson, in *Advances in Ion Chromatography*, Vol. 2, (P. Jandik and R. M. Cassidy, eds), 353–372, (1990). Copyright 1990 Century International.

for the gradient separation of 17 aminoacids present in a protein hydrolyzate. Detection was by IPAD at a Au electrode with a glass-membrane reference electrode. Detection limits (S/N = 3) by IPAD were determined to be ~3–5 pmol. It is especially significant that the sensitivity of IPAD is approximately the same for primary and secondary aminoacids.

Recently LaCourse and Owens [18] demonstrated the superiority of IPAD over PAD for the determination of thiocompounds using standard reversed-phase conditions. Interestingly, IPAD enables the direct determination of thio redox couples [i.e., —SH/—S—S—] and numerous other sulfur moieties at a single Au electrode. In addition, the kinetics for detection of adsorbed S compounds is quite favorable at pH 5, and it is not necessary to perform IPAD under alkaline conditions for thiocompounds. Figure 5.12 shows the separation and determination of several bioactive compounds including the reduced (GSH) and meth-

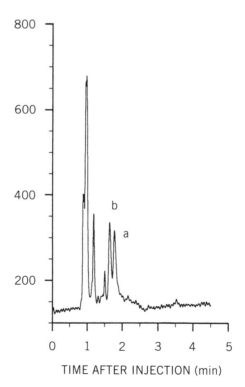

Figure 5.12. Detection of thiocompounds in watermelon by HPLC–IPAD. Peaks: (*a*) GSH; (*b*) methionine. Reprinted from W. R. LaCourse and G. S. Owens, Pulsed electrochemical detection of thiocompounds following microchromatographic separations, *Anal. Chim. Acta.* **307**, 301–319 (1995) with kind permission of Elsevier Science–NL, Sara Burgerhartstraat 25, 1055 KV Amsterdam, the Netherlands.

ionine in watermelon. The IPAD waveform gives lower LODs and more stable baselines, and eliminates oxide-induced artifacts.

The advantage of IPAD compared with PAD relates to maximization of the analyte signal and minimization of baseline magnitude and drift for oxide-catalyzed detections (mode II). Figure 5.13 compares IPAD with PAD for carbohydrates at oxide-free surfaces (mode I), and, as expected, there is no significant difference in S/N between the two techniques. In fact, the additional processing

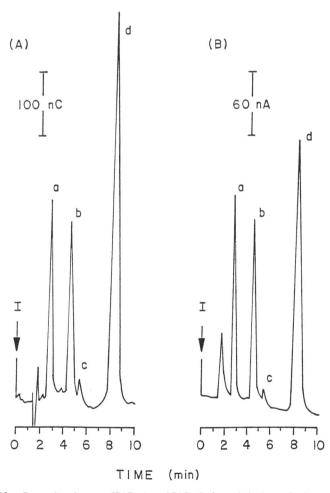

Figure 5.13. Comparison between IPAD (*A*) and PAD (*B*) for carbohydrates. Peaks: (*a*) fucose, 60 pmol; (*b*) glucose, 75 pmol; (*c*) unknown; (*d*) sucrose, 200 pmol. Reprinted from W. R. LaCourse, Pulsed electrochemical detection at noble metal electrodes in high performance liquid chromatography, *Analusis* **21**, 181–195 (1993) with kind permission of Elsevier Science.

steps in IPAD may lead to increased noise and less signal than PAD. Hence, the continued use of PAD for carbohydrate and alcohol detections is recommended.

MULTICYCLE WAVEFORMS

Numerous variations of PED waveforms (i.e., PAD, RPAD, APAD, and IPAD) have been tested, and some offer unique advantages. The use of even more complex waveforms, including multicycle waveforms, have been employed to a limited degree to either afford greater sensitivity or enhanced selectivity.

Dasenbrock et al. [19] focused on HPLC–IPAD analysis of sulfur-containing antibiotics for their determination within milk extracts. This work uses a modified IPAD waveform, which incorporates four cyclic scans in the detection step (see Fig. 5.14). This multicycle detection step has been shown to give lower LODs for all penicillins and cephalosporins (see Chapter 7). Although these studies are in their early stages, this complex waveform more efficiently exploits the formation and utilization of transient oxide intermediates for mode II detections. In addition, the average of the signals from the individual cycles (i.e., $n = 4$ for this waveform) is outputted, which should result in a $n^{1/2}$ increase in S/N. The expected increase in S/N is not totally realized, because the use of fast consecutive cyclic sweeps does not allow for the adsorption of fresh analyte between each cycle. Hence, the contribution to the total signal from the second scan cannot be as great as that from the first scan, and as a consequence, the maximum benefit of multicycle waveforms is often achieved with four or five cycles. It is also important to remember to keep the overall cycle time of the IPAD waveform short enough to maintain chromatographic peak integrity.

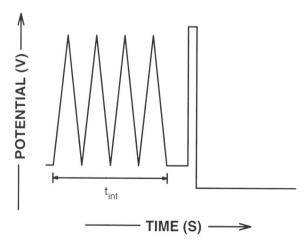

Figure 5.14. Multiple-cycle IPAD waveform.

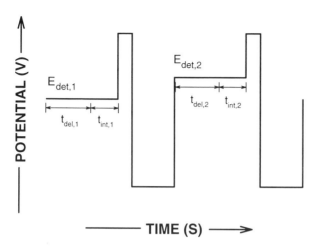

Figure 5.15. Multipex-PAD waveform showing two detection steps in a single cycle.

In order to increase selectivity of electrochemical detection in HPLC, monitoring the current at more than one potential has been implemented using cells with two, or more, working electrodes. The ability to monitor the electrochemical response of a compound at two different potentials is analogous to a "dual-wavelength" uv-absorption detector. Not only can one differentiate on the basis of voltammetric resolution (i.e., different potentials monitored), but qualitative information is provided on a particular analyte in that the ratio of the responses at each electrode will be characteristic of a compound's hydrodynamic voltammogram. Although the hardware to monitor multiple potentials simultaneously is straightforward, in practice, the use of more than two electrodes is rarely done.

An analyte's voltammetric response is dependent on the functional groups being oxidized. Each analyte's response and qualitative information can be derived from a minimum of two potentials. In PAD, dual-potential monitoring can be accomplished easily at a single electrode by using multiplex PAD (MPAD) (see Fig. 5.15). MPAD waveforms incorporate multiple detection parameters in one cycle. Figure 5.16*A* shows the separation of several polar aliphatic compounds. These results were performed at a single Au electrode. The chromatogram at +50 mV (---------) shows the anodic signal for each of the carbohydrate (i.e., alcohol-containing) analogs. As expected from the CV of an amine compound, the lysine signal is weak. The chromatogram at +350 mV (_____) shows a reduced signal for the carbohydrates, due to the effect of inhibition by the forming oxide, and a large signal for lysine, which is oxide-catalyzed. Figure 5.16*B* shows the difference chromatogram of these plots. All aminoacids and peptides give *negative* peaks, and all carbohydrate-based compounds give *positive* peaks. As in dual-wavelength spectroscopy, peak purity can be determined by plotting the ratio of the two wavelengths monitored. Since the response factor

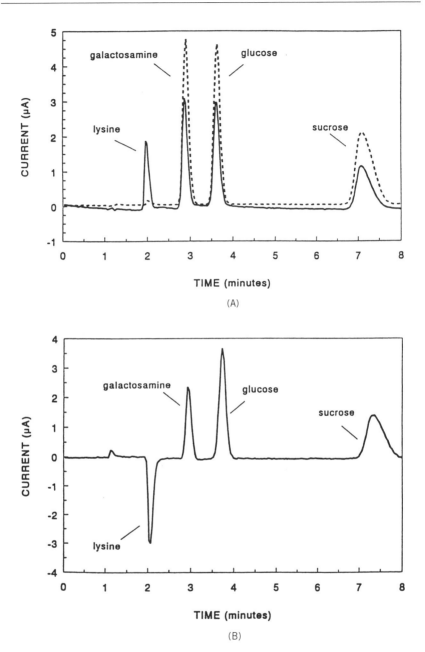

Figure 5.16. Chromatograms (*A*) and difference plot (*B*) using HPLC–MPAD for the separation and detection of a mixture of four compounds. Detection potential: (---------) +50 mV and (_____) +350 mV.

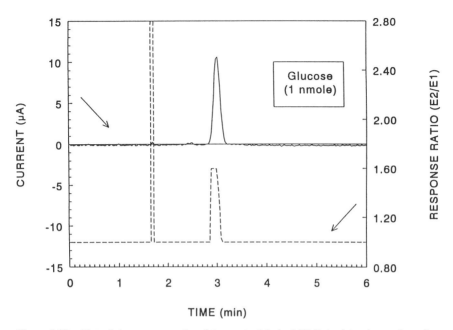

Figure 5.17. Plot of the response ratio of two potentials in MPAD to determine peak purity. Glucose is indicated as being pure by the square response of the ratio output.

at each wavelength would be constant, the ratio of the response factors would also be constant. Hence, the output from the elution of a pure compound will be a square wave. A similar argument can be advanced for two potentials in electrochemical detection, and Figure 5.17 shows the response ratio at two potentials is effective for the determination of peak purity using MPAD. The chromatogram for glucose at a single potential is shown for reference. One major advantage of MPAD is that the same electrode is used for each measurement, and as a consequence, any effects attributable to differences in the history of the electrode are eliminated.

REFERENCES

1. D. Gilroy, *J. Electroanal. Chem.* **71**, 257 (1976).
2. D. Gilroy, *J. Electroanal. Chem.* **83**, 329 (1977).
3. D. Gilroy and B. E. Conway, *Can. J. Chem.* **46**, 875 (1978).
4. H. Angerstein-Kozlowska, B. E. Conway, and W. B. A. Sharp, *J. Electroanal. Chem.* **43**, 9 (1973).
5. S. G. Roscoe and B. E. Conway, *J. Electroanal. Chem.* **224**, 163 (1987).
6. H. Angerstein-Kozlowska, B. E. Conway, K. Tellefsen, and B. Barnett, *Electrochim. Acta* **34**, 1045 (1989).

7. R. Roberts and D. C. Johnson, *Electroanalysis* **4**, 741–749 (1992).

8. J. E. Vitt, L. A. Larew, and D. C. Johnson, *Electroanalysis* **2**, 21 (1990).

9. W. R. LaCourse, W. A. Jackson, and D. C. Johnson, *Anal. Chem.* **61**, 2466–2471 (1989).

10. D. C. Johnson, J. A. Polta, T. Z. Polta, G. G. Neuburger, J. Johnson, A. P-C. Tang, I-H. Yeo, and J. Baur, *J. Chem. Soc., Faraday Trans. 1* **82**, 1081–1090 (1986).

11. T. Z. Polta and D. C. Johnson, *J. Electroanal. Chem.* **209**, 159–169 (1986).

12. S. Altunata, R. L. Earley, D. M. Mossman, and L. E. Welch, **42**, 17–25 (1995).

13. D. G. Williams and D. C. Johnson, *Anal. Chem.* **64**, 1785 (1992).

14. L. E. Welch, W. R. LaCourse, D. A. Mead, Jr., D. C. Johnson, and T. Hu, Anal. Chem. **61**, 555–559 (1989).

15. W. R. LaCourse, *Analysis* **21**, 181–195 (1993).

16. W. R. LaCourse, D. A. Mead, Jr., and D. C. Johnson, *Anal. Chem.* **62**, 220–224 (1990).

17. W. R. LaCourse and D. C. Johnson, in *Advances in Ion Chromatography*, Vol. 2, P. Jandik and R. M. Cassidy, eds., Century International, Medfield, MA, 1990.

18. W. R. LaCourse and G. S. Owens, *Anal. Chim. Acta* **307**, 301–319 (1995).

19. C. O. Dasenbrock, C. M. Zook, and W. R. LaCourse, paper presented at Ohio Valley Symposium, Oxford, OH, June 1996.

6   Pulsed Voltammetry: Waveform Optimization

To detect is human, to optimize, divine.

Pulsed electrochemical detection offers the analyst virtually universal detection of compounds with alcohol, amine, and sulfur-containing moieties, but universal detection does not automatically imply optimal performance. In the development of analytical methodology, it is imperative that all system parameters be either optimized or normalized. Optimization often focuses on maximizing or minimizing the effect of a system parameter. In PED waveform optimization, maximizing the signal-to-noise ratio (S/N) is the main objective. For some system parameters, there may be a wide range of acceptable values, and the analyst is forced to select, or settle, on a single condition, or value. This practice is known as *normalization* of system variables. Occasionally, the optimization of one parameter will adversely affect another system parameter. For example, the S/N of one compound may be optimized at the expense of the S/N of another, or the optimization of the cleaning polarizations for S/N under neat conditions for one analyte may interfere with overall system reproducibility when the compound is determined in the presence of the sample matrix, which may contain strongly adsorbing or fouling interferents. Under these conditions, one must be open to compromise for the purpose of *optimal overall performance*, which must take into consideration all aspects of the analytical method. Optimization and normalization of an analytical technique offer the analyst excellent performance and reproducibility.

In dc amperometry, a constant potential is applied to an electrode over time. Selection of the applied potential can be accomplished by one of two methods. The first method uses the LCEC system, and repetitive injections of the analyte are made while incrementally changing the detection potential. Hence, a hydrodynamic voltammogram is produced from which the optimal detection potential can be educed. This method is both time- and material-intensive. A second method uses a rotating-disk electrode (RDE) to simulate the hydrodynamics of the flowing eluent stream over a fixed electrode, and linear sweep voltammetry to scan a specified range of applied potentials. Scans are performed for the supporting electrolyte–mobile phase and the analyte. A current–potential plot similar to that of the HDV is produced, and the optimal potential can be determined as explained in Chapter 3. This method is quicker and requires less material then the first method. The effect of capacitive currents due to scanning of the applied potential is negligible since both the background and the analyte are scanned over the same range, and the associated charging currents are eliminated on subtraction of the two data files. Since the current is continuously sampled, no time parameters need be optimized. Hence, we are dealing with the optimization of a single detection variable.

At the heart of all pulsed electrochemical detection is the waveform. The waveform in PED represents a pattern of potentials versus time. The most common waveform is that of the PAD waveform, which consists of three potential steps (i.e, E_{det}, E_{oxd}, and E_{red}). One pattern of these potential steps constitutes one cycle of the waveform. In the process of optimizing a PAD waveform, not only must the potentials be optimized, but so must the time periods (i.e., t_{det}, t_{oxd}, and t_{red}) for which they are applied. As described in Chapter 4, the detection step consists of a delay time (t_{del}) and an integration time (t_{int}), both of which should also be optimized. Hence, for the PAD waveform, which is the simplest in PED, eight waveform parameters must be optimized and/or normalized. More complex waveforms (e.g., IPAD or APAD) require the optimization of additional parameters or special considerations.

Whereas PAD has matured as a detection technology applied to various chromatographies, the approach to waveform optimization has not followed the same path. The majority of PAD waveforms in use today are probably based on published waveforms and anecdotal information with some form of empirical optimization. A dilemma that the analyst must confront is how to optimize PAD for a new compound, possibly in limited quantity, for which there is no published waveform. One approach, which mimics the one used in dc amperometry, uses repetitive injections of the analyte into the HPLC–PAD system while incrementally changing the waveform parameter that is being optimized. The settings are typically changed over a specified range, and a stepwise hydrodynamic voltammogram can be plotted for each parameter of the PAD waveform. Figure 6.1shows the plot of current versus E_{det} for glucose at a Au electrode in 0.1 M NaOH using 20-mV increments. The analyst would choose approximately +200 mV as the optimized detection potential. The repetitive injection approach can be

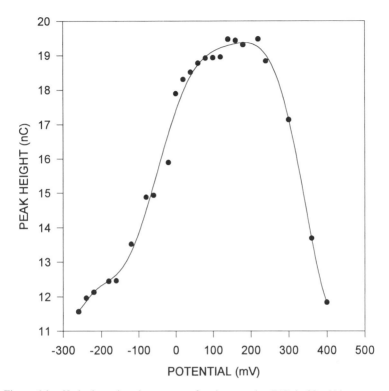

Figure 6.1. Hydrodynamic voltammogram for glucose using PAD in 20-mV increments.

used to optimize all PAD waveform parameters. Since this approach is empirical, it tends to be time-consuming, material-intensive, and tedious to do all eight waveform parameters. For instance, let's assume that we would like to optimize each of the eight waveform parameters for a particular compound over a +800-mV/800-ms range using 20-mV/20-ms increments with a chromatographic run time of ~5 min. The optimization would take ~33 h or, from another viewpoint, taking into consideration coffee breaks, lunch and committee meetings; almost an entire work week. The use of flow injection analysis (FIA) of an unretained peak in HPLC will be affected deleteriously by contributions due to impurities, which need not be PED-active to induce "background-suppression" signals (see Fig. 9.10). Hence, the use of the repetitive injection method necessitates the development of at least a rudimentary separation, which also precludes the quick survey of large numbers of analytes for PED activity.

In addition to the repetitive injection approach, the choice of potential values in PAD, especially E_{det}, has been based on the current–potential responses from cyclic voltammetric experiments in the presence and absence of the analyte at a RDE or, more recently, in the flow-through cell itself. A major drawback of using cyclic voltammetry is that for electrocatalytic detections the data are often inaccurate for selecting E_{det} values. This observation is especially true for electrocatalytic detections where loss of surface activity can occur with adsorption of detection products. In contrast, detection potentials for LCEC can be selected by CV and linear sweep voltammetry because the electrode is not intimately involved in the detection mechanism. As shown in Figure 6.2A, cyclic voltammetric data are generated via a triangular potential–time (E-t) waveform, and, inherently, potential and time are coupled. As a consequence, analyte response at a particular potential is biased by the "history" of the electrode surface at previous potentials in the scan (Fig. 6.2B). This effect is particularly noticeable for electrocatalytic detection mechanisms wherein loss of surface activity can occur as a consequence of the adsorption of detection products (i.e., mode I detection). This bias will often result in an inaccurate estimation of the optimal potential. In addition, the contributions of currents from oxide formation and double-layer charging are significantly different in CV and PAD and can obscure the CV response for a test compound present at low concentrations. Unlike potential scanning voltammetry for dc electroactive compounds, the surface plays an important role and can differ in the presence and absence of analyte, and "background-corrected" CVs will often contain artifacts, which do not reflect faradaic processes. Aside from inappropriate potential selection of PAD waveform parameters from CV data, CV is virtually useless for educing optimal time parameters for PAD waveforms, and their choice has been based largely on the repetitive injection approach.

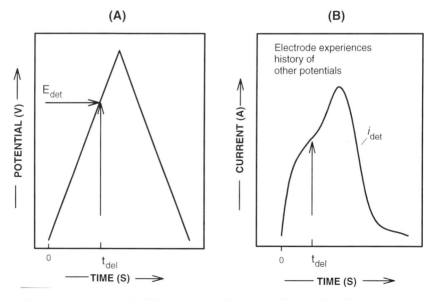

Figure 6.2.   Plots of a (*A*) CV waveform and its potential–time coupled (*B*) *i–t* response.

PULSED VOLTAMMETRY

The choice of optimal PED waveform parameters (e.g., E_{det}) is most logically based on the voltammetric response obtained when a PED waveform is applied to an electrode that is under hydrodynamic control (e.g., RDE). In order to "scan" a particular range of potentials, small incremental changes are made to the parameter being optimized for each cycle of the waveform.

For each cycle in the optimization of E_{det}, the current is sampled after t_{del} (Fig. 6.3A), and the electrode has no "history" at any other potentials (Fig. 6.3B). In other words, the scan of detection potentials is successfully decoupled from the time parameter, and a "true" chronoamperometric signal is obtained only at a single potential. Anodic and cathodic polarizations of the PED waveform serve to clean and reactivate the electrode after each E_{det}, and the electrode is ready for a new E_{det}. The resulting $i–E_{det}$ plot is a true representation of PAD response over that range of E_{det} values. This approach to studying and optimizing E_{det} of a three-step potential waveform at a noble-metal electrode has been called *pulsed voltammetry* (PV) [1–5] and *pulsed voltammetric detection* (PVD) [6]. The technique of PV was first applied by Neuberger and Johnson in 1987 to study the PAD response for carbohydrates [1–3]. More recently, the Johnson group has applied PV to characterize the PAD response of sulfur-containing compounds at both Au and Pt electrodes and the oxide formation process [4,7]. Since PAD combines amperometric detection with continuous cleaning of the electrode surface, PV offers these same features to the analyst who is interested in performing

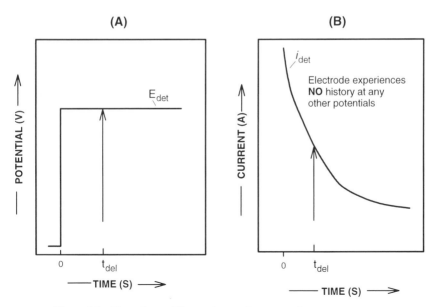

Figure 6.3. Plots of a (*A*) PV waveform and corresponding (*B*) *i–t* response.

voltammetry under electrode-fouling conditions. Ewing et al. used PV to obtain voltammetric information at Pt electrodes in static biological microenvironments [8].

Previous applications of PV focus primarily on the study of electrochemical response mechanisms at noble metal electrodes or on the optimization of E_{det} to be used in HPLC–PAD. LaCourse and Johnson [9] have extended the concept of PV to study the effect of all waveform parameters on PAD response, and PV has been determined to be the definitive method for the optimization of PED waveforms to be applied in HPLC detection.

PV Instrumentation

A typical pulsed voltammetric waveform for the optimization of E_{det} of PAD is shown in Figure 6.4. As in PAD, the current is monitored for a period t_{int} after t_{del}. It is important to remember that this same time period is measured for independent variations of all potential and time parameters. Hence, the effect of changing a particular parameter on detection response is educed. Because the duration of the PV experiment is not limited as in HPLC–PAD, the waveform for each set of parametric values, or step, can be repeated, and each data point then represents an average of the number of cycles sampled. In addition, the first several cycles in each step are used to establish steady-state conditions, and they are rarely collected as part of the final average. By averaging the response of repetitive cycles, the precision and S/N can be enhanced. There appears to be very little advantage in going beyond 25 cycles per step, and often even fewer cycles are sufficient for good analytical information. Since the resultant data are stored in digital files, the background response (no analyte) can be easily subtracted from the sample response in order to show just

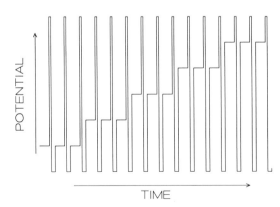

Figure 6.4.   Typical pulsed voltammetry waveform for E_{det}. Reprinted with permission from W. R. LaCourse and D. C. Johnson, *Anal. Chem.* **65**, 50–55 (1993). Copyright 1993 American Chemical Society.

the response of the analyte. The source code for a fundamental PV program is given in Appendix A.

Often the choice of an optimal value is clearly evident from the analyte's response alone. For certain parameters, the S/N is a better plot from which to make a decision. Since repetitive cycles are averaged for each data point, one also has the standard deviation available for manipulation. Hence, the S/N can be calculated as follows:

$$\frac{S}{N} = \frac{i_d}{\sigma_d} = \frac{i_a - i_r}{(\sigma_a^2 + \sigma_r^2)^{0.5}}$$

where i_a, i_r, and i_d are the average values of the analyte, residual, and difference currents, respectively; and σ_a, σ_r, and σ_d are the corresponding standard deviations.

PAD WAVEFORM OPTIMIZATION

Pulsed voltammetry has proved to be useful in the optimization of PAD waveforms, especially for mode I detections. As noted in Chapter 5, mode II detections are best handled with more advanced waveforms. This does not mean that PAD cannot be used for the detection of amine compounds and thiocompounds. It turns out that PV is the only definitive method to guarantee the best performance of PAD for these compounds. Since carbohydrate applications have dominated PED for the past decade, we will use carbohydrates, and glucose in particular, to explain optimization using PV. The overall procedure is to initially develop a separation using a generalized PAD waveform for detection. Once the mobile-phase conditions used for separation are established, this solution can be used as the supporting electrolyte for the PV experiments. Typically, E_{det} is optimized first, followed by t_{del} and t_{int}. Once the detection step has been defined, the cleaning pulses are optimized for effective cleaning and system reproducibility.

Carbohydrates

The optimum value of E_{det}, as well as other waveform parameters, for PAD can be easily and accurately chosen on the basis of PV data obtained during the application of the PAD waveform at an RDE. The experiment is performed in the presence and absence of the compound of interest, and the two electronic data files are subtracted to produce a "background-corrected" response. Figure 6.5 shows the "background-corrected" $i-E_{det}$ response (positive scan direction) at a Au RDE for equimolar concentrations of sorbitol (............), glucose (_____), maltose (---------), fructose (_-_-_) and sucrose (-.-.-.-.-.). As expected for an electrocatalytically driven system, the anodic signals for all the carbohydrates are similar in nature. The anodic signal in the -600- to -200-mV (wave E) region corresponds to the aldehyde group of the reducing sugars (i.e., glucose and

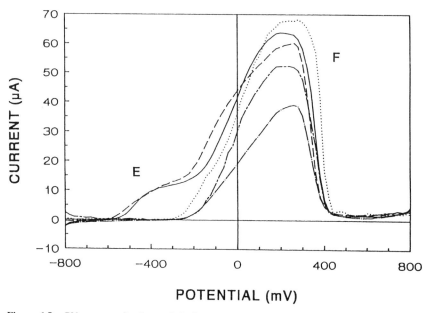

POTENTIAL (mV)

Figure 6.5. PV response for five carbohydrates at the Au RDE in 0.1 M NaOH. Solutions (0.1 mM): (............) sorbitol; (_____) glucose; (_-_-_) fructose; (-.-.-.-.-.) sucrose; (---------) maltose. Reprinted with permission from W. R. LaCourse and D. C. Johnson, *Anal. Chem.* **65**, 50–55 (1993). Copyright 1993 American Chemical Society.

maltose) being converted to the corresponding carboxylic acids. The peak in the region of −200 to +400 mV (wave F) corresponds to the direct oxidation of the alcohol groups. The differences in response are attributable to differences in the diffusion coefficient, the total number of electrons transferred, and the extent to which the detection mechanism is under mass transport or surface control. Table 6.1 shows that the response factors for several different glucose derivatives are equivalent once the number of electrons transferred is considered. Nevertheless, a maximum response is obtained at E_{det} = +200 mV for all carbohydrates at Au electrodes in 0.1 M NaOH. The general and optimal values for all waveform parameters for glucose at a Au electrode in 0.1 M NaOH are listed in Table 4.1. The application of this value enables virtual universal detection of all carbohydrates. Since detection is occurring in the oxide-free region of the electrodes, only a small portion of the anodic signal is attributable to surface oxide formation, and, at +200 mV, there is practically no cathodic response for the reduction of dissolved O_2. S/N plots of carbohydrates generally follow the anodic response curve. Cyclic voltammograms for the same set of carbohydrates (Fig. 6.6), are not as uniform in response because of the fouling of the electrode during the course of the detection scan. Hence, the CV data are more difficult to interpret.

TABLE 6.1 Response Characteristics of Glucose and Related Analogues

Compound	Functional Groups	Response (mA M^{-1})	Number of Electrons Transferred (n)	(mA M^{-1} n^{-1})
Glucose	5(OH)^a 1(CHO)	910 ± 7	~8	114 ± 1
Gluconic acid	5(OH)^a 1(COOH)	670 ± 29	~6	113 ± 5
Glucaric acid	4(OH) 2(COOH)	430 ± 24	~4	108 ± 6
Glucuronic acid	4(OH) 1(CHO) 1(COOH)	480 ± 28	~4	120 ± 7
2-deoxy-D-glucose	4(OH)^a 1(CHO)	218 ± 27	~2	109 ± 14

^{a}Denotes terminal hydroxyl group.

The majority of optimizations and previous applications of PV have focused almost entirely on the detection potential. More recently, experimental evidence indicates that all the parameters are intricately connected. The next step in optimization using PV is to study t_{del} while holding E_{det} at its optimum value. The PV waveform consists of repeating the same cycle of applied potentials, while incrementally changing t_{del}. Figure 6.7 shows the PV response and S/N of glucose and maltose. For any value of $t_{del} > 20$ ms, a significant anodic response is obtained for both sugars. Note that the response for maltose diminishes more rapidly than that for glucose. This fact is attributable to the rapid fouling of the Au surface by the detection products of maltose. Glucose exhibits minimal fouling. The i–t_{del} curves of most monosaccharides and disaccharides will fall off at rates intermediate to those for glucose and maltose. One's first inclination is to select the shortest time possible for t_{del} in order to obtain the maximum signal, but this would be assuming that S/N follows the anodic response curve. This is not the case. At short times, the noise is much more prevalent, and, as a result, the S/N is less. The value $t_{del} = 240$ ms is chosen as optimal for general detection of glucose and maltose.

As with t_{del}, the "background-corrected" i–t_{int} plot is less informative than the S/N plot. Figure 6.8 shows the PV response for glucose and maltose as a result of variation in t_{int}. The anodic response of these compounds is only minimally affected by varying t_{int}. The S/N plot suggests that the $t_{int} > 200$ ms is slightly better for both compounds. A further consideration supporting this choice is that 200 ms is a multiple of 16.667 ms, or 1 60 Hz^{-1}. As discussed in Chapter 4, contributions from noise originating from 60- and 50-Hz power sources can be minimized by using multiples of 16.667 ms. The significance of the optimization of t_{del} and t_{int} is best illustrated with the use of chromatograms. Figure 6.9*A* shows the chromatogram using PAD at the optimized conditions determined thus

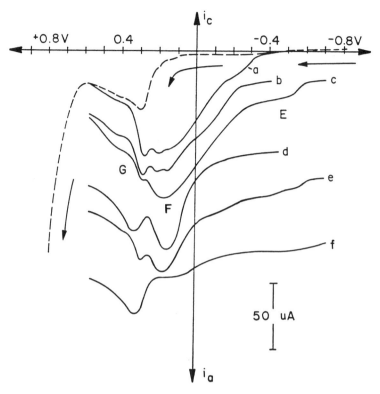

Figure 6.6. Voltammetric response (i–E) for six carbohydrates in deaerated 0.1 M NaOH. Positive scans from -0.8 to 0.8 V and currents greater than the background residual are shown. Each scan has been offset as indicated. Curves: (---------) residual; (———) (*a*) 0.15 mM sorbitol, offset; 0 μA; (*b*) 0.18 mM fructose, offset: 40 μA; (*c*) 0.15 mM glucose, offset: 60 μA; (*d*) 0.29 mM sucrose, offset: 80 μA (*e*) 0.15 mM maltose, offset: 100 μA; (*f*) maltoheptaose, 0.09 mM, offset: 120 mA.

far (i.e., E_{det} = +200 mV, t_{del} = 240 ms, and t_{int} = 200 ms). Figure 6.9*B* shows the effect of reducing t_{del} to 20 ms. The baseline drift is worse, and the S/N has diminished. Figures 6.9*C* and 6.9*D* show the same chromatograms as *A* and *B* except that t_{int} = 20 ms in both cases. Note the dramatic deterioration in S/N and baseline stability. Recently, advances in waveform optimization software from the author's laboratory has resulted in combining the optimization of t_{del} and t_{int} into a single experiment with the use of a deconvolution algorithm. The result is that the optimization of these two parameters can be optimized in <30 s for n = 25 cycles. Under these conditions, extraneous long-term noise and drift are eliminated from the data.

The anodic signal derived from the oxidation of the analyte does not reach a steady-state value for the majority of carbohydrates over the lifetime of the detection period. This attenuation is attributed to fouling of the electrode surface

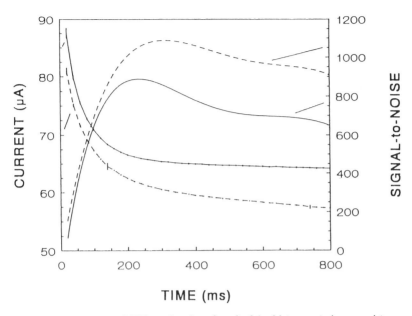

TIME (ms)

Figure 6.7. PV response and S/N as a function of t_{del} for 0.1 mM (———) glucose and (---------) maltose at the Au RDE in 0.1 M NaOH. Reprinted with permission from W. R. LaCourse and D. C. Johnson, *Anal. Chem.* **65**, 50–55 (1993). Copyright 1993 American Chemical Society.

by adsorbed detection products and, depending on the potential, a limited amount of surface oxide. This fouling is reflected in an attenuation of the anodic signal over extended periods of time, and it also accounts for the lack of success of detection of carbohydrates at Au (and Pt) electrodes with constant applied potentials. Fortunately, we can overcome the effects of the fouling by incorporating pulsed potential cleaning in the overall cycle.

The anodic potential pulse induces the formation of stable surface oxide, which results in "oxidative cleaning" of the noble-metal electrode. Fouling products are oxidatively desorbed from the electrode surface. Figure 6.10A shows the PV response of E_{oxd} at $t_{oxd} = 180$ ms. It is important to remember that even though we are changing the anodic pulse, we are observing the effect on the detection. The experiment commences with a clean electrode surface. The once-clean electrode surface becomes progressively fouled during the experiment as E_{oxd} is changed from -800 to $+300$ mV. Over this range no surface oxide is being formed during the anodic pulse, and, as a consequence, no cleaning results. Commencing at approximately $+300$ mV, the amount of surface oxide formed is sufficient to evoke oxidation cleaning of the electrode surface, and the anodic signal begins to increase.

As discussed in Chapter 5, the extent of oxide formation is dependent on the E_{oxd}, and the amount of oxide formation will determine the effectiveness of the

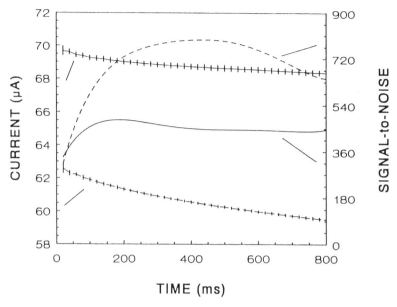

Figure 6.8. PV response and S/N as a function of t_{int} for 0.1 mM (————) glucose and (- - - - - - -) maltose at the Au RDE in 0.1 M NaOH. Reprinted with permission from W. R. LaCourse and D. C. Johnson, *Anal. Chem.* **65**, 50–55 (1993). Copyright 1993 American Chemical Society.

cleaning. Thus, one can use a higher potential for a short duration or a lower potential for a longer period of time. The lower limit of E_{oxd} is determined by the potential at which oxide formation commences, and the upper limit is determined by solvent breakdown and the extent of bubble formation. The limits of t_{oxd} are a function of the minimum time needed to form a monolayer of surface oxide in the presence of the analyte and its contribution to the overall cycle time. As noted in Figure 6.10*B*, the maximum response of glucose at all E_{oxd} potentials between +600 and +1000 mV is achieved in ~150–200 ms. Cleaning of the electrode appears to be most rapidly achieved at 180 ms. Hence, $E_{oxd} = +800$ mV and $t_{oxd} = 180$ ms are chosen for optimal response of carbohydrates.

The third potential pulse has two major functions, which are removal of the surface oxide formed during the application of E_{oxd} and adsorption of analyte to the electrode surface. The oxide formed during E_{oxd} leaves the electrode in an inert state, and E_{red} and t_{red} are chosen to complete reductive dissolution of the surface oxide. Figure 6.11 shows the variation of the PAD response for glucose as a function of E_{red} for $t_{red} = 300$ ms. The response is constant for $E_{red} = -800$ to +100 mV, which indicates that sufficient cleaning is taking place within the 300 ms to completely remove the oxide layer from the electrode surface. At $E_{red} > +200$ mV, reductive dissolution of surface oxide does not occur and carbohydrate response is terminated. Incomplete removal of surface oxide will result in

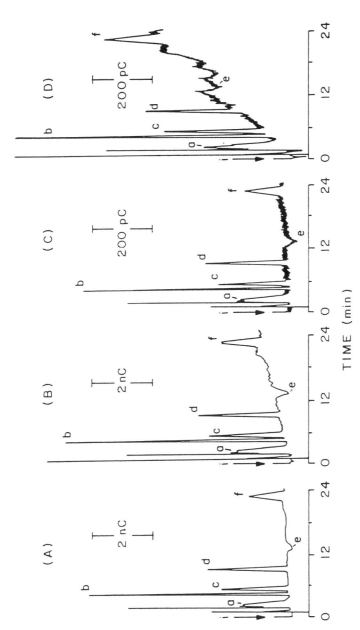

Figure 6.9. Comparison of LC–PAD results with variations of t_{del} and t_{int} in the PAD waveform. Parameters: (A) $t_{del} = 240$ ms, $t_{int} = 200$ ms; (B) $t_{del} = 20$ ms, $t_{int} = 200$ ms; (C) $t_{del} = 240$ ms, $t_{int} = 20$ ms; (D) $t_{del} = 20$ ms, $t_{int} = 20$ ms. Peaks (100 pmol): (a) sorbitol; (b) glucose; (c) fructose; (d) sucrose; (f) maltose. Comment: peak e represents injected dissolved O_2. Reprinted with permission from W. R. LaCourse and D. C. Johnson, *Anal. Chem.* **65**, 50–55 (1993). Copyright 1993 American Chemical Society.

161

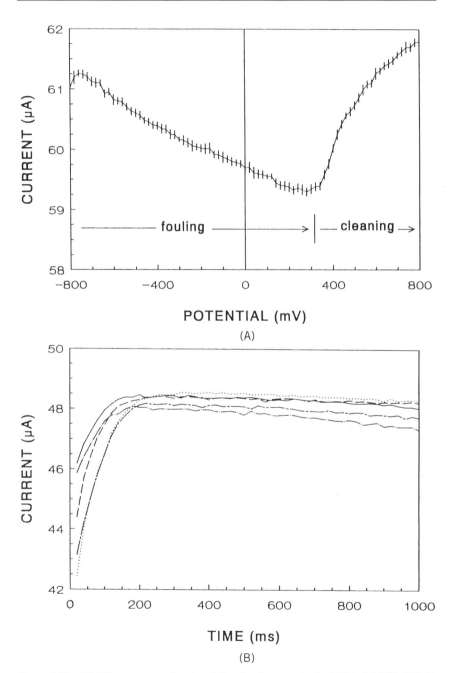

Figure 6.10. (*A*) PV response as a function of E_{oxd} for glucose at the Au RDE in 0.1 M NaOH. (*B*) PV response as a function of t_{oxd} at five E_{oxd} values for glucose. E_{oxd} (mV): (............) 600; (---------) 700; (———) 800; (-..-..-..-..) 900; (——·——) 1000. Reprinted with permission from W. R. LaCourse and D. C. Johnson, *Anal. Chem.* **65**, 50–55 (1993). Copyright 1993 American Chemical Society.

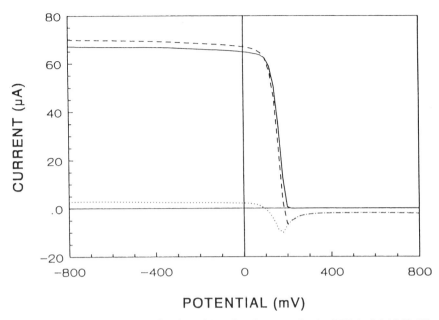

POTENTIAL (mV)

Figure 6.11.   PV response as a function of E_{red} for glucose at the Au RDE in 0.1 M NaOH. Solutions: (............) residual; (--------) 0.1 mM glucose; (_____) difference. Reprinted with permission from W. R. LaCourse and D. C. Johnson, *Anal. Chem.* **65**, 50–55 (1993). Copyright 1993 American Chemical Society.

the reduction of surface oxide during the detection step, which may significantly contribute to the residual response. Since oxide formation and dissolution is very sensitive to fluctuations in temperature, pH, and ionic strength, baselines obtained in HPLC–PAD for waveforms with incomplete oxide reduction will exhibit drifting and noise. Another consequence of inadequate oxide removal is a decreased analytical signal because the electrode surface is not fully available for the detection process. Hence, any value below $+100$ mV could be chosen for E_{red}. Another parameter that can be used to assist the analyst in selecting the optimal potential is the relative standard deviation (RSD) of the response. For the optimization of the PAD waveform for carbohydrates, it was determined that E_{red} $= -300$ mV gave the best signal reproducibility. Thus, $E_{red} = -300$ mV was chosen as the optimal value for this parameter. For complex samples which may contain compounds that strongly foul the electrode surface, an E_{red} value of -800 mV is recommended to promote catalytic–cathodic cleaning of the electrode surface.

   In addition to removal of the oxide during the reduction step, which was formed at E_{oxd}, adsorption of amine- and thio-containing carbohydrates to the "clean," or oxide-free, electrode occurs concomitantly with the removal of surface oxide. Hence, both the potential of maximum adsorption and adsorption time must be considered in the determination of reduction step parameters. On the other hand, no evidence exists to suggest the extensive preadsorption of nonsubstituted carbohydrates under typical PAD conditions. Shown in Figure 6.12 is the PV response of glucose and maltose as a function of t_{red} for E_{red} = -300 mV. The response for glucose reaches a plateau for $t_{red} > 100$ ms, whereas t_{red} over ca. ~ 300 ms is required to reach a maximum response for maltose. A t_{red} of 360 ms is recommended as optimal in the general application of PAD for carbohydrates.

It is important to note that oxide removal and analyte adsorption can continue to occur during the detection step, and, as a consequence, the detection step parameters and reduction adsorption pulse parameters are interdependent. The optimal PAD waveform values for carbohydrate detection at a gold electrode in 0.1 M NaOH are listed in Table 4.1 for a ~ 1-Hz waveform. Limits of detections using the above are routinely 1–10 pmol injected.

For faster waveforms, which are required to maintain peak integrity and system efficiency in microchromatographic and capillary electrophoresis systems, a compromise must often be made in the time of each potential step. The use of microelectrodes in these systems greatly reduces double-layer charging effects, and as a consequence, the time period of the detection step can be shortened to under 100 ms with no adverse effects, especially for mode I detec-

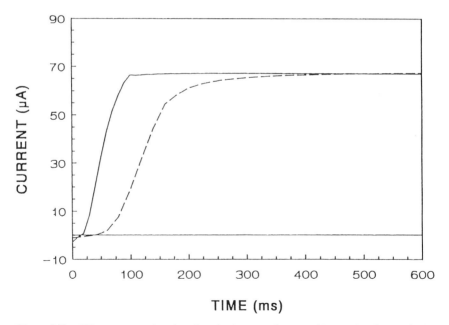

Figure 6.12. PV response as a function of t_{red} for (———) glucose and (---------) maltose at the Au RDE in 0.1 M NaOH. Reprinted with permission from W. R. LaCourse and D. C. Johnson, *Anal. Chem.* **65**, 50–55 (1993). Copyright 1993 American Chemical Society.

tions. On the other hand, fouling may be more serious for microelectrodes because of the higher fluxes. The work of Roberts and Johnson [10,11] clearly demonstrated that the majority of oxide formation and removal occurs in under 50 ms. In addition, the rate of formation is potential-dependent. Thus, shorter time periods for the formation and dissolution of surface oxide can be achieved at higher and lower potentials, respectively. These severe potential excursions may also assist in eliminating fouling species from the electrode surface. Hence, waveforms optimized for frequency and sensitivity are expected to be on the order of 5–10 Hz.

Amines and Amino Acids

As discussed in Chapter 4, amine-based and sulfur-containing compounds are mode II detections, and oxidation is expected to commence with the onset of oxide formation. Figure 6.13 shows the pulsed voltammograms of lysine in 0.1 M NaOH. Note that the response of lysine is concomitant with oxide formation.

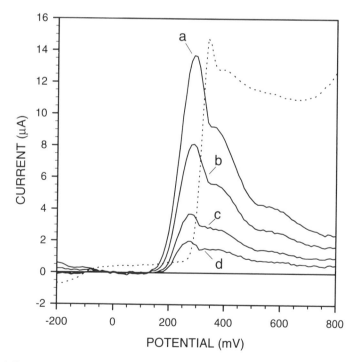

Figure 6.13. Pulsed voltammetric response of lysine at a Au RDE in 0.1 M NaOH. This plot is background-corrected. Solutions: (_____) lysine, (*a*) 100 μM, (*b*) 50 μM, (*c*) 20 μM, (*d*) 10 μM; (............) 0.1 M NaOH, deaerated. Reprinted from W. R. LaCourse in *Electrochemical Detection in Liquid Chromatography and Capillary Electrophoresis* (P. Kissinger, ed.), in press, by courtesy of Marcel Dekker.

The "background-corrected" response of lysine shows a peak at approximately +270 mV. The response of lysine, which is typical of other amines, is prominent in this region because of the presence of a highly catalytic oxide intermediate present at approximately +180 to +270 mV [see residual response (............) in Fig. 6.13].

As with carbohydrates, t_{del} plots for lysine show a large initial response, which quickly diminishes with time. Unlike glucose, the response does not level off. This effect is understandable for two reasons. First, the majority of the signal is derived from the oxidation of the preadsorbed analyte, and to a lesser extent from mass transport to the electrode surface simultaneously with the detection process. This observation is supported by hydrodynamic and scan rate studies, which show that the response is mostly under the control of surface processes. Second, amine compounds foul the electrodes to a greater extent than do carbohydrates. Hence, the combination of these effects will give rise to a chronoamperogram of the detection step that is steadily diminishing. Again, the optimal t_{del} should be gleaned from both response and S/N plots. For lysine, the optimal t_{del} is chosen to be 200 ms (see Table 6.2).

Since the signal is constantly dropping for many amine-based compounds, as one extends t_{int}, more and more attenuated signal is being averaged into the overall response. Hence, at longer t_{int} periods, the magnitude of the response becomes smaller. Again, it is S/N that is of the utmost importance. The combined effects of greater signal early in the detection step due to oxidation of the preadsorbed analyte and the constant diminution of the mass-transport-derived signal can often lead to optimized delay times that are shorter than those for carbohydrates. The optimized t_{int} value for lysine is ~240 ms.

The oxidation, or cleaning potential pulse, for amine oxidations is complicated by the fact that oxidation of the amine compound continues during the cleaning pulse. The adsorbed amine and its oxidation compete with the formation of surface oxide. Hence, the time of the cleaning pulse must often be extended in duration or increased in magnitude to fully form the surface oxide layer at the electrode surface. The magnitude of E_{oxd} is limited by the commencement of bubble formation in the cell. For lysine, E_{oxd} is +800 mV and applied for 180 ms.

Strong evidence exists that amine compounds are reversibly adsorbed at po-

TABLE 6.2 Optimal PAD Waveform Parameters for Lysine in 0.1 M NaOH

Potential (mV vs. Ag/AgCl)			Time (ms)		
Parameter	General	Optimized	Parameter	General	Optimized
E_{det}	+200 to +400	+270	t_{det}	>150	440
			t_{del}	>100	200
			t_{int}	>50	240
E_{oxd}	+300 to +800	+800	t_{oxd}	>100	180
E_{red}	−800 to +100	−400	t_{red}	>80	360

tentials greater than −300 mV at a Au electrode in 0.1 M NaOH. This would mean that if the reduction potential were chosen to be −600 mV, little or no adsorption of the analyte would occur, and the analyte signal derived on going to the detection potential would be derived from analyte adsorption that would take place during the detection step and mass-transport-derived signal. A much more reproducible signal of greater magnitude can be obtained if maximum adsorption of the analyte is accomplished during the reduction pulse. The optimal potential would be chosen to both reduce the surface oxide formed during the oxidation pulse and invoke adsorption of the analyte of interest. Both of these effects are a function of time, and as a consequence, the reduction pulse is often extended as compared to carbohydrates. This affect is even greater for sulfur compounds. The waveform parameters E_{red} and t_{red} are chosen to be −400 mV and 360 ms, respectively. Figure 6.14 highlights the importance of selecting the correct reduction potential and time parameters for amine-based analyte detection. The response for simple alkyl amines is clearly greater for the reduction potential of (*a*) −400 mV versus (*b*) −800 mV for a fixed t_{red} of 200 ms due to greater adsorption. Similarly, longer t_{red} periods at E_{red} of −400 mV also result in greater signals for the same reason. In comparison to carbohydrates, longer t_{red} periods

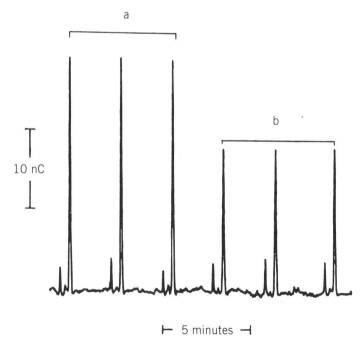

Figure 6.14. Adsorption effect on ethylamine (50 μM) response in PAD at (*a*) E_{red} = −0.4 V and (*b*) E_{red} = −0.8 V. Reprinted with permission from D. A. Dobberpuhl and D. C. Johnson, *Anal. Chem.* **67**, 1254–1258 (1995). Copyright 1995 American Chemical Society.

may be required to allow for greater adsorption of the analyte to the electrode surface. The effect of longer t_{red} periods at the appropriate E_{red} results in greater signals for amine and sulfur-containing compounds. The interdependence of the waveform parameters becomes much more apparent with the optimization of amine and sulfur compounds than for carbohydrates. Since preadsorption of the analyte occurs during the negative-potential pulse, the magnitude of the anodic signal derived during the detection step will be influenced by the parameters of the cathodic pulse. Hence, optimization of t_{del} and t_{int} should be repeated after optimization of the oxidation and reduction steps of the PAD waveform.

Thiocompounds

At $E_{det} = +200$ mV in 0.1 M NaOH at a Au electrode, all carbohydrates, amine-based compounds, and many sulfur compounds are detectable. Hence, if one is concerned with selectivity, a compromise must be made between optimal S/N and simpler chromatograms. For instance, when detecting carbohydrates, a more negative detection potential would be chosen, and the response for amine-based compounds would be greatly attenuated. In fact, carbohydrate detections are often performed at +50 mV.

One of the advantages of PV over HPLC-based methods of optimization is that many solvent conditions can be tested in a relatively short period of time. Although thiocompounds can be detected under alkaline conditions, selective detection of thiocompounds can be performed at Au electrodes under mildly acidic conditions. The use of PV for the proper selection of waveform parameters is even more essential for thiocompounds, due to strong adsorption and the consequential fouling of the electrode by these types of compounds.

The pulsed voltammograms of several thiocompounds in phosphate buffer pH 3.0/ACN (95/5) at a Au electrode are shown in Figure 6.15. The "background-corrected" responses show a maxima at approximately +1300 mV (see Table 6.3). The strong adsorption of sulfur compounds is reflected in the forcing of the peak maxima to more anodic potentials then amine-based and polyhydroxyl compounds due to the suppression of oxide formation at lower potentials. Hence, the background response in the presence of the analyte is different from that in the absence of the analyte, and, when one subtracts residual from the sample voltammogram, a dip occurs in the potential region of +1000 to +1100 mV. This dip is an artifact of the subtraction and is attributed to the fact that we do not know the "true" background in the presence of the analyte. The "background-corrected" response for thiocompounds is legitimized for determining the response in HPLC–PAD, because the peak, or response in HPLC–PAD, is derived from subtracting the baseline (i.e., background signal in the absence of analyte) from the peak response (i.e., signal in the presence of the analyte). In practice, negative peaks are observed at these potentials in HPLC–PAD for thiocompounds.

The parameters E_{oxd}, t_{oxd}, E_{red}, and t_{red} are optimized similarly as for car-

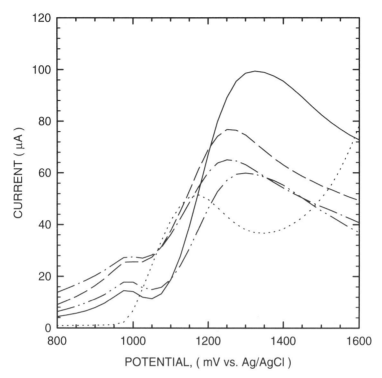

POTENTIAL, (mV vs. Ag/AgCl)

Figure 6.15. Pulsed voltammetric background-corrected response for thiocompounds at a Au RDE. Solutions (98 μM): (-.-.-.-.-..) DTE; (---------) dithiane; (_.._.._.._) AET; (_____) cystamine; and (............) residual. Reprinted from W. R. LaCourse and G. S. Owens, Pulsed electrochemical detection of thiocompounds following microchromatographic separations, *Anal. Chim. Acta.* **307**, 301–319 (1995) with kind permission of Elsevier Science–NL, Sara Burgerhartstraat 25, 1055 KV Amsterdam, the Netherlands.

TABLE 6.3   Optimal PAD Waveform Parameters for Sulfur-Containing Compounds in 95% 100 mM Phosphate Buffer (pH 3)/5% ACN

Potential (mV vs. Ag/AgCl)		Time (ms)	
Parameter	Optimized	Parameter	Optimized
E_{det}	+1300	t_{det}	240
		t_{del}	140
		t_{int}	100
E_{oxd}	+1500	t_{oxd}	180
E_{red}	−600	t_{red}	500

bohydrates and amine-based compounds. The complications encountered with amine-based compounds are even more dramatic for thiocompounds. For instance, Figure 6.16 shows the effect of t_{del} on response. Note how the amplitude of the "background-corrected" response reaches a maximum at ~90 ms, and then quickly and constantly diminishes. As with the amines, the majority of the response is derived from the preadsorbed analyte, which is oxidized early in the chronoamperogram, and the lesser contribution from mass-transport-derived response, which is reflected later on in the plot. Throughout the detection step, fouling of the electrode is occurring and inert oxide is forming. The S/N reaches a plateau at ~260 ms (Fig. 6.16). The importance of preadsorption of thiocompounds is reflected in the plots of current versus t_{red} and various reduction potentials as shown in Figure 6.17. Note the adsorption of the thiocompound, dithiane, is virtually independent of the reduction potential, which must be sufficiently negative to reduce the surface oxide. On the other hand, the adsorption of dithiane requires almost 1500 ms to fully saturate the surface. Since the overall cycle time is limited in order to preserve peak integrity and chromatographic resolution, t_{red} is often chosen to be less than the time required for full coverage of the electrode surface. This may lead to an increase in the RSD of the detection system. For the detection of thiocompounds, the delay and integration times are

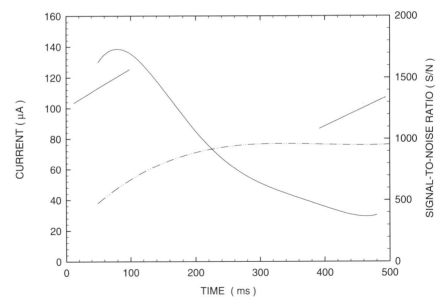

Figure 6.16. Pulsed voltammetric background-corrected response (_____) and S/N (_··_··_) of t_{del} for 98 μM dithiane at a 1-mm Au RDE in 95% 100 mM phosphate buffer/5% ACN. Reprinted from W. R. LaCourse and G. S. Owens, Pulsed electrochemical detection of thiocompounds following microchromatographic separations, *Anal. Chim. Acta.* **307**, 301–319 (1995) with kind permission of Elsevier Science–NL, Sara Burgerhartstraat 25, 1055 KV Amsterdam, the Netherlands.

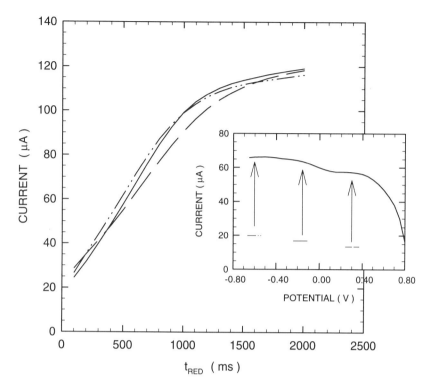

Figure 6.17. Pulsed voltammetric background-corrected response for dithiane (98 μM) as a function of t_{red} and three E_{red} values. *Inset*: pulsed voltammetric response as a function of E_{red} for dithiane at a 1-mm Au RDE in 95% 100 mM phosphate buffer/5% ACN. E_{red} (V): (---------) +.30, (————) −0.15, (−.−.−.−) −0.60. Reprinted from W. R. LaCourse and G. S. Owens, Pulsed electrochemical detection of thiocompounds following microchromatographic separations, *Anal. Chim. Acta.* **307**, 301–319 (1995) with kind permission of Elsevier Science–NL, Sara Burgerhartstraat 25, 1055 KV Amsterdam, the Netherlands.

reduced slightly from their optimized values in favor of extending t_{red}, which gives an overall greater response. If the thiocompounds of interest are of sufficient peak width, then the use of a longer t_{red} period ($\sim$1500 ms) and t_{del} would almost triple the response while also increasing its reproducibility.

Probably the most important question to ask is whether the PV response is truly a reflection of the response one would see in a flow-through electrochemical cell. Figure 6.18 compares the response of the E_{det} plot using the optimized waveform parameters at a RDE and the E_{det} in the flow-through electrochemical cell derived from measuring the peak heights as E_{det} is incrementally changed. The overall comparison of the two techniques is excellent.

As research continues, the time of optimization is being drastically reduced. Presently, a single parameter of the PAD waveform is optimized by incrementing

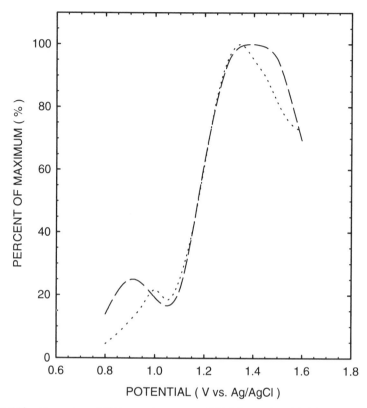

Figure 6.18.   Comparison of PV (...........) response to HPLC (---------) peak heights. Analyte: DTE, 100 μM. Reprinted from W. R. LaCourse and G. S. Owens, Pulsed electrochemical detection of thiocompounds following microchromatographic separations, *Anal. Chim. Acta.* **307**, 301–319 (1995) with kind permission of Elsevier Science–NL, Sara Burgerhartstraat 25, 1055 KV Amsterdam, the Netherlands.

it and holding all other waveform parameters constant. Hence, the experiment consists of applying a series of PAD waveform pulses to the electrochemical cell in both the presence and absence of analyte. For instance, optimization of t_{del} or t_{int} involves using separate PAD cycles to measure sequential values of t_{del} or t_{int}. If we were to optimize both t_{del} and t_{int} ($n = 10$, cycle = 1 s, time step = 2 ms, range = 10–510 ms), the experiment (data collection) by the existing method would take ~2.8 hr.

We have recently implemented an improved data collection and reduction strategy for all time parameters, which results in a significant time savings. For each PAD cycle of an E_{det} optimization waveform, the entire chronoamperogram of the detection step is stored in a matrix, which results in a three-dimensional plot of current versus E_{det} versus t_{det} (see Fig. 6.19). The matrix for each E_{det} value is then deconvoluted to yield a three-dimensional plot of current versus t_{del}

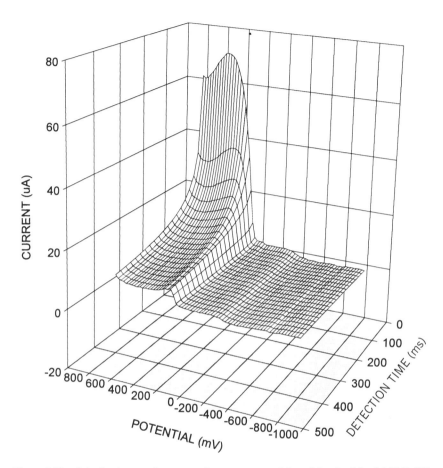

Figure 6.19. Pulsed voltammetric response (current vs. potential vs. delay time) for 0.1 M NaOH.

versus t_{int}. Hence, the data from one PAD cycle may be used to obtain the response at several t_{del} and t_{int} values simultaneously. Data collecion is greatly reduced at the expense of data manipulation, which is easily handled with today's computer systems. Figure 6.20*A* shows the surface plot of S/N for the optimization of t_{del} and t_{int} for glucose. Figure 6.20*B* shows the contour plots for the same data in Figure 6.20*A*. Color plots would do these plots much more justice. The optimized values of t_{del} (240 ms) abd t_{int} (200 ms) found by pulsed voltammetry are in good agreement with the deconvolution approach. In contrast, the experiment described above using the new strategy takes only 20 s. It is important to remember that all the data can be collected in a single experiment. In addition, the dependence of the parameters is better accounted for using this technique. We are presently expanding this technique to the other waveform parameters and on-line optimization.

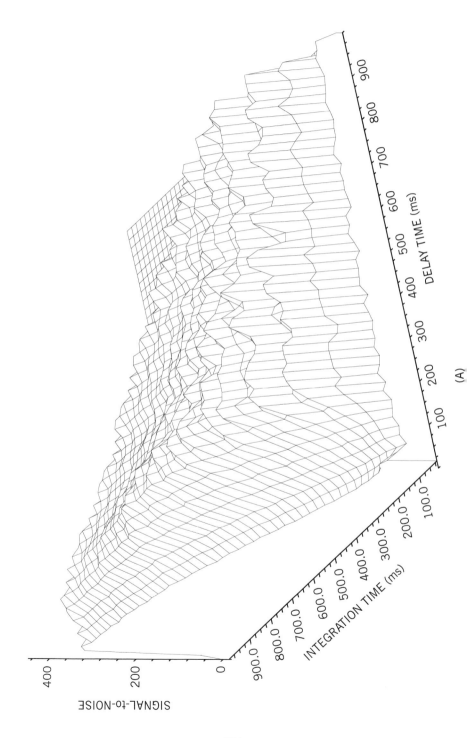

(A)

174

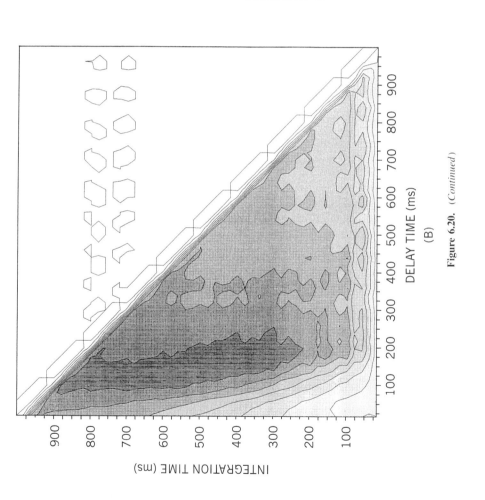

Figure 6.20. (*Continued*)

RPAD AND APAD WAVEFORM OPTIMIZATION

The use of PV for the optimization of RPAD is straightforward in that the E_{oxd} and E_{red} potentials are reversed in their application to the electrode. This may either be accomplished by inputting E_{red} for E_{oxd} and E_{oxd} for E_{red} or altering the order of the steps in the program. Otherwise, the procedure for optimization is the same for PAD.

Activated pulsed amperometric detection (APAD), which is yet to find widespread use, requires the addition of one more potential pulse to program, which is executed after the reduction pulse. Williams and Johnson [6] have performed this technique using "activated pulsed voltammetry." Their work concluded that a high potential pulse of a short duration is optimal. This agrees well with theory.

IPAD WAVEFORM OPTIMIZATION

The IPAD waveform is best perceived as a combination of CV and PV. Presently, work is under way to develop an IPAD optimization program similar to that of PV. Until that work is complete, the cyclic step in the IPAD waveform is best studied using CV, while the cleaning pulses are best optimized using PV. There are several requirements pertaining to the cyclic sweep of E_{det} in IPAD that must be satisfied to achieve maximum success for applications to HPLC. These considerations are as follows:

1. The cyclic scan of E_{det} must begin (E_{dst}) and end (E_{dnd}) at a value for which the electrode is free of surface oxide (i.e., $E <$ ca. $+0.1$ V vs. SCE for Au in 0.1 M NaOH).

2. The value of E_{dnd} should not extend into the region for cathodic detection of dissolved O_2 (i.e., $E < -0.1$ V vs. SCE), if dissolved O_2 is present.

3. The positive scan maximum (E_{dmx}) must not extend beyond the value for anodic solvent breakdown (i.e., ca. $+0.7$ V vs. SCE).

It is to be noted from the residual $i-E$ curve for Au in Figure 5.1B that only a small potential region centered at approximately -0.1 V vs. SCE is appropriate to satisfy criteria (1) and (2) above in 0.1 M NaOH containing dissolved O_2. The purging of solvents with He and use of O_2-impermeable tubing can result in virtually O_2-free conditions, which greatly relaxes constraints resulting from criteria 1 and 2.

Figure 6.21 shows the PV of alanine at a Au electrode in 0.1 M NaOH. Notice that the majority of the analytical signal is in the region of approximately $+100$ to $+500$ mV. Although signal is present over the range of approximately $+500$ to $+800$ mV, it quickly decays as the oxide layer on the electrode surface is formed rapidly and the signal becomes very noisy. Thus, although the third criterion specifies that one should not scan into potentials at which O_2 evolution occurs, I recommend that a greater S/N can be achieved by not scanning to the possible

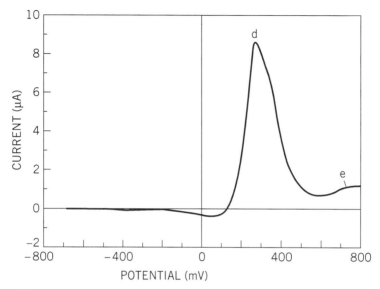

Figure 6.21. Pulsed voltammetric response of 10 mM alanine at a Au RDE. This plot is background-corrected. Waves: d, amine oxidation under mass-transport control; e, amine oxidation under surface control. Reprinted from W. R. LaCourse, Pulsed electrochemical detection at noble metal electrodes in high performance liquid chromatography, *Analusis* **21**, 181–195 (1993) with kind permission of Elsevier Science.

limit of this criterion. As discussed in Chapter 5, it is evident that the analytical signal quickly diminishes in the first 200 ms, especially at oxide-forming potentials where the rate of oxidation is rapid.

At present, the IPAD waveform should be optimized by utilizing the criteria presented above in relation to CV data and using PV to optimize the pulsed potential steps as described for optimizing PAD waveforms. Table 6.4 lists both the general and accepted IPAD waveform parameters under several pH conditions. Since IPAD scans through a range of potentials, all compounds that are PED-active under the experimental conditions are detected. Hence, IPAD offers less selectivity than PAD.

OTHER SYSTEM CONSIDERATIONS

All carbohydrates that lack a formal charge are weakly acidic with pK_a values in the range of ~ 12–14. As a result, carbohydrates are readily separated in highly efficient anion-exchange columns using alkaline mobile phases. The chromatographic k' values increase with decreasing pK_a and increasing molecular weight. In addition, alditols can be separated at even higher pH values. Amine-

TABLE 6.4   Optimized IPAD Waveform Parameters at Different pH Conditions

Parameter Potential, mV; Time, ms	NaOH 16 mM, pH 12	Borate Buffer 20 mM, pH 9.0	Phosphate Buffer 10 mM, pH 8.0	Acetate Buffer 10 mM, pH 5.5
E_{dst}	−350	−250	−300	−350
E_{dmx}	1100	1300	1400	1500
E_{dnd}	−350	−250	−300	−350
E_{oxd}	1200	1400	1500	1700
E_{red}	−1100	−850	−800	−650
t_{dst}	20	20	20	20
t_{dup}	181	194	210	218
t_{dwn}	181	194	210	218
t_{dnd}	0	0	0	0
t_{det}	382	408	440	456
t_{del}	20	20	20	20
t_{int}	362	388	420	436
t_{hld}	0	0	0	0
t_{oxd}	100	100	100	100
t_{red}	200	200	200	200

based compounds are separated under cation-exchange conditions, and thiocompounds are typically separated under reversed-phase conditions. Therefore, it is important to consider the optimization of PED waveforms as a function of all variables normally considered when the various chromatographic separations are optimized (pH, ionic strength, organic modifier concentration, temperature, etc.). Because these parameters can influence the rates for formation and dissolution of surface oxide and the sensitivity of the electrocatalytic response mechanisms at noble-metal electrodes, the requirement of a definitive method for optimization of PED waveforms is well warranted.

The effect of pH variation on the oxide formation process can be explained on the basis of the pH-dependent nature of the oxide formation reaction [i.e., $Au(H_2O)_{ads}$ ---≫ $Au–OH + H^+ + e^-$]. The optimal E_{det} value corresponds approximately to the potential for the onset of oxide formation. Furthermore, optimal values of E_{oxd} and E_{red} are approximately fixed with respect to the potential for the onset of oxide formation. Hence, for a given change in pH (ΔpH), the optimal values of E_{det}, E_{oxd}, and E_{red} must be adjusted by the amount -60 mV per pH unit. Table 6.4 shows that the same basic IPAD waveform is used for many different pH and buffer conditions by merely changing the potentials to compensate for the changes in pH. For pH gradients, the effect of the shifting optimal potential can be overcome with the use of a pH electrode as the reference electrode.

Typical ionic strengths (μ) in HPLC–PED are in the range of 50–100 mM, and variations within this range are of virtually no consequence to the optimization of PED waveform parameters. However, under conditions of ion-gradient elution, baseline drifts have been observed. This effect can be overcome by the postcolumn addition of a solution with high ionic strength to effectively "buffer" the ionic strength in the effluent stream passing through the electrochemical cell.

The presence of organic modifiers to the mobile phase in HPLC–PED can have a much greater consequence on the optimized PED waveform than is caused by changes in μ and pH. This is true especially when the modifiers are strongly adsorbed at the electrode surface and interfere with access by the analyte to specific reaction sites on the electrode surface with corresponding attenuation of the analytical response. Furthermore, adsorbed organic modifiers can suppress the oxide formation process with the necessity of increasing E_{oxd} values.

Large temperature fluctuations can cause significant changes in the sensitivity of PED response, especially for compounds whose detection is under mixed or surface control. Whereas the majority of present-day HPLC systems rely on passive thermostatic control normally found in air-handling systems of well-designed analytical laboratories, PED will be more reproducible with active thermostatic control at $<1°C$ precision.

PULSED VOLTAMMETRY (PV): A QUANTITATIVE AND MECHANISTIC TOOL

Since PV is based on using a train of PED waveforms, the technique should and does have the same quantitative aspects as PED itself. Figure 6.22 shows the calibration curve for glucose at +200 mV. Glucose is linear over four decades with a detection limit of 100 nM. These data agree well with published literature results for HPAEC–PAD. The detection limit in concentration units can be reduced by using a smaller cell volume. Interestingly, since the time of the PV experiment is not time-limited, numerous PAD cycles can be averaged to increase S/N. Earlier in this chapter, the effect of the number of cycles on the response of glucose was discussed, and a fivefold increase in S/N was reported for 25 pulses as compared to $n = 1$. Hence, PV can also be used as an analytical tool. In addition, there is limited selectivity depending on the detection potential chosen.

Another advantage of PV emulating PED is that mechanisms of detection can be studied using PV. Numerous studies have used pulsed voltammetry to study the rate of oxide growth and dissolution at noble-metal electrodes. In addition, the qualitative aspects of the response have been investigated. For instance, Figure 6.23 shows the response of (———) glucose and (---------) fructose. These compounds differ by the presence of a single aldehyde group. The aldehyde functionality should begin to be detected at approximately -600 mV at a Au electrode in 0.1 M NaOH and reach a plateau. Indeed, when the PV for fructose is subtracted from the PV for glucose, an anodic wave (............) commencing at about -600 mV and reaching a plateau is observed (see Fig. 6.23).

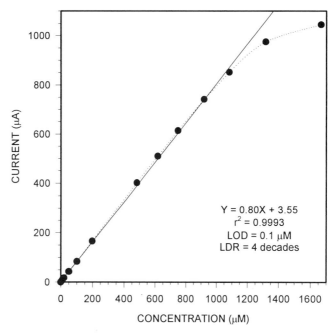

Figure 6.22. Pulsed voltammetric calibration curve for glucose in 0.1 M NaOH.

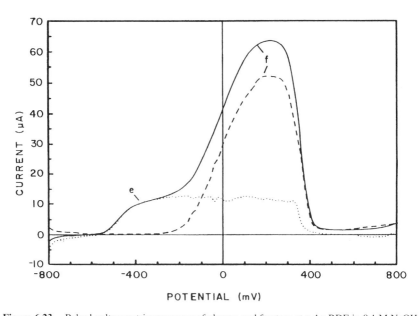

Figure 6.23. Pulsed voltammetric responses of glucose and fructose at a Au RDE in 0.1 M NaOH. These plots are background-corrected. Waves: e, aldehyde oxidation; f - alcohol oxidation. Solutions: (_____) 0.1 mM glucose; (---------) 0.1 mM fructose; and (............) difference or aldehyde response. Reprinted from W. R. LaCourse, Pulsed electrochemical detection at noble metal electrodes in high performance liquid chromatography, *Analusis* **21**, 181–195 (1993) with kind permission of Elsevier Science.

Hence, the quantitative and qualitative nature of PV can be used to investigate mechanisms in PED.

CONCLUSIONS

All the parameters of any PED waveform can be optimized definitively using PV and CV in a batch cell for a variety of possible chromatographic conditions. Because the effects of variations of the waveform parameters are not mutually exclusive, it is important to reexamine the effects of variation of each parameter after approximate optimal values have been determined. PV can ultimately benefit from the use of automated simplex or other optimization protocols. The results from PV in the batch cell agree well with results determined in the flow-thorough electrochemical cell, and the overall goal is to perform PED optimization on-line. PV also can be used as a quantitative tool as well as for mechanistic studies.

REFERENCES

1. G. G. Neuberger and D. C. Johnson, *Anal. Chim. Acta* **192**, 205–213 (1987).
2. G. G. Neuberger and D. C. Johnson, *Anal. Chem.* **59**, 150–154 (1987).
3. G. G. Neuberger and D. C. Johnson, *Anal. Chem.* **59**, 203–204 (1987).
4. P. J. Vandeberg, J. L. Kowagoe, and D. C. Johnson, *Anal. Chim. Acta* **260**, 1–11 (1992).
5. L. A. Larew and D. C. Johnson, *J. Electroanal. Chem.* **262**, 167–182 (1989).
6. D. G. Williams and D. C. Johnson, *Anal. Chem.* **64**, 1785–1789 (1992).
7. P. J. Vandeberg, J. L. Kowagoe, and D. C. Johnson, *Anal. Chim. Acta* **260**, 1–11 (1992).
8. T. K. Chem, Y. Y. Lau, D. K. Y. Wong, and A. G. Ewing, *Anal. Chem.* **64**, 1264–1268 (1992).
9. W. R. LaCourse and D. C. Johnson, *Anal. Chem.* **65**, 50–55 (1993).
10. R. Roberts and D. C. Johnson, Electroanalysis **4**, 741–749 (1992).
11. R. E. Roberts and D. C. Johnson, *Electroanalysis* **6**, 193–199 (1994).

7 Applications of PED

A separation! A separation! My kingdom for a separation!

Up to this point we have focused on the nuts and bolts of pulsed electrochemical detection, but the true test of any analytical technique is how well it works for real-world applications. The number of papers involving PED has increased significantly since its inception, and the tables in Appendix B list virtually all the PED publications to the present time. Many other papers mention PED only in their experimental sections, and this observation strongly reflects the routine use and maturation of this technique. The routine application of PED is especially pertinent for carbohydrate determinations, which encompass nearly two-thirds of all applications (see Fig. 7.1). Rather than attempt to review the individual papers, I will instead focus on the fundamental aspects of PED applications in relation to classes of compounds (e.g., carbohydrates, amine-based compounds, thiocompounds). In this manner, you will be better able to apply PED to your own application.

CARBOHYDRATES AND ALDITOLS

Pulsed amperometric detection has gained prominence since its inception as a sensitive detection system for all carbohydrates and derivatives following their separation by high-performance liquid chromatography. This popularity can be traced to the ease of executing PAD for oxide-free (mode I) detections for

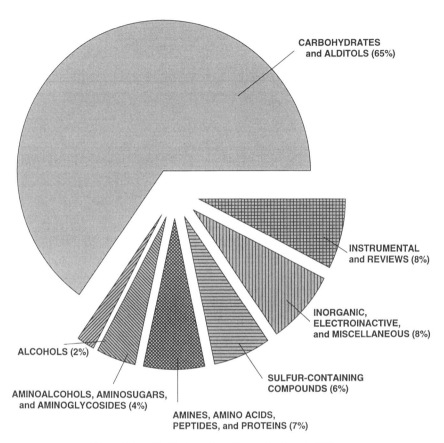

Figure 7.1. Distribution of specific applications in PED.

aldehyde and alcohol-containing compounds at Au and Pt electrodes in alkaline media (pH 12). As discussed earlier, the use of Au electrodes has the distinct advantage that detection can be achieved without simultaneous reduction of oxygen. The anodic signal for polyalcohols and carbohydrates in 0.1 M NaOH is obtained at $E_{det} = +0.0$ to $+0.20$ V versus Ag/AgCl, and a value in this range is typically chosen for E_{det} in the PAD waveform.

Early efforts focused on existing carbohydrate separation modes with post-column addition of NaOH to achieve the high pH required for PAD. Hence, separation of mixtures of carbohydrates was achieved using a Ca(II)-loaded cation-exchange column with a deionized water mobile phase operated at elevated temperature [1,2]. Although the feasibility of PAD was unequivocally demonstrated, transition-metal-loaded columns offered inadequate resolution. Most other separation approaches (e.g., gel filtration, reversed-phase, amino/cyano bonded silica, normal phase, affinity chromatography, borate-assisted anion exchange) for carbohydrates were found to have significant compatibility problems with PAD, inadequate resolution, or limited utility. Nevertheless, some of these separation techniques offer unique advantages to a limited number of applications, and they have been used successfully.

The most versatile and complementary approach to HPLC–PAD is realized when the stationary phase can tolerate the conditions of pH and ionic strength required for optimum PAD detection with no postcolumn addition. This mode of separation was made possible with the advent of alkaline-tolerant polymeric, anion-exchange phases. Presently, these columns offer high performance, high efficiency, and excellent pH tolerance.

Carbohydrates are typically regarded as nonionic, but under highly alkaline conditions they behave as weak acids. As noted in Table 7.1, carbohydrates have pK_a values in the range of 12–14 [3,4]. Hence, chromatographic separations of complex mixtures can be achieved using *high-performance anion-exchange chromatography* (HPAEC) with alkaline mobile phases [5]. Figure 7.2 shows the separation of five common carbohydrates under isocratic conditions of 0.1 M NaOH. Note that the order of separation of glucose and fructose agrees quite well with the pK_a values of the individual compounds. Many other separations are performed at a hydroxide-ion concentration of 10–20 mM, and under these conditions the separation of isomeric monosaccharides (glucose and galactose) are possible. For separations that are not performed under adequately alkaline conditions, the detection requirement of high pH can be satisfied by postcolumn addition of alkaline reagents.

Shown in Figure 7.3 is the separation of alditols and aldoses in 480 mM NaOH, and the order of elution generally correlates with the pK_a values of the carbohydrates. Since sugar alcohols (i.e., reduced carbohydrates) are weaker acids than their nonreduced counter parts (i.e., sorbitol and glucose or ducitol and galactose; Table 7.1), a higher pH is required to induce greater ionization and columns with higher ion-exchange capacity are required for efficient separations. In general, aldoses elute later than their reduced alditol forms. It is important to

TABLE 7.1 Dissociation Constants of Common Carbohydrates in Water at 25°C

Carbohydrate	pK_a
Fructose	12.03
Mannose	12.08
Xylose	12.15
Glucose	12.28
Galactose	12.39
Dulcitol	13.43
Sorbitol	13.60
α-Methyl glucoside	13.71

Source: Data were taken from *Lange's Handbook of Chemistry*, 13th edition.

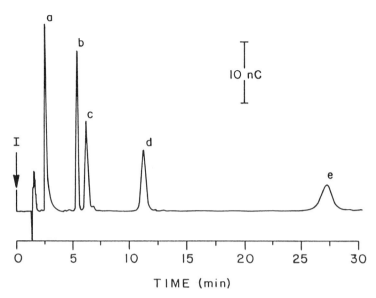

Figure 7.2.   Pulsed amperometric detection of five carbohydrates following isocration separation in HPAEC. Peaks: (*a*) glucitol, (*b*) glucose, (*c*) fructose, (*d*) sucrose, (*e*) maltose. Reprinted from W. R. LaCourse and D. C. Johnson, Optimization of waveforms for pulsed amperometric detection (p.a.d.) of carbohydrates following separation by liquid chromatography, *Carbohydr. Res.* **215**, 159–178 (1991) with kind permission of Elsevier Science–NL, Sara Burgerhartstraat 25, 1055 KV Amsterdam, the Netherlands.

remember that, in addition to the degree of ionization of the sugar, other factors (e.g., charge site accessibility, partitioning, and molecular size) play a role in the separation process, and, as evidenced in Figure 7.3, mannose elutes earlier than expected.

Paskach et al. [4] determined the retention times for 93 sugars and sugar alcohols with one to four residues using HPAEC under isocratic elution with 0.1 M NaOH. Capacity factors for monosaccharides were lowest for sugar alcohols and higher for analogous aldoses and ketoses. In general, capacity factors increased with increasing number of carbon atoms. Both increasing number of hydroxyl groups on a structure and increasing chain length of a homologous series resulted in a corresponding increase in capacity factor.

Since carbohydrates can undergo a number of well-documented reactions (e.g., transformations, eliminations, deacetylation), a major concern with the use of high-pH mobile phases is sample stability. Table 7.2 lists the reactions that may occur in alkaline solutions and observations as to the extent of their effect in conditions pertinent to HPAEC. Except for the decomposition of carbohydrates with strong leaving groups (e.g., D-glucose-3-sulfate), the majority of these reactions are slow and do not affect the chromatography of carbohydrates adversely

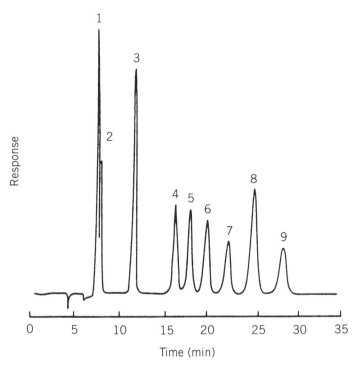

Figure 7.3. HPLC–PAD of several alditols and aldoses. Electrode: Au. Column: Dionex CarboPac MA1. Elution: isocratic with 480 mM NaOH. Peaks: 1, myo-inositol; 2, glycerol; 3, xylitol; 4, sorbitol; 5, dulcitol; 6, mannitol; 7, mannose; 8, glucose; and 9, galactose. Reprinted from D. C. Johnson and W. R. LaCourse/*Carbohydrate Analysis: High Performance Liquid Chromatography and Capillary Electrophoresis* (Z. ElRassi, ed), 1994, with kind permission of Elsevier Science–NL, Sara Burgerhartstraat 25, 1055 KV Amsterdam, the Netherlands.

over the time period of an analytical separation, especially at ambient temperatures.

As noted above for homologous series of oligosaccharides, capacity factors increase in a regular manner with chain length. Since the separation is based on an anion-exchange mechanism, the retention time of oligosaccharides and carbohydrates in general can be decreased by the addition of simple anions (e.g., hydroxide, nitrate, oxalate, acetate ions). Figure 7.4 shows the separation of a homologous series of glucopolymers from glucose (g1) to maltoheptaose (g7) with a mobile phase of 100 mM NaOH/30 mM $NaNO_3$. Although nitrate is a satisfactory "pusher" anion under isocratic conditions, equilibration–regeneration time for the column is excessive (i.e., > 1 h). Hence, isocratic separations of oligosaccharides using strong "pusher" anions are possible, but they are of limited use for gradient systems.

Complex mixtures of carbohydrates are typically achieved using gradient

TABLE 7.2 Reactions Specific to Carbohydrates at High pH

Reaction	Comments/Observations
Lowbry de Bruyn, van Ekenstein transformations (epimerization and tautomerization)	No evidence for the formation of D-glucose or D-mannose from D-fructose Glucose in 150 mM NaOH for 4 days at ambient temperature—no evidence of mannose or fructose *N*-Acetyl glucosamine (GlcNAc) in 100 mM NaOH—epimierization to *N*-acetyl mannosamine (ManNAc) after 2–3 h. Oligosaccharides exhibit 0–5% epimerization; reduced oligosaccharides and sugar alcohols do not epimerize
Deacetylation of *N*-acetylated sugars	GlcNAc in 150 mM NaOH overnight at room temperature—20% hydrolysis to glucosamine GlcNAc and other *N*-acetylated sugars give only one peak by HPAEC
β-Elimination or peeling of 3-*O*-substituents on reducing sugars	Glucopyranosyl-β-1-3 glucopyranose in 150 mM NaOH for 4 h—80% destruction of disaccharide to glucose and an unidentified peak; only one peak by HPAEC D-glucose-3-sulfate decomposes rapidly during HPAEC

Source: Compiled from *Technical Note 20*, Dionex Corp., Sunnyvale, CA.

chromatography. Increased concentrations of OH⁻ above the minimum alkaline conditions required to induce ionization of the carbohydrates results in decreased retention times due to competition of OH^- for anion-exchange sites. Hence, complex mixtures can be effectively separated using NaOH-gradient elution. In addition, the use of OH^- gradients minimizes equilibration time after a chromatographic run. Unfortunately, increases in OH^- concentration results in a corresponding and unacceptable increase in the baseline, which is deleterious to high-sensitivity applications. Baseline changes occur because the optimum potential for carbohydrate detection is very near to the potential for oxide formation. As discussed previously, the onset for oxide formation is pH-dependent, and it shifts by the amount of approximately −60 mV per pH unit. Therefore, increases in OH^- concentration result in a large increase in the background signal because the rate of oxide formation is increasing at the chosen E_{det} value. Figure 7.5*A* shows the separation of seven carbohydrates using a linear gradient from 2 mM NaOH to 200 mM NaOH (0–15 min) followed by isocratic elution with 200 mM NaOH (15–30 min). Note the significant increase in the background signal over the chromatographic run. In order to maintain adequate ionic strength for electrochemical detection, supporting electrolyte (i.e., 0.20 M KNO_3) was added postcolumn.

There are several approaches to overcoming the shift in baseline due to NaOH gradients. The earliest method was to mitigate the effect by the postcolumn addition of concentrated NaOH (e.g., 600 mM), which can reduce the pH change from 2 pH units to less than 0.1 pH unit. This method greatly reduces baseline changes, but high-sensitivity work is still limited. A more sophisticated approach

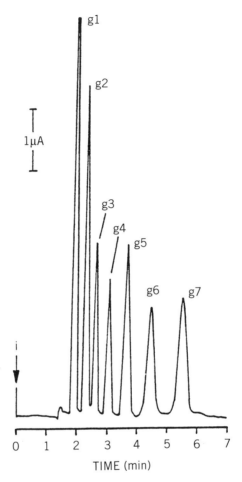

Figure 7.4. Isocratic separation of seven maltooligosaccharides using HPAEC–PAD. Peaks: (g1) glucose, 1.7 nmol; (g2) maltose, 1.2 nmol; (g3) maltotriose, 0.5 nmol; (g4) maltotetraose, 0.4 nmol; (g5) maltopentaose, 0.5 nmol; (g6) maltohexaose, 0.3 nmol; and (g7) maltoheptaose, 0.4 nmol. Reprinted from W. R. LaCourse, Pulsed electrochemical detection at noble metal electrodes in high performance liquid chromatography, *Analusis* **21**, 181–195 (1993) with kind permission of Elsevier Science.

to the rising baseline in Figure 7.5*A* is to substitute a pH reference electrode for the more common fixed-potential Ag/AgCl reference electrode [6]. The potential of the pH reference electrode shifts as a function of pH (i.e., $-60 \, \text{mV} \, \text{pH}^{-1}$) in a fashion similar to that for the onset of oxide formation, and thereby E_{det}, which remains constant relative to the pH reference electrode, automatically shifts by $-60 \, \text{mV} \, \text{pH}^{-1}$. Figure 7.5*B* clearly illustrates the benefit of substituting the pH

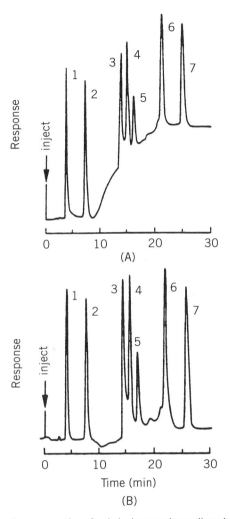

Figure 7.5. Anion-exchange separation of carbohydrates under gradient elution with pulsed amperometric detection using a SCE (*A*) and a glass, pH (*B*) reference electrode. Sugars (ppm): (1) sorbitol (13), (2) fucose (15), (3) glucose (25), (4) fructose (44), (5) sucrose (25), (6) turanose (50), (7) maltose (25). Reprinted with permission from W. R. LaCourse, D. A. Mead, and D. C. Johnson, *Anal. Chem.* **62**, 220–224 (1990). Copyright 1990 American Chemical Society.

reference electrode for the Ag/AgCl reference electrode with respect to eliminating the effect of a NaOH gradient.

Probably the simplest approach to the gradient separations of complex mixtures of carbohydrates is to use a PED-inert anionic species, such as acetate ion, to reduce retention times by competing for anion-exchange sites [7]. The application of acetate gradients in the presence of constant and often dilute concentrations of NaOH has little or no effect on pH, and as a consequence, a Ag/AgCl reference electrode can be used in the PED cell. Figure 7.6 shows the separation using an acetate gradient and the detection of cornstarch hydrolyzates at various degrees of enzymatic degradation, as denoted by the sample's corresponding dextrose equivalent (D.E.) value [8]. Figure 7.6D shows a D.E. = 20, and represents corn syrup. Hence, HPAEC–PAD can be used to "fingerprint" the extent of starch hydrolysis. Numerous examples of the separation of homologous series of carbohydrates, including amylose, inulins, colominic acids, and glactouronic acids, have been published (see Table B.1 in Appendix B).

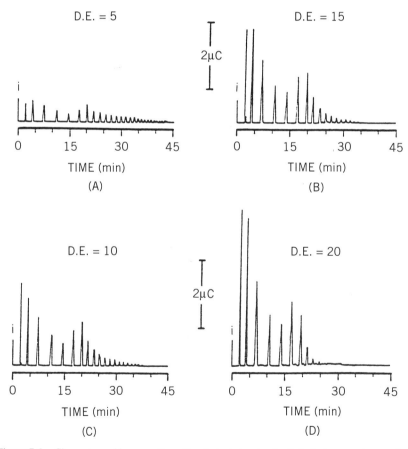

Figure 7.6. Chromatographic separation of carbohydrates in starch hydrolyzate using gradient elution. Solutions: (*A*) DE = 5; (*B*) DE = 10; (*C*) DE = 15; (*D*) DE = 20. Reprinted from W. R. LaCourse, Pulsed electrochemical detection at noble metal electrodes in high performance liquid chromatography, *Analusis* **21**, 181–195 (1993) with kind permission of Elsevier Science.

Under gradient conditions, acetate ion is typically the preferred "pusher" ion in that it offers rapid equilibration, is PED-inactive, and allows for near-maximal resolution of carbohydrate moieties. Recently Wong and Jane compared the effects of acetate and nitrate as pushing agents in HPAEC–PAD for the separation and detection of debranched amylopectins [9]. They found that in comparison with the commonly used acetate ion, nitrate offered greater reproducibility, accuracy, and lower limits of detection. Figure 7.7 shows the HPAEC–PAD chromatographic profiles of the enzymatic debranched tapioca and wheat amylopectins using a nitrate gradient. Impressively, baseline resolution of peaks up to a degree of polymerization (DP) of 66 was achieved. The separations were designed to be completed within 100 min.

In addition to homologous series, complex mixtures of charged sugars can be readily separated using acetate gradients and detected with PAD. A separation of 15 mono- and diphosphorylated monosaccharides is shown in Figure 7.8. This particular separation was performed using a stepped acetate gradient.

Probably the most prominent and significant impact to the pharmaceutical and biotechnology industries of HPAEC–PAD is its application to the separation and quantitation of neutral and aminomonosaccharides [10], as well as positional isomers of glycopeptides [11,12] derived from glycoproteins. Although sequencing and compositional analysis of the protein portion of glycoproteins has been performed for decades, the ability to analyze the glyco portion was nearly impossible. The introduction of HPAEC–PAD facilitates the elucidation of the complex antennary structures of glycoconjugates without the need of any preinjection or postcolumn derivatization chemistry. A chromatogram of the six monosaccharides typically found in the hydrolyzates of glycoproteins is shown in Figure 7.9A. These six sugars are fucose, galactosamine, glucosamine, galactose, glucose, and mannose, and they are separated isocratically using ~16 mM NaOH and a CarboPAC-PA1 column (Dionex). Often 2-deoxyglucose, which is not naturally found in mammalian systems, is added as an internal standard. Aminosugars, or basic sugars, typically elute earlier than their nonsubstituted analogs. Figure 7.9B illustrates the trace-level determination (i.e., 1 pmol injected of each) of these same compounds using HPAEC–PAD. This assay allows for the compositional analysis of a glycoconjugate.

Separations of high mannose, hybrid, complex oligosaccharides, as well as polysaccharides are readily separated and detected using HPAEC–PAD. Figures 7.10A and 7.10B show the structures of synthetic complex oligosaccharides and the separation of this mixture, respectively. Undoubtedly, the most impressive aspect of HPAEC is its ability to resolve closely related structures of glycopeptides. Figure 7.11A shows the structure of two closely related asialylated fetuin glycopeptides. Even though they differ in the linkage of only one carbohydrate, the two compounds are well resolved from one another in HPAEC (Fig. 7.11B). The application of HPAEC–PAD to the characterization of glycoproteins has

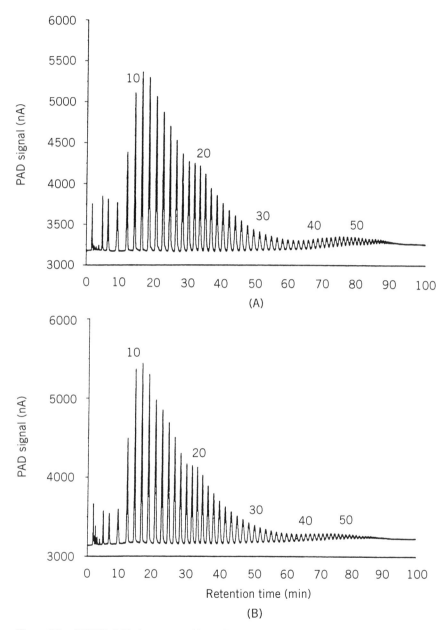

Figure 7.7. HPAEC–PAD chromatographic profiles of the enzymatic debranched (*a*) tapioca amylopectin and (*b*) wheat amylopectin. Peak numbers indicate the degree of polymerization. Reprinted from Ref. (9), p. (73) by courtesy of Marcel Dekker, Inc.

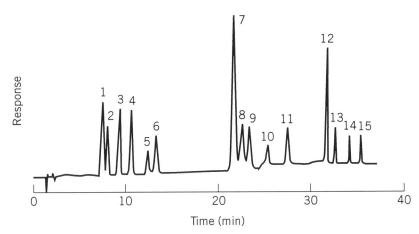

Figure 7.8. Analysis of mono- and diphosphorylated monosaccharides. Column: Dionex CarboPac PA1. Electrode: Au. Eluents: (A) 100 mM NaOH, (B) 100 mM NaOH/1.0 M NaOAc. Elution: isocratic at 90% A/10% B (0-20 min), isocratic at 80% A/20% B (20-30 min), isocratic at 50% A/50% B (30-40 min.) Peaks: 1, α-D-galactosamine-1-P; 2, α-D-glucosamine-1-P; 3, α-D-galactose-1-P; 4, α-D-glucose-1-P; 5, α-D-ribose-1-P; 6, β-D-glucose-1-P; 7, D-glucosamine-6-P; 8, D-galactose-6-P; 9, D-glucose-6-P; 10, D-fructose-1-P; 11, D-fructose-6-P; 12, α-D-glucuronic acid-1-P; 13, α-D-glucose-diP; 14, β-D-fructose-2,6-diP; and 15, D-fructose-1,6-diP (P = phosphate). See Dionex Technical Note TN20 for details. Reprinted with permission from Technical Note TN20, Dionex, March 1989. Copyright 1989 Dionex Corporation.

been reviewed by Townsend [13], and Table B.1 in Appendix B lists additional references.

Bioapplications include the determination of xylitol in serum and saliva [14], sugars in red blood cells [15,16], bile acids [17], and polysaccharides from influenza virus conjugates and polysaccharide vaccines [18]. HPAEC–PAD has been used for the assessment of carbohydrates in HIV patients [19] and as a diagnostic tool for lysosomal diseases [20] and diabetes [21]. Food assays include monosaccharides in grated cheese [22], citrus juices [23,24], dairy products [25], soluble coffee [26], wines [27], and pectic substances in fruits and vegetables [28]. HPAEC–PAD has been invaluable in forensic applications of food authenticity, which includes the determination of adulteration of honey [29], orange juice [30,31] grapefruit juice [32], and maple syrup [33]. Hence, the application of HPAEC–PAD to carbohydrate analysis in real-world samples has been extensive, which reflects the analytical utility of this technique. A complete listing of applications can be found in Appendix B.

Quantitation

A single PAD waveform at a Au electrode can be used to detect virtually all carbohydrates under alkaline conditions. However, electrochemical studies indicate significant differences in reaction dynamics even for very similar carbohy-

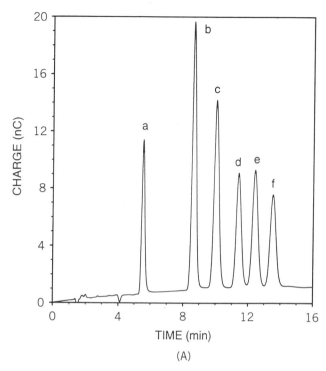

(A)

Figure 7.9. Anion-exchange separation of common monosaccharides found typically in mammalian glycoproteins. Peaks (*A*) 100 pmol each and (*B*) 1 pmol each: (*a*) fucose; (*b*) galactosamine; (*c*) glucosamine; (*d*) galactose; (*e*) glucose; (*f*) mannose. Reprinted from W. R. LaCourse in *Electrochemical Detection in Liquid Chromatography and Capillary Electrophoresis* (P. Kissinger, ed.), in press, by courtesy of Marcel Dekker.

drates. From a mechanistic point of view, the response of the alcohol group in D-glucitol and D-glucose is primarily under mass-transport control. On the other hand, the response of sucrose and maltose is under surface control in that increases in linear velocity of the mobile phase which corresponds to a thinner diffusion layer over the electrode, has a minimal effect on the observed current. The response of fructose is under mixed control. The observations are for experimental classifications of the response for each carbohydrate, and these results should not be extrapolated to infer that the groups of the carbohydrate with which the response is associated is responsible for its mechanism of detection.

One ramification of the differences in controlling mechanisms for carbohydrate response is reflected in the shape of i–C^b plots. Carbohydrates that undergo transport-controlled reactions tend to produce linear plots over larger ranges of C^b than do those reactions under surface control. Calibration plots for five common carbohydrates are shown in Figure 4.10. These data were obtained at a Au RDE in 0.1 M NaOH. In agreement with the electrochemical characterization findings, the plots for D-glucitol (▲) and D-glucose (●) are more linear

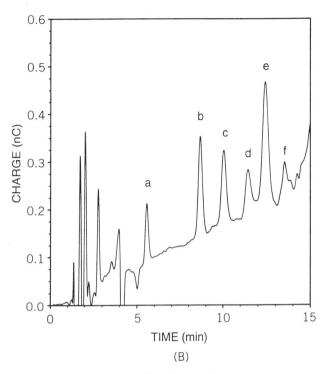

Figure 7.9. (*Continued*)

than those for sucrose (○) and maltose (△). The response for fructose (■) is intermediate to the two extremes. Two tentative explanations are offered for the nonlinear i–C^b plots for reactions characterized as being under surface control: (1), the reactant molecules can be strongly adsorbed, and, therefore, response is controlled by the adsorption isotherm for the reactant; and (2), if detection products are strongly adsorbed with the result of surface fouling, the current response will be attenuated more abruptly during the detection step for large values of C^b for which full coverage of the surface is more quickly achieved.

The changes in hydroxide-ion concentration, which are needed to effect various separations in HPAEC, will undoubtedly result in an observed change in peak response in PED. Table 7.3 shows that the linear range of the response for PAD is relatively unaffected by small changes in pH. The range of linear response for D-glucose by HPAEC–PAD is ~10–10,000 pmol for mobile-phase conditions of 20 mM (pH 12.3) to 200 mM (pH 13.3) NaOH. The enhanced sensitivity is typically offset by an increase in system noise, and the S/N is more or less constant as reflected in similar detection limit values. It is important to remember that significant deviation from linear response can be expected when using a mobile phase having a low buffer capacity (e.g., low NaOH concentration). This

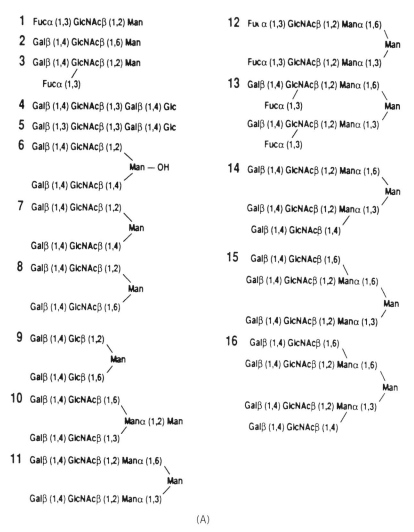

(A)

Figure 7.10. (*A*) Structures and (*B*) analysis of synthetic complex oligosaccharides. Reprinted with permission from Technical Note TN20, Dionex, March 1989. Copyright 1989 Dionex Corporation.

effect is attributable to the consumption of the OH^- ion by the H^+ generated in the detection reaction. Therefore, the linear dynamic range for 20 mM NaOH is not expected to be as large as that for 200 mM NaOH.

Using optimized waveform conditions, Table 7.4 lists the quantitative aspects of the carbohydrates separated in Figure 7.2. Regression analysis of the calibra-

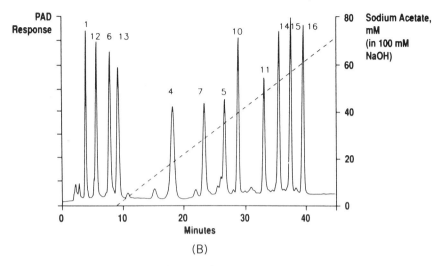

Figure 7.10. (*Continued*)

tion data for these carbohydrates indicates that they give linear response from their limit of detection up to 1000 pmol. The >1000-pmol range was not tested in this particular study, but glucose has been reported to be linear over five decades. The limits of detection range from 1 pmol (i.e., 40 nM or 10 ppb) for D-glucitol to 5 pmol (i.e., 200 nM or 70 ppb) for sucrose. With present PED technology using optimized waveforms and data-smoothing algorithms, detection limits for monosaccharides are now less than 1 pmol injected.

Quantitative determination of individual oligosaccharides in a polysaccharide separation is still quite challenging. Larew et al. [34,35] approached the problem by attempting to convert each eluting peak of a polyglucose separation to an equivalent amount of glucose via a postcolumn immobilized glucoamylase reactor. Hence, for 100% conversion of the components of each eluting peak, a single glucose calibration curve would suffice for peak area quantitation. Unfortunately, conversion efficiencies for oligosaccharides of DP = 2–7 were neither 100% nor similar. More recently, LaCourse [8] has studied the PAD response characteristics of individual oligosaccharides. Figure 7.12 shows the response factors of glucopolymers of DPs from 1 to 7. Interestingly, the response increases from DP = 1 to DP = 5 and then remains fairly constant. Because the molar flux of these carbohydrates is a function of the diffusion coefficient of the reactant, which decreases with increased molecular weight, PED sensitivity for linear oligosaccharides is expected to decrease with increasing DP value. Hence, all quantitative applications of LC–PED must be based on careful calibration plots using standard solutions for each sample component.

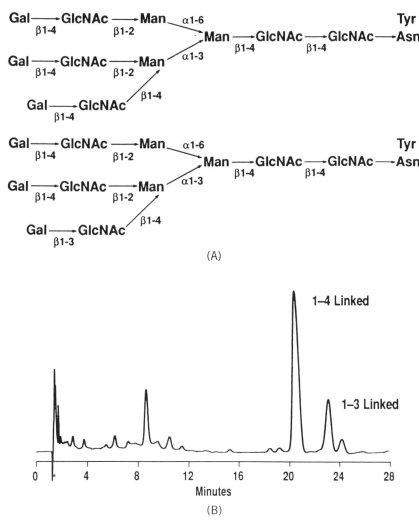

(A)

(B)

Figure 7.11. (*A*) Structures and (*B*) analysis of asialyted fetuin glycopeptides. Reprinted with permission from Technical Note TN20, Dionex, March 1989. Copyright 1989 Dionex Corporation.

ALIPHATIC ALCOHOLS AND POLYALCOHOLS

Aliphatic alcohols are volatile, and their separation and detection are typically performed using gas chromatography with flame ionization detection (GC–FID). Unfortunately, polyalcohols are highly polar and are prone to pyrolysis, which makes them less amenable to GC. Although the separation of aliphatic alcohols including glycols and polyalcohols is easily achieved in liquid chromatography,

TABLE 7.3 Linearity of D-Glucose Response at Various NaOH Concentrations

NaOH (mM)	Linear Range (pmol)	a (nC pmol)	b (nC)	R^2
		Response: nC = $a \cdot$ (pmol) + b		
20	10–10,000	0.0958	−0.2997	0.9999
50	10–10,000	0.1638	+0.367	1.0000
100	10–10,000	0.2456	+1.379	1.0000
200	10–10,000	0.3258	+9.377	0.9999

their quantitative determination is hindered by the lack of a sensitive detector. Aliphatic alcohols have no inherent chromophore or fluorophore and are considered to be electroinactive under constant applied potentials. Hence, derivatization to benzoate esters is often used to improve the detection properties of aliphatic alcohols [36,37].

PAD was introduced in 1981 for the detection of monoalcohols at Pt electrodes [38,39]. This work demonstrated the viability of aliphatic alcohol determination by using PAD with flow injection analysis. With the advance of pH-stable polymer-based columns, the determination of underivatized aliphatic alcohols with PAD has proved to be sensitive and simple. PAD has been applied to the determination of alcohols at both Pt and Au working electrodes under acidic and alkaline conditions, respectively [40].

In HPLC, aliphatic alcohols are generally separated under reversed-phase conditions. Polymer-based stationary phases with mixed-mode separation capabilities are effective for the separation of simple alcohols in the reversed-phase mode with the advantages of a second mode of separation (i.e., ion-exclusion or ion-exchange chromatography) and tolerance for a wide range of pH conditions. Ion-exclusion columns contain styrene-based, fully sulfonated resin, which is normally used for ion-exclusion separations of weak acids. Separations that use these types of columns are accomplished via Donnan exclusion (a function of acid strength and degree of ionization), steric exclusion (size), hydrogen bonding

TABLE 7.4 Quantitation Aspects of Monosaccharides in Figure 7.2

Compound	Linear Range (pmol-LOD)[a]	$2a$ (nC pmol)	b (nC)	R^2	%RSD (pmol, n)
		Response: nC = $a \cdot$ (pmol) + b			
Glucitol	1000, 1	0.249	−1.01	1.0000	12.7 (100, 5)
Glucose	1000, 2	0.247	−0.45	0.9999	3.6 (100, 5)
Fructose	1000, 2	0.137	0.12	1.0000	2.8 (100, 5)
Sucrose	1000, 2	0.097	0.19	1.0000	2.2 (100, 5)
Maltose	1000, 5	0.040	0.11	1.0000	1.7 (100, 5)

[a]LOD for S/N = 3 from a 5-pmol injection. Average noise is ±0.1 nC.

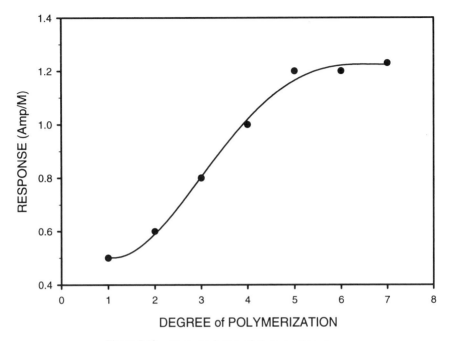

Figure 7.12.   Response factors of glucose polymers.

(attraction to the negatively charged sulfonate group), and adsorption–partitioning (hydrophobicity). Since alcohols and polyalcohols are not ionizable, the retention of smaller alcohols (e.g., ethanol) is based mainly on hydrogen bonding and size exclusion, whereas larger alcohols are retained by adsorptive interactions with the resin polymer. Separations of very hydrophobic alcohols on this type of resin are impractical because of excessively long retention times. The capacity factor (k') for ethanol is nearly independent of pH ($HClO_4$ concentration) and ionic strength ($NaNO_3$ concentration).

   In addition to offering the greatest selectivity for these columns, acidic mobile phases are ideally suited for PAD at a Pt working electrode. Platinum electrodes show a good response for aliphatic alcohols even in 1 M $HClO_4$, whereas Au electrodes show an attenuated response for monoalcohols, glycols, and glycerol even in highly alkaline media. A possible contributing factor is the fact that the —CH_2OH group is substantially more hydrophilic than the alkyl chain, and, hence, the alcohol molecule at the electrode surface is preferentially oriented with the —CH_2OH group toward the bulk of the solution. The HPLC–PAD separation of 13 poly-, primary, and secondary alcohols is shown in Figure 7.13 to illustrate the range and applicability of 100% aqueous separation of aliphatic alcohols and polyalcohols. This type of separation is especially useful for more hydrophilic alcohols such as glycols.

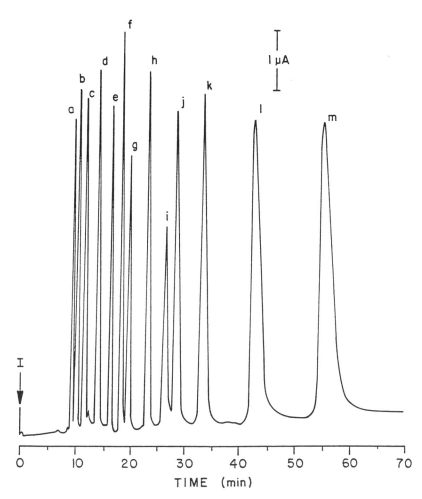

Figure 7.13. LC–PAD for 13 aliphatic alcohols at a Pt working electrode. Samples: (*a*) adonitol, 45 ppm; (*b*) erythritol, 36 ppm; (*c*) glycerol, 9 ppm; (*d*) ethylene glycol, 10 ppm; (*e*) methanol, 30 ppm; (*f*) ethanol, 45 ppm; (*g*) 2-propanol, 177 ppm; (*h*) 1-propanol, 202 ppm; (*i*) 2-butanol, 202 ppm; (*j*) 2-methyl-1-propanol, 120 ppm; (*k*) 1-butanol, 122 ppm; (*l*) 3-methyl-1-butanol, 364 ppm; (*m*) 1-pentanol, 365 ppm. Reprinted with permission from W. R. LaCourse, D. C. Johnson, M. A. Rey, and R. W. Slingsby, *Anal. Chem.* **63**, 134–139 (1991). Copyright 1991 American Chemical Society.

The elution of larger, less polar alcohols (e.g., 3-phenyl-1-propanol and 1-tetradecanol) from polymeric phases requires the addition of an organic modifier to the mobile phase to increase the solvent strength. In addition, the wide range of hydrophobicities of alcohols and polyalcohols is best accomplished using gradient chromatography. The 13 alcohols shown in Figure 7.14 are separated using a

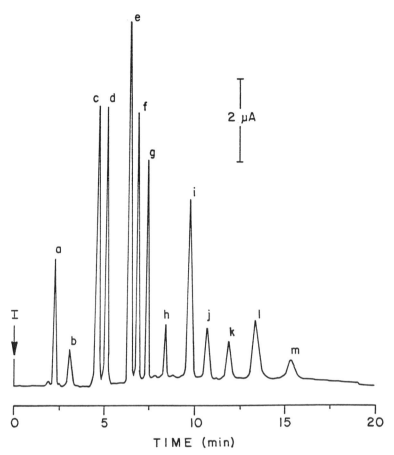

Figure 7.14. LC–PAD of 13 aliphatic alcohols at a Au working electrode. Samples: (*a*) ethanol, 1840 ppm; (*b*) 1-propanol, 460 ppm; (*c*) 2-methyl-2-propen-1-ol, 46 ppm; (*d*) cyclopentanol, 460 ppm; (*e*) phenylmethanol, 69 ppm; (*f*) 1-phenylethanol, 115 ppm; (*g*) 3-phenyl-1-propanol, 115 ppm; (*h*) 2-ethyl-1-hexanol, 460 ppm; (*i*) 1-decanol, 460 ppm; (*j*) 1-undecanol, 920 ppm; (*k*) 1-dodecanol, 920 ppm; (*l*) 1-tridecanol, 920 ppm; (*m*) 1-tetradecanol, 920 ppm. Reprinted with permission from W. R. LaCourse, D. C. Johnson, M. A. Rey, and R. W. Slingsby, *Anal. Chem.* **63**, 134–139 (1991). Copyright 1991 American Chemical Society.

linear acetonitrile gradient. These alcohols are separated solely by a reversed-phase mechanism, and, as expected, the capacity factor for each alcohol decreases as the concentration of organic modifier is increased. Although the use of a Pt electrode in place of a Au electrode offers more sensitive detection, detection in Figure 7.14 is performed at a Au electrode under alkaline conditions, which is more compatible with the presence of organic modifiers. Postcolumn addition of

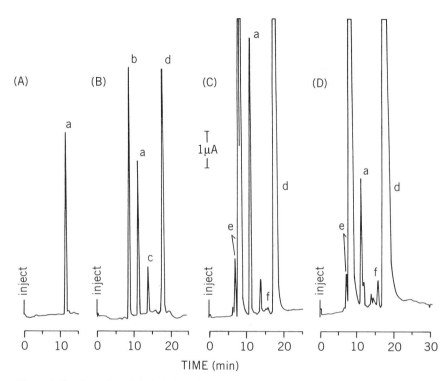

Figure 7.15. Detection of alcohols in various samples by LC–PAD. Samples were (*A*) toothpaste, (*B*) liquid cold formula, (*C*) brandy, and (*D*) wine cooler. Compounds detected were (*a*) glycerol, (*b*) sorbitol, (*c*) propylene glycol, (*d*) ethanol, (*e*) sugars, and (*f*) methanol. Reprinted with permission from W. R. LaCourse, D. C. Johnson, M. A. Rey, and R. W. Slingsby, *Anal. Chem.* **63**, 134–139 (1991). Copyright 1991 American Chemical Society.

NaOH is used to produce the high-alkaline conditions necessary for PED response at a Au electrode.

Figure 7.15 shows the separation on an ion-exclusion column with direct detection for various alcohols and polyalcohols in (*A*) toothpaste, (*B*) liquid cold formula, (*C*) brandy, and (*D*) wine cooler [40]. The selectivity for alcohols in acidic media at a Pt electrode produces simplified chromatograms and contributes to decreased time for sample preparation. Other applications include the determination of ethanol in human breath [41], flavor-active alcohols [42], and free glycerols in biofuels [43], see Table B.2 in Appendix B.

Although the assay for alcohols has been applied to a variety of matrices, the power of PAD is best illustrated by the combination of a chromophoric (uv–vis) detector and a nonchromophoric (PAD) detector. The versatility of separations on mixed-mode ion-exchange columns with selective detection is illustrated in Figure 7.16 by the simultaneous detection of ionic and neutral species in a phar-

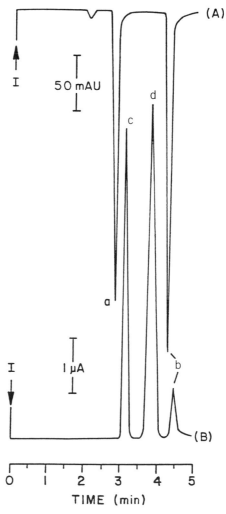

Figure 7.16. Cephalosporin antibacterial assay by cation-exchange/reversed-phase chromatography with (A) uv–vis detection and (B) PAD in series. Compounds detected were (a) p-toluenesulfonic acid, (b) cefazolin, (c) 1,6-hexanediol, and (d) 1,4-cyclohexanediol. Reprinted with permission from W. R. LaCourse, D. C. Johnson, M. A. Rey, and R. W. Slingsby, *Anal. Chem.* **63**, 134–139 (1991). Copyright 1991 American Chemical Society.

maceutical preparation. This experiment utilizes a uv detector and PAD in series after a Dionex PCX-500 column. Under acidic conditions, the cephalosporin-based antibacterial formulation consists of a hydrophobic cation (i.e., cefazolin) and neutral and anionic compounds (i.e., 1–6-hexanediol, 1,4-cyclohexanediol, p-toluenesulfonic acid). The neutral and anionic compounds are separated by the reversed-phase character of the column, while the cationic compounds are separated by a combination of cation exchange and a reversed-phase mechanism. Figure 7.16 shows that the *p*-toluenesulfonic acid and the cefazolin are best detected by uv photometry at 254 nm (*A*), and the two diols, which do not have a chromophore, are easily detected by (*B*) PAD. In addition, the cefazolin has a PAD signal, which has a higher limit of detection than uv, but may be utilized for added selectivity. On the other hand, the PAD waveform was optimized for detection of the diols. This assay could easily be optimized for the sensitive and general detection of cephalosporins.

Quantitation

The limits of detections of simple alcohols at Pt electrodes in acidic media are high in contrast to carbohydrates at Au electrodes in alkaline media due to the reduced number of oxidizable groups and the large background currents from the simultaneous reduction of dissolved O_2. Limits of detection range from 0.1 ppm (160 nmol per 50-μL injection) for methanol to 2 ppm (1100 nmol per 50-μL injection) for pentanol with linearity from the detection limit extending over two to three orders of magnitude. Repeatabilities range of 0.7–6.7% RSD (10 ppm each). Under these conditions, as the number of alcohol groups on the molecule increases, the molar response reaches a plateau after the first three. This may be attributable to temporal limitations of the molecules at the electrode surface and steric hindrance. In addition, as the number of methylene groups increases (methanol to pentanol), the molar (255 ± 1.9 to 38 ± 1.8 nA/nmol) and mass (7.95 ± 0.06 to 0.26 ± 0.02 nA/ng) response factors decrease. As an aside, the detection limit for ethylene glycol has been reported to be ~ 10 ppb (S/N = 3) using the latest equipment with postrun data smoothing.

Even though the Au electrode system (1.3 nA peak-to-peak noise) is an order of magnitude less noisy than the Pt system (13 nA peak-to-peak noise), the sensitivity of simple alcohol detection is drastically reduced at Au in alkaline media in the presence of acetonitrile as compared to the responses of alcohol detection at a Pt electrode in acidic media. The loss in sensitivity is attributable to the presence of ACN, which is strongly adsorbed to the Au electrode surface. The adsorbed acetonitrile blocks surface sites usually available for the anodic detection of the alcohol. Signal attenuation attributed to the presence of ACN results in poorer detectability; the LOD for ethanol is 10 ppm for the Au/base system and 0.1 ppm for the Pt/acid system. Although the more hydrophobic alcohols (e.g., decanol, undecanol, dodecanol, tridecanol, tetradecanol) are separated and detected, quantitation is hindered by poor solubility and the formation of micelles.

AMINOALCOHOLS, AMINOSUGARS, AND AMINOGLYCOSIDES

In PED, the mechanism of detection requires preadsorption of the analyte to the electrode (i.e., intermediate stabilization via free-radical adsorption). Hence, detection of weakly adsorbing compounds (e.g., alcohols) is often inhibited by the presence of organic modifiers (e.g., ACN), which adsorb strongly to the electrode surface. On the other hand, amine- and sulfur-containing functional groups are strongly adsorbed to the electrode surface, and their response is virtually unaffected by the presence of organic modifiers in the mobile phase. Therefore, the adsorption characteristics of multifunctional analytes can be exploited to enhance PED reactivity in both the presence and absence of organic modifiers in the mobile phase.

In PAD, the anodic detection of the alcohol groups (mode I) of aminoalcohols, aminosugars, and aminoglycosides benefit as a consequence of amine adsorption that brings the associated alcohol group close to the electrode surface. The enhanced reactivity is attributed to an increased residence time of the analyte at the electrode surface via the adsorption of the amine group(s). In agreement with this line of thinking, virtually no interference from acetonitrile is observed for the detection of these multifunctional compounds.

Aminoalcohols

Aminoalcohols (i.e., alkanolamines) are important in the chemical and pharmaceutical industries for production of emulsifying agents, corrosion inhibitors, laundry additives, and dyes, and for purifying gases. These compounds lack natural chromophores and fluorophores for photometric and fluorometric detection, and, furthermore, their high polarity virtually eliminates use of GC for quantitative determinations.

Aminoalcohols have been separated using ion-pair chromatography. Alkanesulfonate salts are effective ion-pairing reagents in that they are PAD inactive, can be obtained readily, and are available in a variety of alkyl chain lengths. Baseline separation of five alkanolamines (Fig. 7.17) is achieved with a mobile phase of 2 mM SDS/20 % ACN [44]. The capacity factor (k') decreases as the concentration of ACN is increased, and k' increases as the concentration of ion-pairing reagent is increased.

Linear, branched, and complex alkanolamines have been separated by reversed-phase ion-pair chromatography with mode I detection. Under the conditions used, monoethanolamine and triethanolamine are well resolved; however, monoethanolamine and diethanolamine coeluted. Postcolumn addition of 0.2 M NaOH is used to ensure the high pH needed for optimum sensitivity for PAD at a Au electrode. The detection limit (S/N = 3) for ethanolamine was determined to be 40 ppb (8 ng per 200-μL injection) with linear response from 40 ppb to ~10 ppm.

The high sensitivity and high selectivity of HPLC–PAD contribute to decreased time for sample preparation and simplified chromatograms. Figure 7.18 shows chromatograms for the detection of ethanolamine, triethanolamine, and aminomethylpropanol in a (*A*) cold tablet, (*B*) hand lotion, and (*C*) hair-spray formulation [44]. These samples were prepared by diluting a weighed portion of sample, diluting to the proper level with mobile phase, and filtering.

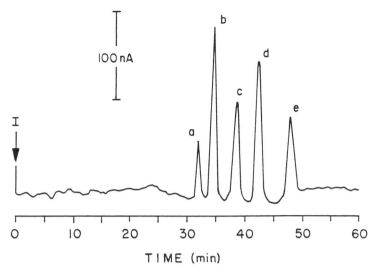

Figure 7.17. Alkanolamine separation using sodium dodecanesulfonate as paired-ion reagent. Samples (10 ppm): (*a*) 2-amino-1-ethanol, (*b*) 4-amino-1-butanol, (*c*) 5-amino-1-pentanol, (*d*) 2-amino-1-butanol, (*e*) 6-amino-1-hexanol. Reprinted with permission from W. R. LaCourse, W. A. Jackson, and D. C. Johnson, *Anal. Chem.* **61**, 2466–2471 (1989). Copyright 1989 American Chemical Society.

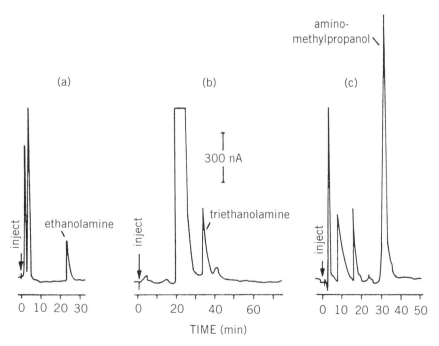

Figure 7.18. Detection of alkanolamines in various samples by LC with PAD. Samples: (*a*) generic cold tablet, (*b*) hand lotion, (*c*) hairspray. Reprinted with permission from W. R. LaCourse, W. A. Jackson, and D. C. Johnson, *Anal. Chem.* **61**, 2466–2471 (1989). Copyright 1989 American Chemical Society.

207

The advance of pH-tolerant, mixed-mode (i.e., ion-exchange and reversed-phase) columns have been applied to the separation of alkanolamines as cations in the presence of ACN (see Fig. 7.19) [8]. The mobile phase is 5% acetonitrile/0.1 M $HClO_4$. With the postcolumn addition of NaOH, detection of the alcohol group is easily performed at $E_{det} = +0.2$ V without interference from acetonitrile due to the presence of the amine group. The mixture of five alkanolamines is similar to the mixture separated using ion-pair chromatography (Fig. 7.17), which required over 50 min for elution of all the compounds. The obvious advantage is a two-thirds reduction in analysis time by using mixed-mode chromatography, accompanied by a decrease in the limit of detection as a result of less band broadening.

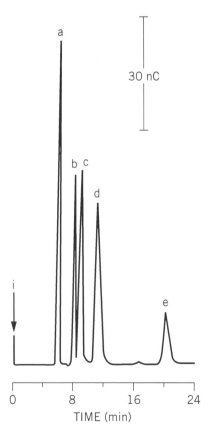

Figure 7.19. HPLC–PAD of alkanolamines using mixed-mode chromatography. Peaks (10 ppm each): (*a*) 2-amino-1-ethanol, (*b*) 4-amino-1-butanol, (*c*) 5-amino-1-pentanol, (*d*) 2-amino-1-butanol, (*e*) 6-amino-1-hexanol. Reprinted from W. R. LaCourse, Pulsed electrochemical detection at noble metal electrodes in high performance liquid chromatography, *Analusis* **21**, 181–195 (1993) with kind permission of Elsevier Science.

Aminosugars, Glycoconjugates, and Aminoglycosides

As with aminoalcohols, the anodic detection of aminosugars, glycoconjugates, and aminoglycosides can benefit from the same surface adsorption interactions observed for alkanolamines. Figure 7.20 shows the separation of (*a*) glucosamine, (*b*) glucose, and (*c*) *N*-acetylglucosamine. Glucosamine shows the strongest signal, which is attributable to strong adsorption of the analyte to the electrode surface via the amine group. Substitution on the amine group reduces its ability to adsorb to the electrode surface, as evidenced by the reduced signal of *N*-acetylglucosamine as compared to glucosamine. Glucose gives the smallest signal for this equimolar mixture.

An interesting effect results from the postcolumn addition of ACN. HPLC–PAD results are shown in Figure 7.21 for the separation of a carbohydrate mixture containing sorbitol (*a*), fucose (*b*), galactosamine (*c*), glucose (*d*), *N*-acetyl-galactosamine (*e*), fructose (*f*), and sucrose (*g*). Chromatogram *A* represents the postcolumn addition of only 0.2 M NaOH, whereas chromatogram *B* shows postcolumn addition of 0.2 M NaOH/20% ACN which effectively makes the mobile phase of the detector 10% ACN. The response for the majority of the sugars is severely attenuated by the presence of the ACN, with a decrease as large as 97% for sucrose. The single exception for this mixture is galactosamine whose signal was least attenuated by the ACN. The persistence of the signal for galactosamine is the beneficial result of the ability of the amine group to adsorb in spite of the presence of ACN on the oxide-free Au surface. In the case of *N*-acetylgalactosamine, the loss of signal in the presence of ACN is indicative of the weakness of amine adsorption as a result of steric hindrance from the acetyl group. This phenomenon is considered to have the same origins as in the case of the ACN effect on alcohol and alkanolamine detection; specifically, the strong adsorption of the amine groups prevents interference by adsorbed ACN in the surface-catalyzed detection mechanism of the alcohol groups.

Applications of HPAEC–PAD pertaining to the determination of aminosugars and glycoconjugates, which were discussed earlier in this chapter, have far-reaching biochemical and medical ramifications. Of no less importance has been the growing number of papers involving aminoglycosidic antibiotics. The ability to quantitate aminoglycosides in a variety of matrices (e.g., fermentation broths, formulations, foodstuffs, biological matrices) has been hindered by poor optical detection properties and problems with chemical derivatization as a consequence of their multifunctionality (i.e., forms multiple product species). On the other hand, the multifunctional nature of aminoglycosides is ideally suited to PAD.

One of the earliest applications by Polta and Johnson [45] was aminoglycoside antibiotics by PAD at a Pt electrode. They focused their efforts on nebramycin factors, which are produced by fermentation of *Streptomyces tenebrarius*. Samples were loaded onto a preconcentrator column in 10 mM phosphate buffer, pH 5.2, and then backflushed with NaOH onto a neutral polystyrene analytical column. Detection limits were ~80 ppb for tobramycin. Tobramycin

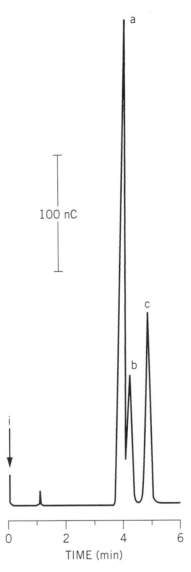

Figure 7.20. HPLC–PAD of aminosugars using anion-exchange chromatography. Peaks (0.1 μM each): (*a*) glucosamine, (*b*) glucose, (*c*) *N*-acetylglucosamine. Reprinted from W. R. LaCourse, Pulsed electrochemical detection at noble metal electrodes in high performance liquid chromatography, *Analusis* **21**, 181–195 (1993) with kind permission of Elsevier Science.

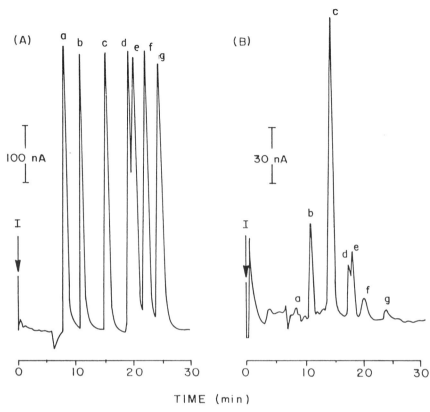

Figure 7.21. LC–PAD peaks for seven carbohydrates with detection made (*A*) in the absence and (*B*) in the presence of 10% acetonitrile in the mobile phase. Peaks: (*a*) sorbitol, 1.6 nmol; (*b*) L-fucose, 1.9 nmol; (*c*) D-galactosamine, 1.5 nmol; (*d*) D-glucose, 1.9 nmol; (*e*) N-acetyl D-galactosamine, 2.5 nmol; (*f*) D-fructose, 3.1 nmol; (*g*) sucrose, 2.3 nmol. Reprinted from W. R. LaCourse and D. C. Johnson, Optimization of waveforms for pulsed amperometric detection (p.a.d.) of carbohydrates following separation by liquid chromatography, *Carbohydr. Res.* **215**, 159–178 (1991) with kind permission of Elsevier Science–NL, Sara Burgerhartstraat 25, 1055 KV Amsterdam, the Netherlands.

and apramycin were determined in fermentation broth and spiked blood serum by HPLC–PAD using this technique (see Fig. 7.22). This work was extended by Statler in 1990 [46], and tobramycin was determined by anion-exchange chromatography followed by PAD at a Au electrode under alkaline conditions, which was added postcolumn. The detection limit for tobramycin was improved to 0.2 ppb. Chromatographic resolution was improved largely because of improvements in column technology, and detection sensitivity was improved by using a Au electrode under alkaline conditions.

McLaughlin and Henion [47] used PAD with ion-spray mass-spectrometry (MS) detection for spectinomycin, hygromycin B, streptomycin, and dihydro-

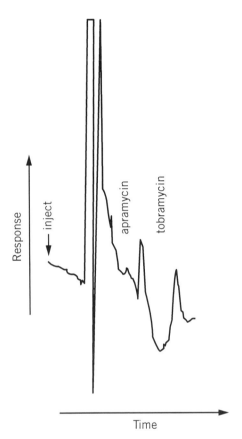

Figure 7.22. Chromatogram of blood serum spiked with 0.6 ppm each tobramycin and apramycin. Reprinted from J. A. Polta, D. C. Johnson, and K. T. Merkel, Liquid chromatographic separation of aminoglycosides with pulsed amperometric detection, *J. Chromatogr.* **324**, 407–414 (1985) with kind permission of Elsevier Science–NL, Sara Burgerhartstraat 25, 1055 KV Amsterdam, the Netherlands.

streptomycin. The target compounds in an extract of a control bovine kidney fortified at the 20-ppm level are shown in the chromatogram in Figure 7.23. McLaughlin and Henion found that PAD detection was compatible with the ion-pairing agent and more versatile than MS detection. The MS detection was more sensitive than PAD, but was incompatible with the ion-pairing agent. Other antibiotic applications include spectinomycin and its degradation [48], the four major components of gentamicin to aid in the identification of the source of pharmaceutical preparations of gentamicin [49], and neomycin A, B, and C in topical lotions [50]. Figure 7.24 shows a chromatogram of a topical neomycin preparation in which neomycin B and C are clearly identifiable. The detection limit for this assay was 0.2 ppm for each compound. The determination of aminoglycoside antibiotics by HPLC–PAD is presently of significant research

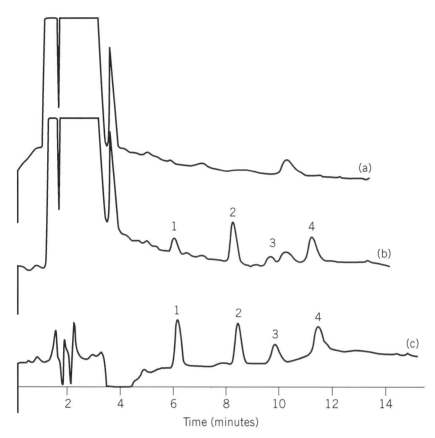

Figure 7.23. HPLC–PAD of (*a*) extract of control bovine kidney, (*b*) extract of bovine kidney fortified at the 20-ppm level, and (*c*) synthetic mixture of standards at levels of 15 ng per component injected, representative of 100% recovery. Peaks: (1) spectinomycin; (2) hygromycin B; (3) streptomycin; (4) dihydrostreptomycin. Reprinted from L. G. McLaughlin and J. D. Henion, Determination of aminoglycoside antibiotics by reversed-phase ion-pair high-performance liquid chromatography coupled with pulsed amperometry and ion spray mass spectrometry, *J. Chromatogr.* **591**, 195–206 (1992) with kind permission of Elsevier Science–NL, Sara Burgerhartstraat 25, 1055 KV Amsterdam, the Netherlands.

interest, and antibiotic assays by HPLC–PED have recently been reviewed (see Tables B.3 and B.7).

AMINES, AMINOACIDS, PEPTIDES, AND PROTEINS

Numerous amine-based compounds can be detected directly at Pt [51] and Au [44,52] electrodes in alkaline solutions by the oxide-catalyzed mechanism of mode II using PED. The criterion for detectability is the existence of a non-

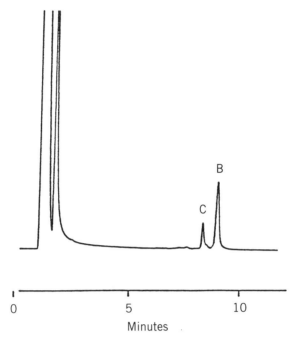

Figure 7.24. Neomycins B and C in topical lotion using HPLC–PAD. Reprinted with permission from Application Note 66R, Dionex, June 1991. Copyright 1991 Dionex Corporation.

bonded electron pair on the N atom of the amine to facilitate adsorption at the electrode surfaces. Therefore, primary, secondary, and tertiary amines are detected, and quaternary amines are not.

Amines

As with simple alcohols, the quantitative determination of aliphatic amines has been hindered by the lack of a sensitive detector. Aliphatic amines have no inherent chromophore or fluorophore and are considered to be electroinactive under constant applied potential. In addition, amines are highly polar and, as a result, are not amenable to gas chromatography (GC), and liquid chromatographic separations are often plagued by peak tailing. Preinjection (e.g., hydrazines) and postseparation (e.g., ninhydrin) derivatizations have been used to improve the detection and chromatographic properties of aliphatic amines [53].

Several studies have focused on the detection of aliphatic monoamines and diamines by HPLC–PAD [54–56], but amine detections are oxide-catalyzed and are best performed using IPAD. A fine example involves the use of HPLC–IPAD to directly determine biogenic amines as indicators of seafood spoilage was published by Draisci et al. [54]. They separated putrescine, histidine, cadaverine, and histamine using a cation-exchange column with an ACN gradient. Figure

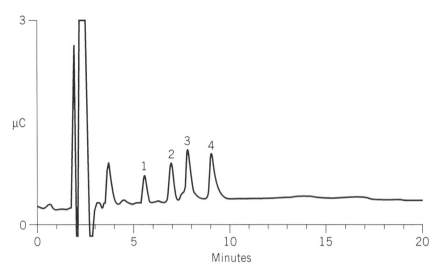

Figure 7.25. HPLC–PAD analysis of biogenic amines in fish using HPLC–PAD. Peaks: (1) putrescine, (2) histidine, (3) cadarverine, (4) histamine. Reprinted from R. Draisci, S. Cavalli, L. Lucentini, and A. Stacchini, *Chromatographia* **35**, 584–590 (1993).

7.25 shows the chromatograms of amines extracted from spoiled canned herrings. Detection was performed by IPAD following postcolumn addition of NaOH. Other aliphatic amine applications are listed in Table B.4 in Appendix B.

Aminoacids

Most aminoacids in biological materials are aliphatic and have been perceived as not electroactive [57,58]. The conclusions of these researchers is biased by the observations that a persistent signal is not obtained at common solid anodes (i.e., Au, Pt, C) using dc amperometry. Hence, their detection in HPLC has commonly been achieved by photometric absorbance after derivatization using ninhydrin reagent, by fluorescence using *o*-phthaldehyde reagent, or by amperometric detection of the phenyl- or methylthiohydantion adducts at Ag and Hg electrodes.

Voltammetrically, aminoacids behave similarly to amines, and aminoacids can be detected directly at Pt and Au electrodes in alkaline solutions by the oxide-catalyzed mechanism of mode II. The Au electrode under alkaline conditions is preferred over Pt to minimize interference from dissolved O_2. Although PAD has been successfully used for the determination of aminoacids, IPAD is preferred for mode II detections.

Simple mixtures of aminoacids can be separated isocratically using anion-exchange chromatography [59]. For complex mixtures, it is essential to perform separations with gradient-elution chromatography. Figure 5.11 shows the chromatogram of a protein hydrolyzate containing 17 aminoacid residues using

HPLC–IPAD with a pH reference electrode [60]. The aminoacid mixture was separated using an anion-exchange column with a quaternary gradient, which incorporated both a pH and an organic modifier gradient. The use of IPAD with a Ag/AgCl reference electrode results in a significant decrease in the baseline shift for small changes in pH (i.e., pH <2). However, optimum potential values for the IPAD waveform also shift with the change in pH, and the negative consequence of this fact can be decreased significantly by use of a glass membrane, whose pH-dependent potential changes similarly to that for the onset of oxide formation. An improved separation of 21 aminoacids was performed by Martens and Frankenberger [61] using the gradient program given in Table 7.5 (see also Fig. 7.26). Presently, detection limits (i.e., S/N = 3) for HPLC–IPAD for aminoacids are typically 1–50 pmol injected with comparable sensitivities for primary and secondary aminoacids.

Peptides and Proteins

As discussed earlier in this chapter, HPAEC–PAD is used routinely for the detection of glycopeptides. HPLC–IPAD has also been applied to the separation and detection of peptides and glycopeptides from bovine fetuin hydrolyzates using typical peptide separation conditions (i.e., reversed-phase chromatography with 0.1% TFA and an ACN gradient). Postcolumn addition of NaOH was required to increase ionic strength and the pH to improve the sensitivity of the analysis. Figure 7.27 compares uv detection (*A*) and IPAD (*B*) for a bovine fetuin hydrolyzate [8]. The results are quite similar except for the superior sensitivity of IPAD for the early-eluting compounds. Other applications have focused on tyrosine and tryptophan-containing peptides [62] as well as sulfur-containing peptides [62,63]. Protein applications have been limited to pulsed amperometric detection using conducting polymer-based electrodes [64–66].

TABLE 7.5 Quadratic Gradient Program for the Amino Acid Separation in Figure 7.26 [61]

	Percent Eluent			
Time (min)	*A* 23 mM NaOH/ 7 mM NaH$_2$BO$_3$	*B* 80 mM NaOH/ 23 mM NaH$_2$BO$_3$	*C* 0.65 M NaOAc	*D* 1 mM NaOH/ 0.3 M H$_3$BO$_3$
0	100	0	0	0
10	100	0	0	0
16.5	0	100	0	0
20.0	0	100	0	0
32–50	0	0	100	0
50–60	0	0	0	100
60–70	100	0	0	0

Note: All gradients are linear.

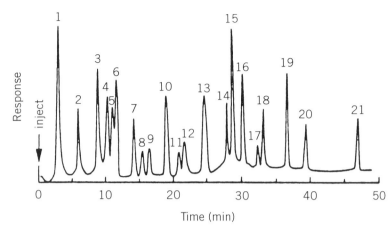

Figure 7.26. LC–PED results for mixture of twenty amino acids. Peaks: (1) arginine, (2) lysine, (3) glutamine, (4) asparagine, (5) threonine, (6) alanine, (7) glycine, (8) serine, (9) valine, (10) proline, (11) isoleucine, (12) leucine, (13) methionine, (14) system, (15) histidine, (16) phenylalanine, (17) glutamic acid, (18) aspartic acid, (19) cysteine + cystine, (20) tyrosine, (21) tryptophan. Aminoacid separation. Reprinted from Ref. (61), p. (430) courtesy of Marcel Dekker, Inc.

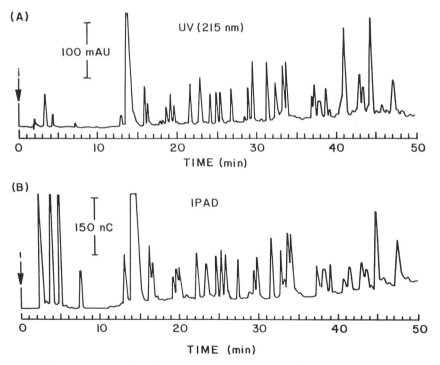

Figure 7.27. Separation and detection using (*A*) UV and (*B*) IPAD of bovine fetuin hydrolyzate. Reprinted from W. R. LaCourse, Pulsed electrochemical detection at noble metal electrodes in high performance liquid chromatography, *Analusis* **21**, 181–195 (1993) with kind permission of Elsevier Science.

SULFUR-CONTAINING COMPOUNDS

Numerous organic and inorganic sulfur compounds are adsorbed at the oxide-free surfaces of Au and Pt electrodes and can be detected by mode II [67–71]. These compounds include sulfides, disulfides, thiols, thioethers, thiophenes, thiocarbamates, organic thiophosphates, and numerous inorganic compounds. Adsorption is prerequisite to detection and, therefore, at least one nonbonded electron pair must reside on the S atom. Hence, sulfonic acids and sulfones are not detected. Electrode fouling is more prevalent when detecting sulfur compounds than either alcohol or amine compounds. Hence, much work has centered on understanding the detection mechanisms of sulfur-containing compounds and overcoming the adverse effects of electrode fouling.

It is evident from our discussion that sulfur-containing aminoacids (e.g., cysteine, cystine, methionine) can be detected with PED [72–74]. Both the amine

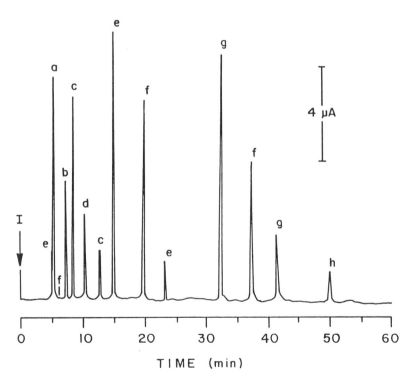

Figure 7.28.   Chromatogram of sulfur-containing pesticides separated on a C_{18} reverse-phase column with PAD. Peaks (250 ng): (*a*) Ethion, (*b*) Methomyl, (*c*) Dimethoate, (*d*) Aldecarb, (*e*) Phorate, (*f*) Thiometon, (*g*) Guthion, (*h*) Sulprotos. Reprinted from A. Ngoviwatchai and D. C. Johnson, Pulsed amperometric detection of sulfur-containing pesticides in reversed-phase liquid chromatography, *Anal. Chim. Acta.* **215**, 1–12 (1989) with kind permission of Elsevier Science–NL, Sara Burgerhartstraat 25, 1055 KV Amsterdam, the Netherlands.

and sulfur moieties of sulfur-containing aminoacids are detected at Au electrodes in alkaline conditions. Fortuitously, the selective detection of virtually all sulfur-containing compounds can be achieved under mildly acidic conditions, which eliminates the need for postcolumn addition of NaOH and allows for the separations using silca-based phases. HPLC–PAD has been demonstrated for a mixture of sulfur-containing pesticides [71]. The separation of eight pesticides was achieved on a C_{18} column using a mobile phase consisting of 50% (v/v) acetonitrile in an acetate buffer (pH 5.0) (see Fig. 7.28). The presence of the acetonitrile in the mobile phase did not attenuate the PAD response for the pesticides because of the preferential and strong adsorption of the sulfur atoms at the Au surface, and postcolumn addition of NaOH was not required.

Recently LaCourse and Owens [75] demonstrated the superiority of IPAD over PAD for the determination of thiocompounds using standard reversed-phase conditions. These results were as expected since sulfur detections are mode II. Figure 7.29 shows separation of several bioactive compounds including the

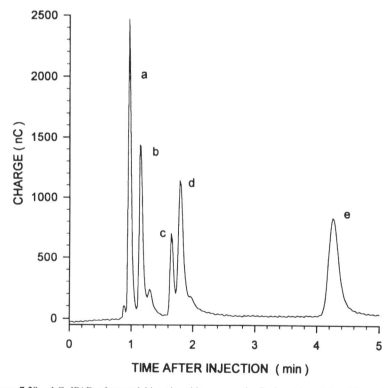

Figure 7.29. LC–IPAD of several bioactive thiocompounds. Peaks: (*a*) cysteine, 50 pmol; (*b*) homocysteine, 50 pmol; (*c*) methionine, 50 pmol; (*d*) GSH, 50 pmol; (*e*) GSSG, 50 pmol. Reprinted from W. R. LaCourse and G. S. Owens, Pulsed electrochemical detection of thiocompounds following microchromatographic separations, *Anal. Chim. Acta.* **307**, 301–319 (1995) with kind permission of Elsevier Science–NL, Sara Burgerhartstraat 25, 1055 KV Amsterdam, the Netherlands.

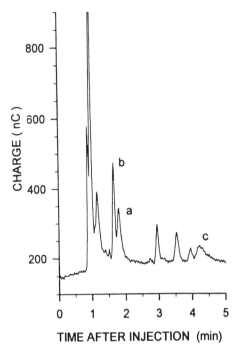

Figure 7.30. Detection of thiocompounds in chicken liver. Peaks: (*a*) GSH, (*b*) methionine, (*c*) GSSG. Reprinted from W. R. LaCourse and G. S. Owens, Pulsed electrochemical detection of thiocompounds following microchromatographic separations, *Anal. Chim. Acta.* **307**, 301–319 (1995) with kind permission of Elsevier Science–NL, Sara Burgerhartstraat 25, 1055 KV Amsterdam, the Netherlands.

reduced (GSH) and oxidized (GSSG) forms of glutathione. Interestingly, IPAD enables the direct determination of thio redox couples (i.e., —SH/—S—S—) and numerous other sulfur moieties at a single Au electrode. The high selectivity of PED for thiocompounds under mildly acidic conditions reduces sample preparation and produces simpler chromatograms of complex mixtures. Figure 7.30 shows the determination of thiocompounds in chicken liver. The IPAD waveform gives lower LODs and more stable baselines, and eliminates oxide-induced artifacts.

Because of their ubiquitous use in modern medical practice, analysis of penicillin and cephalosporin antibiotics in biological matrices and preparations has been the focus of much research. PED is a good choice for the analysis of these sulfur-containing compounds. Welch and co-workers [76–79] have been prominent in the development of HPLC–PAD methods for the separation and quantitation of penicillin antibiotics. The separation of mixtures of common penicillins was accomplished using a C_{18} reversed-phase column with a 0.2 M acetate buffer (pH 4.75) and an ACN/MeOH organic modifier mixtures. Since PAD detection has limited compatability with organic modifier gradients, a traditional gradient could not be used. Instead, they kept the total amount of organic constant, but varied the composition of the organic portion of the mobile phase. This served to preserve the baseline stability while also allowing for better separation of the early-eluting compounds. Using this approach, Kirchmann et al. [78] were able to separate eight penicillins within a milk extract matrix in less than 30 min (see Fig. 7.31). Their work focused on the comparison and optimization of direct and indirect modes of PAD and related advanced PED waveforms. Detection limits for PAD ranged from 2 to 0.7 ppm and an order of magnitude lower when using an on-line preconcentration scheme. Altunata et al. [79] attempted to improve on earlier work with penicillin through the investigation of various three- and four-step waveforms in both direct and indirect detection modes. The best detection limits of 0.2 and 0.3 ppm are accomplished with three-step indirect PAD and four-step indirect PAD, respectively.

LaCourse and Owens [75] report on the use of PED following microbore separation of biologically important compounds. The analysis of the sulfur-containing antibiotics, lincomycin, penicillin G, and cephalexin using IPAD following reversed-phase liquid chromatography is shown in Figure 7.32. This work

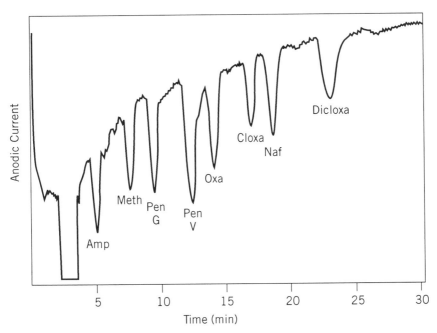

Figure 7.31. The separation of eight penicillins in a milk extract. All penicillins are present at 3–4 ppm. Reprinted from Ref. (78), p. (1769) by courtesy of Marcel Dekker, Inc.

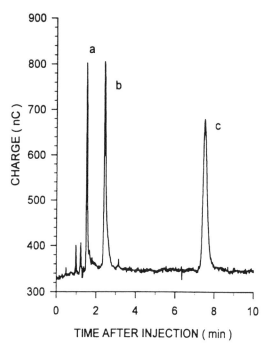

Figure 7.32. Application of LC–IPAD to sulfur-containing antibiotics. Peaks: (*a*) cephalexin, 25 pmol; (*b*) lincomycin, 50 pmol; (*c*) penicillin-G, 150 pmol. Reprinted from W. R. LaCourse and G. S. Owens, Pulsed electrochemical detection of thiocompounds following microchromatographic separations, *Anal. Chim. Acta.* **307**, 301–319 (1995) with kind permission of Elsevier Science–NL, Sara Burgerhartstraat 25, 1055 KV Amsterdam, the Netherlands.

showed that PED detection is applicable to microbore separations, see Chapter 9, and that good mass detection is achievable with this technique.

Recent studies by Dasenbrock, Zook, and LaCourse [80] have focused on the HPLC–IPAD analysis of five sulfur-containing antibiotics for their determination within milk extracts. The compounds are separated on a C_8 column with a mobile phase of 20% 500 mM acetate buffer (pH 3.75)/5% acetonitrile/75% water. Table 7.6 shows the detection limits achievable with single injections of each compound. All of these detection limits are at or below the target levels for antibiotic residues in milk set by the FDA. The IPAD analysis was also compared with a uv detection scheme by adding a uv detector in line. In all cases, the IPAD detection was more sensitive and selective for the compounds of interest than uv detection. Work was also done on milk extracts spiked with selected target compounds (see Fig. 7.33). The analytes of interest were detectable within the matrix at the 20–50-ppb level for cephapirin and ampicillin, respectively. Ongoing work has focused on improving the extraction and concentration of the target compounds from milk.

TABLE 7.6   Summary of the Analytical Figures of Merit Achievable with HPLC–IPAD[a]

Compound	Target Level (ppb)	LOD (ppb)	LOL (ppb)	CV at 50 ppb (%)
Amoxicillin	10	1	50	1.4
Ampicillin	10	2	500	1.9
Cephapirin	20	1	200	1.7
Penicillin G	5	5	500	3.6
Cloxacillin	10	5	500	2.4

[a]These data were collected by injecting each compound separately with a mobile phase chosen to yield similar k' values for each compound [80]. *Abbreviations:* LOD—limit of detection (S/N = 3); LOL—limit of linearity; CV—coefficient of variation.

MISCELLANEOUS APPLICATIONS AND REVIEWS

Many of the early efforts to discern the mechanism of anodic oxygen transfer utilized inorganic species as test species. It only follows that these compounds are also applicable to PED following ion chromatography. Inorganic analytes include H^+ and OH^- [81], heavy-metal ions [82], triorganotin species [83],

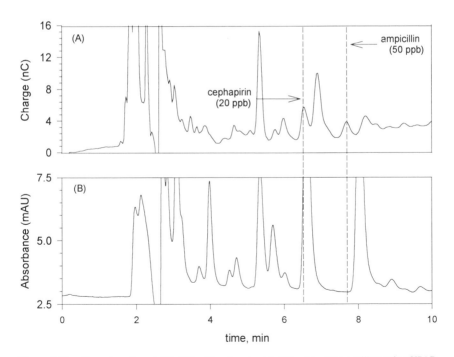

Figure 7.33. Detection of ampicillin (50 ppb) and cephapirin (20 ppb) within a milk matrix of IPAD (*A*) and uv detection at 254 nm (*B*). Reprinted from Ref. (93) by courtesy of Marcel Dekker, Inc.

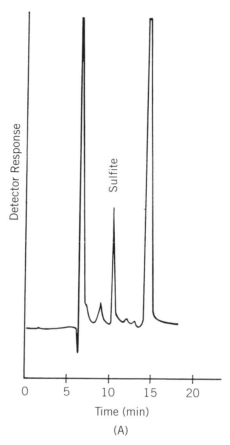

Figure 7.34. Chromatograms of (*A*) commercial 5% v/v alcohol beer containing 5.5 mg/L sulfite diluted 1:20 with pH 9.0 buffer and (*B*) inorganic mercury compounds in the 100–200-ppb range. (*A*) Reprinted with permission from H. P. Wagner and M. J. McGarrity, *J. Am. Soc. Brew. Chem.* **50**, 1–3 (1992). Copyright 1992 American Society of Brewing Chemists. (*B*) Reprinted from T. S. Hsi, J. S. Tsai, and J. Chin, *Chem. Soc.* (*Taipei*) **41**, 315–322 (1994) with permission from the author.

arsenic(III) [84,85], mercury species [86], sulfite and other inorganic sulfur-containing species [87–89], and cyanide [89]. Figure 7.34*A* shows the determination of sulfite in beer by ion exclusion [87]. Figure 7.34*B* shows mercury speciation [86]. Along with these applications, others are listed in Table B.6 in Appendix B.

Throughout this text the instrumental developments listed in Table B.7 in Appendix B have been discussed in detail. Individuals interested in designing their own systems are encouraged to read papers on the construction of a micro-computer-controlled pulsed amperometric detector system [90], electrochemical cell designs with respect to PAD waveforms [84], and S/N optimization in direct

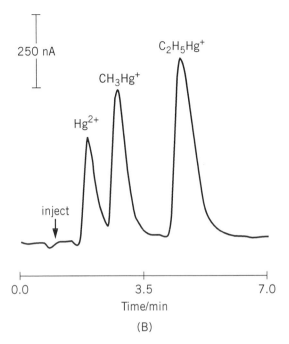

Figure 7.34. (*Continued*)

curent and pulsed amperometric detection [91]. In addition, the most recent developments are discussed in Chapter 9. Numerous reviews have been written that cover either specific areas, such as carbohydrates [92] or antibiotics [93], or more general topics. A compilation of instrumental developments and review papers can be found in Table B.7 (in Appendix B).

REFERENCES

1. S. Hughes and D. C. Johnson, *J. Agric. Food Chem.* **30**, 712 (1982).
2. S. Hughes and D. C. Johnson, *Anal. Chim. Acta* **149**, 1 (1983).
3. J. A. W. Beenackers, B. F. M. Juster, and H. S. van der Baan, *Carbohydr. Res.* **140**, 169 (1985).
4. T. J. Paskach, H. P. Lieker, P. J. Reilly, and K. Thielecke, *Carbohydr. Res.* **215**, 1 (1991).
5. R. D. Rocklin and C. A. Pohl, *J. Liq. Chromatogr.* **6**, 1577 (1983).
6. W. R. LaCourse, D. A. Mead, Jr., and D. C. Johnson, *Anal. Chem.* **62**, 220 (1990).
7. J. D. Olechno, S. R. Carter, W. T. Edwards, and D. G. Gillen, *Am. Biotech. Lab.* **5**, 38 (1987).
8. W. R. LaCourse, *Analysis* **21**, 181–195 (1993).
9. K. S. Wong and J. Jane, *J. Liq. Chromatogr.* **18**, 63 (1995).
10. M. R. Hardy, R. R. Townsend, and Y. C. Lee, *Anal. Biochem.* **170**, 54 (1988).
11. M. R. Hardy and R. R. Townsend, *Proc. Natl. Acad. Sci.* (USA) **85**, 459 (1988).

12. R. R. Townsend, M. R. Hardy, O. Hindsgaul, and Y. C. Lee, *Anal. Biochem.* **174**, 459 (1988).

13. R. R. Townsend, in *Carbohydrate Analysis: High Performance Liquid Chromatography and Capillary Electrophoresis*, A. El Rassi, ed., Elsevier, Amsterdam, 1995.

14. K. Ohsawa, Y. Yoshimura, S. Watanabe, H. Tanaka, A. Yokota, K. Tamura, and K. Imaeda, *Anal. Sci.* **2**, 165–168 (1986).

15. N. Kuroda, S. Taka, T. Kajikawa, M. Niimi, T. Ishida, and K. Kawanishi, *Tonyobyo* (Tokyo) **37**, 695–698 (1994).

16. N. Kuroda, S. Tada, T. Kajikawa, M. Niimi, T. Ishida, and K. Kawanishi, *Tonyobyo* (Tokyo) **38**, 979–983 (1995).

17. M. F. Chaplin, *J. Chromatogr.* **664**, 431–434 (1995).

18. C. M. Tsai, X. X. Gu, and R. A. Byrd, Vaccine, **12**, 700–706 (1994).

19. S. C. Fleming, J. A. Kynaston, M. F. Laker, A. D. J. Pearson, M. S. Kapembwa, and G. E. Griffin, *J. Chromatogr.* **640**, 293–297 (1993).

20. H. J. Blom, H. C. Andersson, D. M. Krasnewich, and W. A. Gahl, *J. Chromatogr.* **533**, 11–21 (1990).

21. S. Tanaka, K. Nakamori, H. Akanuma, and M Yabuuchi, *Biomed. Chromatogr.* **6**, 63–66 (1992).

22. R. M. Pollman, *J. Assoc. Off. Anal. Chem.* **72**, 425–428 (1989).

23. D. R. White, Jr. and W. W. Widmer, *J. Agric. Food Chem.* **38**, 1918–1921 (1990).

24. C. M. Zook and W. R. LaCourse, *Curr. Seps.* **14**, 48–52 (1995).

25. J. van Riel and C. Olieman, *Carbohydr. Res.* **215**, 39–46 (1991).

26. J. Prodolliet, M. B. Blanc, M. Bruelhart, L. Obert, and J. M. Parchet, *Colloq. Sci. Int. Cafe* **14**, 211–219 (1992).

27. J. L. Bernarl, M. J. Del Nozal, L. Toribio, and M. Del Alamo, *J. Agric. Food Chem.* **44**, 507–511 (1996).

28. K. A. Garleb, L. D. Bourquin, and G. C. Fahey, Jr., *J. Food Sci.* **56**, 423–426 (1991).

29. K. W. Swallow and N. H. Low, *J. Agric. Food Chem.* **38**, 1828–1832 (1990).

30. K. W. Swallow, N. H. Low, and D. R. Petrus, *J. Assoc. Off. Anal. Chem.* **74**, 341–345 (1991).

31. D. R. White, Jr. and P. F. Cancalon, *J. AOAC Int.* **75**, 584–587 (1992).

32. N. H. Low and G. G. Wudrich, *J. Agric. Food Chem.* **41**, 902–909 (1993).

33. J. G. Stuckel and N. H. Low, *J. Agric. Food Chem.* **43**, 3046–3051 (1995).

34. L. A. Larew, D. A. Mead, Jr., and D. C. Johnson, *Anal. Chim. Acta* **204**, 43 (1988).

35. L. A. Larew and D. C. Johnson, *J. Electroanal. Chem.* **60**, 1867 (1988).

36. F. Nachtmann and K. W. Budna, *J. Chromatogr.* **136**, 279 (1977).

37. T. Jupille, *J. Chromatogr. Sci.* **17**, 160 (1979).

38. S. Hughes, P. L. Meschi, and D. C. Johnson, *Anal. Chim. Acta* **132**, 11 (1981).

39. S. Hughes and D. C. Johnson, *Anal. Chim. Acta* **132**, 11 (1981).

40. W. R. LaCourse, D. C. Johnson, M. A. Rey, and R. W. Slingsby, *Anal. Chem.* **63**, 134 (1991).

41. G. Schiavon, N. Comisso, R. Toniolo, and G. Bontempelli, *Electroanalysis* **8**, 544–548 (1996).

42. E. Le Fur, P. X. Etievant, and J. M. Meunier, *J. Agric. Food Chem.* **42**, 320–326 (1994).

43. P. Lozano, N. Chirat, J. Graille, and D. Pioch, *Fresenius' J. Anal. Chem.* **354**, 319–322 (1996).

44. W. R. LaCourse, W. A. Jackson, Jr., and D. C. Johnson, *Anal. Chem.* **61**, 2466 (1989).

45. J. A. Polta and D. C. Johnson, *J. Chromatogr.* **324**, 407 (1985).

46. J. A. Statler, *J. Chromatogr.* **527**, 244 (1990).

47. L. G. McLaughlin and J. D. Henion, *J. Chromatogr.* **591**, 195 (1992).

48. J. G. Phillips and C. Simmonds, *J. Chromatogr.* **675**, 123 (1994).

49. L. A. Kaine and K. A. Wolnik, *J. Chromatogr.* **674**, 255 (1994).

50. Application Note 66R, *Neomycin in Topical Lotions*, Dionex, Sunnyvale, CA, 1991.

51. J. A. Polta and D. C. Johnson, *J. Liq. Chromatogr.* **6**, 1727 (1983).

52. W. A. Jackson, Jr., W. R. LaCourse, D. A. Dobberpuhl, and D. C. Johnson, *Electroanalysis* **3**, 607 (1991).

53. L. R. Snyder and J. J. Kirkland, in *Introduction to Modern Liquid Chromatography*, Wiley, New York, 1979.

54. R. Draisci, S. Cavalli, L. Lucentini, and A. Stacchini, *Chromatographia* **35**, 9–12 (1993).

55. D. A. Dobberpuhl and D. C. Johnson, *Anal. Chem.* **67**, 1254–1258 (1995).

56. D. A. Dobberpuhl, J. C. Hoekstra, and D. C. Johnson, *Anal. Chim. Acta* **322**, 55–62 (1996).

57. M. Malfoy and J. A. Reynaud, *J. Electroanal. Chem.* **114**, 213 (1980).

58. H. M. Joseph and P. Davies, *Curr. Seps.* **4**, 62 (1982).

59. L. E. Welch, W. R. LaCourse, D. A. Mead, Jr., and D. C. Johnson, *Talanta* **37**(4), 377 (1990).

60. L. E. Welch, W. R. LaCourse, D. A. Mead, Jr., and D. C. Johnson, *Anal. Chem.* **61**, 555 (1989).

61. D. A. Martens and W. T. Frankenberger, Jr., *J. Liq. Chromatogr.* **15**, 423–439 (1992).

62. J. A. van Riel and C. Olieman, *Anal. Chem.* **67**, 3911–3915 (1995).

63. M. J. Donaldson and M. W. Adlard, *J. Chromatogr.* **509**, 347–356 (1990).

64. O. A. Sadik and G. G. Wallace, *Anal. Chim. Acta* **279**, 209–212 (1993).

65. O. A. Sadik and G. G. Wallace, *Anal. Chim. Acta* **302**, 131 (1995).

66. W. Lu, H. Zhao, and G. G. Wallace, *Anal. Chim. Acta* **315**, 27–32 (1995).

67. P. J. Vandeberg, J. L. Kowagoe, and D. C. Johnson, *Anal. Chim. Acta* **260**, 1 (1992).

68. T. Z. Polta and D. C. Johnson, *J. Electroanal. Chem.* **209**, 159 (1986).

69. T. Z. Polta, D. C. Johnson, and G. R. Luecke, *J. Electroanal. Chem.* **209**, 171 (1986).

70. D. C. Johnson and T. Z. Polta, *Chromatogr. Forum* **1**, 37 (1986).

71. A. Ngoviwatchai and D. C. Johnson, *Anal. Chim. Acta* **215**, 1 (1988).

72. P. J. Vandeberg and D. C. Johnson, *Anal. Chem.* **65**, 2713–2718 (1993).

73. A. J. Tudos and D. C. Johnson, **67**, 557–560 (1995).

74. A. Liu, T. Li, and E. Wang, *Anal. Sci.* **11**, 597–603 (1995).

75. W. R. LaCourse and G. S. Owens, *Anal. Chim. Acta* **307**, 301 (1995).

76. L. Koprowski, E. Kirchmann, and L. E. Welch, *Electroanalysis* **5**, 473 (1993).

77. E. Kirchmann and L. E. Welch, *J. Chromatogr.* **633**, 111 (1993).

78. E. Krichmann, R. L. Earley, and L. E. Welch, *J. Liq. Chromatogr.* **17**, 1755 (1994).

79. S. Altunata, R. L. Earley, D. M. Mossman, and L. E. Welch, *Talanta* **42**, 17 (1995).

80. C. O. Dasenbrock, C. M. Zook, and W. R. LaCourse, paper presented at Ohio Valley Chromatography Symposium, June 1996.

81. J. A. Polta, I. H. Yeo, and D. C. Johnson, *Anal. Chem.* **57**, 563–564 (1985).

82. M. Hara, Y. Saeki, and N. Nomura, *Toyama Daigaku Kyoikugakubu Kiyo* **40**, 43–47 (1992).

83. C. W. Whang and W. L. Tsai, *J. Chin. Chem. Soc.* **36**, 179–186 (1989).

84. R. S. Stojanovic, A. M. Bond, and E. C. V. Bulter, *Electroanalysis* **4**, 453–461 (1992).

85. D. G. Williams and D. C. Johnson, *Anal. Chem.* **64**, 1785–1789 (1992).

86. T. S. Hsi and J. S. Tsai, *J. Clin. Chem. Soc.* (Taipei), **41**, 315–322 (1994).

87. H. P. Wagner and M. J. McGarrity, *J. Am. Soc. Brew. Chem.* **50**, 1–3 (1992).

88. H. P. Wagner, *J. Am. Soc. Brew. Chem.* **53**, 82–84 (1995).

89. S.-W. Park, S.-W. Hong, and J. H. You, *Bull. Korean Chem. Soc.* **17**, 143–147 (1996).

90. J. G. Pentari and C. E. Efstathiou, *Anal. Instrum.* **15**, 329–345 (1986).

91. R. D. Rocklin, T. R. Tullsen, and M. G. Marucco, *J. Chromatogr.* **671**, 109–114 (1994).
92. D. C. Johnson and W. R. LaCourse, in *Carbohydrate Analysis: High Performance Liquid Chromatography and Capillary Electrophoresis*, A. El Rassi, ed., Elsevier, Amsterdam, 1995.
93. W. R. LaCourse and C. O. Dasenbrock, in *Advances in Chromatography*, P. R. Brown and Eli Gruska, eds., Marcel Dekker, New York, in press.

8 Instrumental Considerations

An analyst hath no better thing under the sun, than to separate, and to detect, and to be merry.

In the preceding chapters, we have focused on the concepts and basic tenets of PED. Literally hundreds of assays have been developed for HPLC–PED, and many of these applications are used routinely each day. Unfortunately, many analysts spend a significant amount of time dealing with various problems and unexpected difficulties. These problems include equipment malfunctions and failures as well as poor chromatograms, excessive noise, and baseline anomalies. Nothing is more vexing than the inconvenience and/or wasted expense of service calls that could have been avoided or that we could have easily handled ourselves. Unfortunately, the lessons learned in another laboratory rarely diffuse elsewhere. Only the working assay and/or positive results are often published, and many of us are left to rediscover the wheel. We know this for a fact because after telling colleagues about how we addressed this random and onerous problem, they often respond, "I solved that same problem months ago."

The use of electrochemical detection (ED) often exacerbates the perception of problems because many users are unfamiliar with ED, and when a problem does arise, an unsureness or bias against the detector is often present. The purpose of this chapter is to discuss the overall instrumental setup in regard to PED in order to avoid potential difficulties. It will also address the most common problems associated with PED, and how to overcome them using an organized approach. It is important to remember that any instrument works best when the data are not needed or when any data will do. (Just kidding!)

THE INSTRUMENTAL SETUP

Figure 8.1 shows a generic high-performance chromatography system with electrochemical detection. Our discussion will focus mainly on optimal setup of the system with respect to PED. It is assumed that all instrumental components are working properly in accordance with manufacturers' specifications.

Since the most common application of PED is for mode I detections, especially carbohydrates, the overall HPLC system should be alkaline-tolerant. This means that all flow paths throughout the HPLC system should be polymer-based materials (e.g., PEEK, Teflon). Since glass is etched by NaOH, any hydroxide-containing eluants should be maintained in reservoirs made of plastic. All mobile phases and solvents should be freshly filtered and possibly degassed. Sparging control and the ability to blanket the solvents with N_2 or He is highly recommended to eliminate carbonate formation in alkaline solvents and to maintain low levels of dissolved O_2. If on-line degassing is used, inert-gas blanketing is still necessary to avoid carbonate buildup. Wide-diameter Teflon should be used to connect the reservoir to the solvent delivery pump.

If possible, the solvent delivery pump should use inert pump heads, which are typically made from machined PEEK, and check valves. Sapphire balls in the check valves and piston are sufficiently inert. It is important that the piston seals be made of a material that is resistant to the types of mobile phases being used. An inherent aspect of the plunger mechanism is that with each stroke a thin film of fluid is exposed outside the fluid path on the plunger. Evaporation leads to an accumulation of buffer salts and, in the case of high-pH mobile phases, to carbonate salts. Hence, a mechanism to wash the back of the piston is useful to extend the life of the piston and piston seals. This washing procedure should be done daily.

The effects of band broadening are not important prior to injection; therefore, the inner diameter of the tubing from the pump to the injector–autosampler should be as large as possible (i.e., to reduce system backpressure) while maintaining high-pressure strength. The injection system, especially rotor seals, should be made of alkaline-tolerant materials.

Next in the flow path is a guard column, which is optional, followed by the analytical column. A guard column is recommended to protect the integrity of the analytical column, which is more expensive, and to extend its life. It is important to read any and all of the manufacturer's literature relating to the maintenance, handling, and limitations of the column. For example, silica-based columns are compatible only with pH 2–7. By putting a packed-silica column before the injector, one can slightly extend the pH compatibility. Hence, high-pH condi-

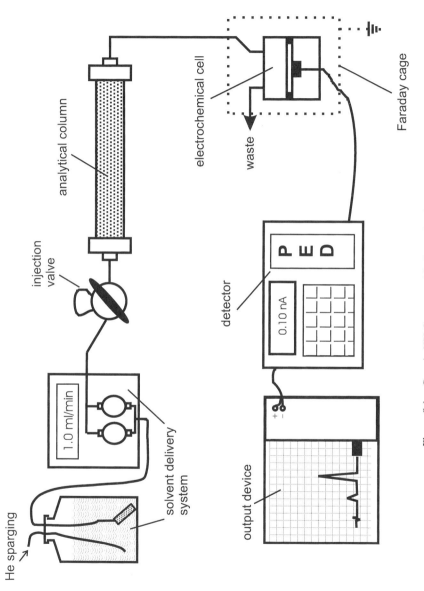

He sparging

injection
valve

analytical column

1.0 ml/min

solvent delivery
system

electrochemical cell

waste

Faraday cage

detector

0.10 nA

P
E
D

output device

Figure 8.1. Generic HPLC system with electrochemical detection.

tions, which are necessary for PED of carbohydrates and amine-based compounds, must be carried out using a polymeric, alkaline-tolerant stationary phase.
The column outlet is then connected to the electrochemical cell inlet. The vast
majority of electrochemical cells are solvent-compatible and tolerant of a wide
range of pH conditions. It is good practice to position the detection cell with the
internal surface of the thin-layer cavity oriented vertically with fluid entering at
the bottom and exiting from the top. Independent of the cell design, this arrangement tends to minimize the collection of small bubbles in the detection cell. After
the electrochemical cell, a waste line should be run to a suitable collection
reservoir. This tubing inner diameter and length can vary to add backpressure
[i.e., usually between 10 and 50 psi (lb/in.^{2})] to the cell to also reduce bubble
formation. If bubble formation is not a concern, then large-inner-diameter tubing
is desirable.

From the injector to the column to the detector, all tubing should be as short as
possible and of the narrowest diameter available to minimize extracolumn effects. Tubing should be cut flat and fit flush, and all fittings should be zero dead
volume. Any extracolumn volume compromises separation efficiency. I highly
recommend the use of PEEK tubing and fittings throughout the chromatographic
flow path. PEEK tubing is available in a wide range of sizes, is color-coded for
ease of size selection, is easy to work with, is tolerant to a wide range of buffer
and solvent conditions, and is impermeable to O_2.

In contrast to carbohydrates and amine-based compounds, which require high-
pH conditions for PED, thiocompounds can be detected over the entire pH range,
including conditions compatible with silica-based reverse-phase chromatography. Hence, stainless-steel- and titanium-based chromatography systems can be
used with PED without any modification to your system. If high-pH conditions
are required for detection only, postcolumn addition of NaOH can be used
without altering the basic chromatographic system. It is recommended that you
perform general system cleaning and pacification of a metal system at least once
a month. The technique of pacification involves removing any columns, flushing
the system with water to remove any buffers and solvents, washing the system
with 20% nitric acid (6 N), and flushing the system again with water until neutral
pH is reached. The pH is best checked by testing the waste outlet using pH paper
of the proper range. Pacification has been shown to protect metal systems from
damage by eluant high in salt (e.g., NaCl) concentration and harsh pH conditions.

The problem with stainless-steel-based chromatography systems is mainly the
intolerance, or corrosion, of the metal surfaces under extreme pH conditions.
Interestingly, in carbohydrate applications, it is the separation or column integrity
that is compromised by metal contaminants that build up on the column. PED is
not incompatible in any way with stainless-steel-based systems. As discussed
above, thiocompounds are readily separated under reversed-phase conditions,
and the electrochemical cell can be connected directly after the column. Depending on a detector's electronic configuration (i.e., working electrode at ground or
auxiliary electrode at ground), the electrochemical cell may need to be isolated

from the chromatography by using plastic or PEEK tubing to connect it to the column outlet.

If one is planning to use a metal-based system for high-pH applications, many problems can be minimized or avoided by simply converting as much of the system as possible to PEEK by replacing all inlet lines, fittings, tubing, and possibly check valves. This conversion will greatly reduce the amount of metal that is in contact with the metal path. Remember to change any pump seals and injector rotors to alkaline-tolerant materials.

The combination of optical detection (i.e., uv–vis or fluorescence) and PED for nonchromophoric compounds is a powerful combination. Unfortunately, optical windows can be etched by hydroxide eluants, and optical detectors cannot be placed in high-pH eluant streams. For low-pH applications, the optical detector can be placed before or after the PED unit. Placing the optical detector downstream of PED requires that the electrochemical cell be of low dead volume. If alkaline conditions are needed only for detection, the optical detector is placed after the analytical column followed by postcolumn addition of NaOH and on to the electrochemical cell. Figure 8.2 illustrates the different instrumental configurations.

If active temperature control is used, I recommend a system that will encompass the column, any postcolumn addition apparatus, and the electrochemical cell. Active temperature control can enhance chromatographic reproducibility and afford opportunities to improve separation efficiency. As noted in this book, the precision of quantitation in PED is also improved.

PED has been shown to be compatible with microbore chromatography, capillary liquid chromatography, and capillary electrophoresis. The concerns discussed above are all directly applicable to microchromatographic systems. Most importantly, microchromatographic systems require smaller extracolumn volumes than do normal-bore systems. Hence, great pains should be taken to use the smallest-inner-diameter tubing available, zero-dead-volume fittings, smaller injection volumes, and smaller detection cell volumes. Electrochemical detection cells are easily adaptable to miniaturization, and this aspect of PED will be discussed in Chapter 9. For the most part, follow the manufacturer's guidelines to obtain the best performance of your system.

POSTCOLUMN REAGENT ADDITION

A significant number of applications involving PED have employed postcolumn reagent (PCR) addition to enhance the compatibility between the chromatographic and detection systems. Most often a NaOH reagent (i.e., > 200 mM) is added postcolumn to increase the alkalinity of the eluant to be compatible with PED. As an example, the separation of xylose and mannose and the determination of some aminoglycosides requires the use of neutral eluants, and PCR

(A) Separation - pH < 8 and Detection - pH < 8. [Detectors are interchangeable.]

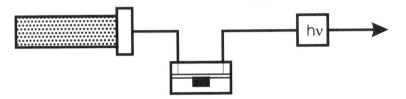

(B) Separation - pH < 8 and Detection - pH > 8.

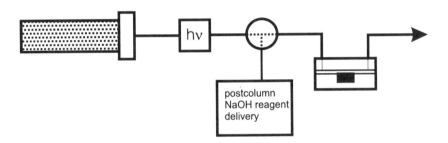

(C) Separation - pH > 8 and Detection - pH < 8.

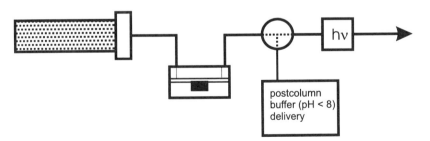

Figure 8.2. Configurations for combining optical and electrochemical detection in series depending on the pH of the separation and detection requirements.

addition of NaOH is used to make PED a viable detection system for these compounds.

During performance of gradient-based chromatography, a deleterious baseline shift occurs with PED as a consequence of the pH dependence of oxide formation at noble-metal electrodes. The addition of a highly concentrated solution of NaOH postcolumn can effectively mitigate the baseline shift by diluting out the pH gradient effect. Similar arguments can be made for ionic-strength gradients, which also affect the baseline signal, with the postcolumn addition of a high-concentration ionic-strength buffer. Hence, PCR can be used to improve gradient compatibility and detection.

A postcolumn addition system consists simply of a reagent delivery pump, a mixing tee, and a mixing coil (see Fig. 8.3). Since electrochemical detection is very sensitive to pressure and flow changes, a pressurized reservoir is commonly used to deliver a pulseless flow of the reagent. The vessel is typically outfitted with a check valve to prevent reagent backup. The major drawback of using a pressure-based delivery system is that it cannot handle a great deal of system backpressure. Unfortunately, the deck seems to be stacked against pressure-based delivery systems in that the postcolumn mixing coil, thin-layer flow-through electrochemical cell, and narrow-bore waste line (i.e., to prevent bubble formation) typically offer significant backpressure. In addition, the high viscosity of

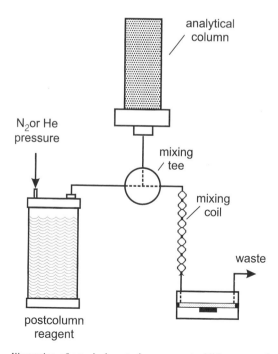

Figure 8.3. Illustration of a typical postcolumn reagent addition system in HPLC–ED.

concentrated NaOH solutions tends to exacerbate this problem. Hence, close attention must be paid to reduce postcolumn backpressure sources to effectively use PCR. Any single-piston pump, even with extensive pulse dampening, usually is inadequate for high-sensitivity work. A high-quality, nearly pulse-free (i.e., expensive) pump is essential for high-quality postcolumn addition.

Delivery of the postcolumn reagent to the chromatographic eluant flow is accomplished via a mixing tee. The mixing tee should be a low-dead-volume fitting. Probably the most crucial component in the PCR system is the mixing coil, which connects the mixing tee to the detector. It is essential that the mixing coil produce a homogeneous solution in the most efficient manner, in other words, with minimal band broadening. I have found that the best mixing coil is a woven or knitted reaction coil. The three-dimensional weave achieves efficient mixing and effectively reduces band-broadening effects by preventing laminar flow patterns to be established. In addition, their open-tubular nature produces less backpressure than do packed-bed reactors, and woven reactors are easy to make using commercially available Teflon tubing. In fact, the first knitted reactor I ever used was crocheted by my mother, and it worked every bit as well as a much more expensive commercial unit.

It is important to remember that postcolumn addition always results in dilution of the eluting peaks. Therefore, a decision must always be made with respect to the flow rate and concentration of the postcolumn reagent. These two factors work against one another in that the same pH can be obtained in the detection cell by a high flow rate of low-concentration reagent and a low flow rate of high-concentration reagent. Often the latter is preferred because it results in less dilution, although, for pressurized delivery systems, the former may produce less backpressure by lowering the viscosity of the reagent as happens with NaOH. A detailed discussion of all aspects of postcolumn addition systems has been published [1].

MAINTAINING HPLC–PED SYSTEMS

Many problems can be avoided with proper maintenance of your PED system. In addition, an understanding of the peculiarities of PED will assist you in troubleshooting more obdurate situations. Our approach will be to start with the working electrode and build up to the complete HPLC–PED system discussing potential sources of problems.

At the heart of PED is the working electrode, which is composed of either Au or Pt. It is advisable to polish fouled electrodes and any new electrodes before use to remove manufacturing remnants. A fouled working electrode typically gives high-background currents and/or a decrease in sensitivity. Fouling can be determined by comparing the analytical signal for a known standard against its expected response.

Polishing any working electrode involves the use of a series of wet, polishing compounds from coarse to fine (e.g., alumina or diamond dust) in conjunction with a polishing pad (i.e., a leathery material) to remove fouling and reactivate the electrode surface. Always follow the manufacturer's instructions closely. Important points to remember are as follows:

1. The working electrode must be held flat against the polishing pad to avoid rounding of the block in which the electrode is embedded. Otherwise, the working electrode block will fit poorly in the assembled cell and leakage will result.

2. Always use different polishing pads for each electrode material to avoid cross-contamination of the electrode surfaces.

3. Rinse off all traces of polishing compound from the working electrode with copious amounts of deionized water. Trace particulates on the surface will alter electrode response.

4. Follow the water rinse with a methanol rinse to remove any oils deposited from the polishing compounds and rinse again with deionized water.

5. Allow the electrode to air-dry. The use of cloths and paper towels may leave residual fibers or cellulose on the electrode surface. *Do not* blow-dry with hair dryers, which may change the integrity of the block or electrode via thermal expansion; or compressed gases, which will more than likely leave oily deposits on the electrode surface.

The polished electrode surface should be shiny and without pits or scratches. A newly polished electrode may take up to 12 h to stabilize, as the surface is reconstructed (i.e., changes effected in surface area) by the application of the pulsed potential waveform. The baseline may drift and peak RSDs will be higher than normal. A milder cleaning procedure involves using a clean white eraser in lieu of the polishing compound/pad setup to remove light fouling from the electrode surface. Rinsing and drying should be performed as described above. This procedure may reduce the amount of equilibration time significantly. This procedure does not always work, and it may leave you taking apart the cell and doing a wet polish.

In dc amperometry, the working electrode must be polished daily to remove foulants. In contrast, the working electrode in PED is continuously cleaned by the application of the pulsed potential waveforms, and the working electrode should be polished only when it's new or fouled. Aged Au and Pt electrodes with no fouling are yellowish-brown and grayish-black in color, respectively, and may have a powdery appearance. This discoloration, also known as *gold black* or *platinum black*, is a result of fine particulate structure on the electrode surface, which may have taken weeks to form. *Do not* polish the electrode to remove this discoloration. It is normal, and in our laboratory, 6 months between polishing is not unusual. In PED, polishing routinely will result in poor detection reproducibility.

If one loses one's focus in life, all that persons's efforts may be for naught. In an analogous manner, the reference electrode is usually a fixed-potential source (i.e., electrochemical half-cell), which serves to maintain the proper working electrode potential. If the reference electrode's inherent potential changes, so does the working electrode potential, and the PED response and reproducibility is compromised. The construction and operation of several reference electrodes were discussed in Chapter 2. Probably the most common reference electrode is Ag/AgCl, and we shall use it for our discussion. Reference electrode problems can typically be traced to either loss of electrolyte over time or blockage of its frit or junction. Restoration of the electrolyte solution can be simply done by refilling the electrode with new electrolyte solution, if possible, or by soaking in a solution of saturated KCl whenever the electrochemical cell is not in use. Storage and soaking of the reference electrode should be carried out in a soaker bottle that suspends the electrode in the soaking solution and prevents corrosion of the electrode contacts (see Fig. 8.4). Over time even with infrequent use, the reference electrode may change color and darken. This change in color is not a problem, and it should not affect PED operation. On the other hand, if the electrode becomes cloudy, changing to a new electrode is advised. Blockage of the reference electrode junction is often visible as a dark band in or at the tip of the frit. The first approach should be to attempt to regenerate the electrode by soaking it in 1 M KCl and 1 M HCl. For a more intractable blockage, the electrode tip, which is typically a frit embedded in glass, can be ground down using either a file or emery paper followed by soaking in 1 M KCl and 1 M HCl. *Be careful* not to cut yourself when filing or polishing glass! If mechanical removal of the blockage does not work, replace the reference electrode. Reference electrodes typically last 3–6 months in our laboratory.

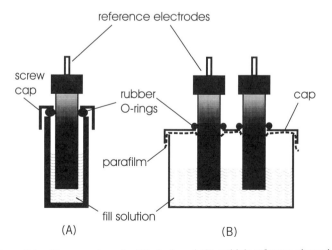

Figure 8.4. Storage systems for (*A*) single and (*B*) multiple reference electrodes.

As with any other electrical device, the auxiliary electrode completes the circuit. If an oxidation reaction occurs at the working electrode, a reduction reaction occurs at the auxiliary electrode and vice versa. Whereas the potential at the working electrode is controlled in order to oxidize the analyte of interest, the potential of the auxiliary electrode will float to any potential to induce a reaction to complete the electron flow. Often the inertness of the auxiliary electrode and the prevalence of water result in the anodic or cathodic breakdown of water. In order to maintain a low current density and prevent bubble formation, the surface area of the auxiliary electrode is typically very large. Occasionally, the auxiliary electrode will darken or become pitted near the working electrode. These conditions are a problem only if the pitting is so extensive that laminar flow cannot be established or the darkening greatly increases resistance of the auxiliary electrode surface. Discoloration can be removed with careful polishing, but severe pitting requires replacement of the auxiliary electrode.

The three electrodes are most commonly configured in a thin-layer channel design, which is depicted in Figure 8.5. The working electrode is separated from the auxiliary electrode by a thin gasket that forms the thin-layer channel. The reference electrode is placed either downstream or directly across from the working electrode. It is important to keep each electrode isolated from the other except via the thin-layer channel. Hence, it is imperative that no leaks occur at the gasket that forms the thin-layer channel or around the reference electrode. These leaks

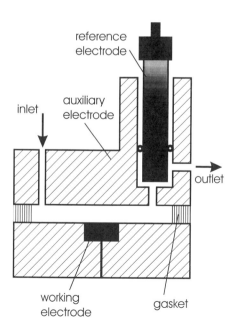

Figure 8.5. Depiction of a generic thin-layer electrochemical cell.

will serve to short-circuit the electrodes external to the desired thin-layer channel, to induce corrosion currents, and to allow O_2 to leak into the cell. Cell leaks result in drifting baselines and increased background noise. Visible drops of fluid at cell interfaces are easy to see, but often the leak is more subtle. Usually signs of leakage are corrosion (i.e., a blackened appearance) under the gasket at the auxiliary electrode surface or a buildup of salts at the edge of the cell gasket. The electrochemical cell should be assembled dry as follows:

1. Since the gasket forms the thin-layer channel and separates the working electrode from the auxiliary electrode, it is of vital importance that the gasket not leak. Inspect the gasket for any creases or deformities. If there is any doubt as to its integrity, replace it with a new gasket, and carefully place it on the auxiliary electrode.
2. The working electrode, which is dry and free of lint, is placed carefully on the gasket and tightened into place.
3. With the cell connected to the eluant inlet, start the fluid flow to expunge any trapped air from the thin-layer channel.
4. Fill the reference electrode chamber, insert the reference electrode causing the excess fluid to overflow, and tighten the reference electrode assembly. Attach the waste line and check that it is flowing properly.
5. Inspect the cell for leaks.

Do not turn on the cell until you are sure that mobile phase containing supporting electrolyte is flowing through the cell. In the case of PED for carbohydrates, test the solution to ensure that it is alkaline (i.e., use a pH test strip to test the waste eluant) *before* turning on the cell. The PED waveform is operating at the limits of solvent breakdown, and operating the cell at other than intended conditions may result in excessive bubble formation.

In connecting the electrochemical cell to the detector, the wires should be as short as possible and not twisted or crossed to minimize electromagnetic noise pickup. After all, the assembled electrochemical cell is an electrical transducer, and it is important to isolate it from electromagnetic interferences. Hence, proper shielding and grounding of the entire HPLC–PED system per the manufacturer's instructions should be followed conscientiously. At a minimum, the cell should be placed inside a Faraday cage to shield it from noise from fluorescent lights, external pumps, and other equipment. If active temperature control is used to lower system RSDs, the temperature controller should be shielded properly so as to avoid adding electrical noise from thermostat contacts. One of the most overlooked possible sources of noise is the waste line, which can act as either an antenna or a conduit to the plumbing system of the building. If more than one HPLC system is using the same waste receptacle, the two instruments will experience cross-talk (i.e., noise) if the eluant streams cross or are submerged under the waste solution. Hence, I highly recommend that the waste-line outlet be placed above the waste reservoir and allowed to drip into the container.

Major problems with HPLC often originate from wear or fluid restriction (e.g., filters) points in the system. Inlet filters should be checked regularly for blockage; pump seals can wear out, causing leakage; and check valves can stick from deposits from buffer crystallization. These problems typically lead to drifting and/or erratic baselines. The rotor and needle seals can become worn and scored, resulting in sample carryover, split peaks, and sample leakage. Particulates, matrix precipitation, and fines can block the column, resulting in higher-than-normal backpressures, which is why the use of a guard column is highly recommended in order to protect the analytical column. In general, the overall system should be checked for visible leaks, bent or crimped tubing, leaky seals, uneven pump-head pressure, and higher-than-normal system pressure. Salt build-up at joints and connections is evidence of a slow leak.

Dissolved O_2 is electrochemically active, and its presence in a sample is typically noted as a negative peak in PED. Dissolved O_2 behaves as an organic molecule, and it is often retained significantly on polymeric columns. The negative peak is of no consequence unless it interferes with an analyte of interest. In alcohol separations using a polymeric column, it typically elutes with ethanol; and in monosaccharide analysis, it routinely elutes near glucose. There are several possible solutions to this problem:

1. Eliminate all dissolved O_2 from the sample. This approach is difficult to employ efficiently while maintaining sample integrity. It is also not amenable to most autoinjectors.
2. Change to another stationary phase with a different selectivity for dissolved O_2 versus the analyte of interest. For instance, monosaccharide analysis is routinely done using a CarboPac-PA1 (Dionex), and dissolved O_2 elutes with glucose. This problem has been overcome by the development of the CarboPac-PA10. Retention of dissolved O_2 has been increased relative to the monosaccharides, and it is no longer a problem.
3. A postcolumn weave made of Teflon tubing can be used to reduce the dissolved-O_2 peak by smearing out the peak and/or equilibrating the amount of dissolved O_2 with the atmosphere [2].

Dissolved O_2 in the mobile phase can cause drifting, especially at Pt electrodes, unless it is maintained at constant levels. The lower concentration of dissolved O_2 is desired to relax constraints on IPAD waveforms.

Preventing system leaks and blanketing solvents with an inert gas is essential to avoid the introduction of atmospheric oxygen into the HPLC system, which can be deleterious to electrochemical detection, and the formation of carbonate in alkaline solvents from CO_2. Monosaccharide analysis, which is one of the most popular applications of PED, sometimes suffers from peaks that elute early. Carbohydrates are separated using anion exchange using hydroxide ion as a weak pushing ion to release the carbohydrate from the column. Carbonate anion has a stronger affinity for the ion-exchange sites at low concentrations of hydroxide, and if it is present in the mobile phase, it can build up on the column. The loss of sites results in reduced retention of the monosaccharides. At concentration of hydroxides above ~50 mM, the carbonate does not build up on the column.

TROUBLESHOOTING HPLC–PED SYSTEMS

In keeping with the goals of this book, it will be assumed that the chromatographic portion of the HPLC–PED system is working correctly, and only problems specifically related to PED will be discussed in detail. Should a HPLC problem arise, several sources could help you identify and solve the problem:

1. The operator's manual and service literature supplied with your instrument. Included in these items is typically a troubleshooting chart, exploded diagrams, and part numbers for replacement parts.
2. Co-workers and senior colleagues are often an invaluable source of information, and many may have experience at solving the same problem you are encountering. Don't be afraid to ask.
3. Often the manufacturer of your instrument can help you solve a problem over the phone. Recently, some manufacturers have set up World Wide Web sites on the Internet to assist customers troubleshoot and solve problems.
4. Because of the maturity of HPLC, there are many books, reviews, and journal columns (e.g., *LC/GC* magazine) describing maintenance and troubleshooting of HPLC systems. Many of the books listed in the bibliography at the end of this chapter contain excellent troubleshooting information.
5. If all else fails, make a service call, especially if it's under warranty. In order to save time and possibly money, make a list of everything you have already tried and give it to the repair person.

It is important to remember that the majority of HPLC problems can be prevented if routine maintenance is practiced.

Essential to troubleshooting PED systems is the establishment of a quality-control regimen. In other words, what does the system do when it is doing it properly? For a particular system, this regimen should include documentation of the system pressure, the capacity factor and efficiency of a mixture of standards, baseline drift over 10 min, and the typical response and background noise for those same standards. A new, hydrated reference electrode should be on hand for comparison to the reference electrode in use. With a high-impedance voltmeter, the voltage difference between the two electrodes when submersed in a conductive solution should be less than ~30 mV. The quality-control regimen will be worth the effort as a reference in troubleshooting your system.

When experiencing a problem, a systematic approach is the best approach to save time and frustration. In PED, the most common problems encountered in order of occurrence are baseline drift, a noisy baseline, and reduced analyte signal. The following is the approach I often use to identify PED problems in my laboratory. It is assumed that the proper method, waveform, mobile phase, column, and electrode material are being used.

If the system pressure and detector response is normal, baselines that drift up and down are almost always a consequence of peaks coming off the column from previous injections. Since these compounds may have been on the column for a long time, they often elute over an extended period of time and at a low signal strength. Since PED is a heterogeneous technique, these low-level interferent peaks can affect analyte response by competing or blocking analyte adsorption. The first step is to remove the column and replace it with a blank column. A blank column can simply be constructed from narrow-bore tubing of sufficient length to produce enough backpressure for proper pump operation. If the baseline drifting goes away, the problem is diagnosed. Otherwise, scrutiny of the remaining system, especially the detector, is necessary. Begin with a careful inspection of the entire chromatographic system for leaks, bad seals, and/or poorly set fittings. Remember to check the solvent reservoirs; changing dissolved-O_2 levels in the mobile phase will often cause the baseline to drift. At this point, examine and test the reference electrode. Nine times out of ten the reference electrode is the source of the problem in electrochemical systems. If necessary, regenerate, grind down, or change the reference electrode. Reinsert the reference electrode and test the baseline by comparing to the normal values determined by your quality-control regimen. If the problem still exists, dismantle the electrochemical cell and inspect for signs of leaks, a faulty gasket, and/or corrosion. Reassemble the electrochemical cell and test the baseline. If all else fails, suspect the electronics of the detector.

If the *baseline is noisy*, it typically is not a consequence of the column. Again begin with a careful inspection of the entire chromatographic system for leaks, bad seals, and/or poorly set fittings. As with baseline drift, check the reference electrode first, followed by an inspection of the cell for leaks. When noise is a concern, the working electrode should be examined for fouling, pitting, and/or signs of receding into the block. If necessary, the working electrode should be polished or changed and the baseline noise compared against the expected values of the quality-control regimen. If no problem is found, the possibility exists that the mobile phase has been contaminated or aged. Change all the solvents with freshly filtered solutions. Remember to wait until the old solvent has flushed out of the system before retesting the baseline noise. If all else fails, suspect the electronics of the detector.

If *reduced analyte signal* is observed, the first step is to test the response of the system using a standard test mixture, which should be a part of your quality-control regimen. Check the flow rate and pH of the eluant at the waste line to ensure that the proper conditions are in use. Carefully inspect the injector and postinjector fluid path for leaks. The injector is most easily tested by switching to an alternate unit, but I usually reserve this as a last resort. Injector problems typically give reduced analyte signal due to leaks from rotor wear, but this loss of sample tends to be highly irreproducible. For weakly adsorbing analytes, reduced analyte signals sometimes result from the coelution of an ultra-late-eluting contaminant, which is broad and possibly undetected directly. Insert a blank column into the system and test the response of a standard compound. If the analyte response is normal as compared to your quality-control mechanism, attempt to wash and regenerate your column using manufacturer's guidelines. Otherwise, the working electrode should be inspected for fouling, pitting, and/or signs of receding into the block. If necessary, polish or change the working electrode, and while the cell is dismantled check the reference electrode. Reassemble the cell and compare the response of your standard against the expected values of the quality-control regimen.

For PED of weakly adsorbing compounds, the remote possibility exists that the mobile phase has been contaminated by a strongly adsorbing compound (e.g., ACN or a thiocompound), which may cause reduced analyte signal. Hence, change all the solvents with freshly filtered solutions and retest the baseline. *Always* use the highest-quality reagents available. As before, if all else fails, it is time to question the electronics of the detector.

In all cases, the chances that the detector electronics are bad is not very high, and as a consequence, most of the problems you encounter should be resolvable. The data acquisition is best tested by simply replacing it with an old-fashioned chart recorder, if you still have one lying around.

REFERENCES

1. B. Lillig and H. Engelhardt, in *Reaction Detection in Liquid Chromatography*, I. S. Krull, ed., Marcel Dekker, New York, 1986.
2. W. R. LaCourse, D. C. Johnson, M. A. Rey, and R. W. Slingsby, *Anal. Chem.* **63**, 134–139 (1991).

BIBLIOGRAPHY

Karger, B. L., L. R. Snyder, and C. Horvath, *An Introduction to Separation Science*, Wiley, New York, 1973.

McMaster, M. C., *HPLC: A Practical User's Guide*, VCH Publishers, New York, 1994.

Snyder, L. R., and J. J. Kirkland, *Introduction to Modern Liquid Chromatography*, 2nd ed., Wiley, New York, 1979.

Snyder, L. R., J. L. Glajch, and J. J. Kirkland, *Practical HPLC Method Development*, Wiley, New York, 1988.

A User's Guide To HPLC Troubleshooting, Technical Notes 1, 3rd ed., Phenomenex, Torrance, CA, 1994.

9 Future Aspects of PED

All's well that detects well!

At present, the multistep potential–time (E–t) waveforms of PED can be applied to the *direct* (i.e., requiring no derivatization), *sensitive*, and *reproducible* detection of a large variety of polar aliphatic compounds including alcohols, glycols, carbohydrates, alkanolamines, aminoacids, and numerous sulfur-containing compounds. These compounds, many of which have biological significance, typically have been classified as nonelectroactive for detection under constant applied potentials and have poor optical detection properties. As shown previously for the determination of diols in a pharmaceutical formulation (Chapter 7), the combination of a chromophoric (uv–vis) detector and a nonchromophoric (PED) detector following a chromatographic separation is a powerful technique.

Present research is being directed toward the application of PED at microelectrodes for use in miniaturized separation systems, indirect detection techniques for macromolecules, and enhanced selectivity with on-line pulsed voltammetry. A brief description of the latest advances in these areas is given below.

MICROELECTRODE APPLICATIONS IN PED

The response in optical detection methods based on Beer's law is dependent on the pathlength of the detector cell. In order to maintain the efficiency of micro-chromatographic and capillary-based separation systems, it is crucial that detection cell volumes also be miniaturized. Since the pathlength of a cell is directly proportional to cell volume, miniaturization is often at odds with optical detection techniques. The loss of response on miniaturization is further exacerbated by compounds with poor optical detection properties. In contrast, electrochemical detection is based on a reaction at an electrode surface. Present technology allows us to make electrodes very small, and, consequently, detector cell volumes can be made similarly small with no decrease in sensitivity. Figure 9.1 illustrates the fundamental difference in response characteristics between optical and electrochemical detection. The combination of electrochemical detection systems with microchromatographic and capillary-based separation techniques, which require detection cells of limited volume, offers increased mass sensitivity, higher chromatographic efficiencies, less solvent consumption, and, in particular, the ability to analyze samples of limited quantity. PED affords these same advantages to virtually all polar aliphatic compounds.

LaCourse and Owens [1] have applied PED following microbore (i.e., 1-mm i.d. column) and capillary (i.e., 180-μm-i.d. column) liquid chromatography to the determination of thiocompounds (e.g., methionine, cystamine, homocysteine, coenzyme A and derivatives) under typical reversed-phase conditions. Interestingly, PED enables the direct determination of thio redox couples [i.e., —SH/—S—S—] and numerous other sulfur moieties at a single Au electrode without derivatization. In a comparative study of PAD and IPAD, IPAD was determined to be better suited for management of the large oxide formation (background) currents, which accompany the faradaic signal of the analyte. IPAD resulted in more stable baselines, eliminated oxide-induced artifacts, and yielded lower limits of detection. Figure 9.2 shows the separation of several thio redox (i.e., —SH/ —SS—) couples using microbore chromatography. The separation of other bioactive compounds was shown in Chapter 7. Table 9.1 lists the analytical figures of merit for the bioactive compounds in Figure 7.29. Mass limits of

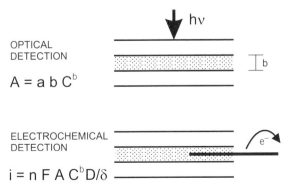

Figure 9.1. Comparison of cell response characteristics for optical and electrochemical detection in microchromatographic and capillary-based systems.

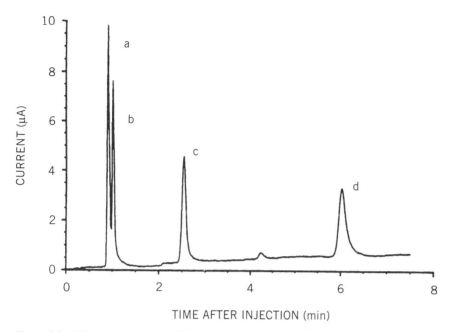

TIME AFTER INJECTION (min)

Figure 9.2. Microbore separation of four model thiocompounds. Peaks (100 pmol each): (*a*) di-thioerythritol; (*b*) *trans*-1,2-dithiane-4,5-diol; (*c*) 2-aminoethanethiol; (*d*) cystamine. Reprinted from W. R. LaCourse and G. S. Owens, Pulsed electrochemical detection of thiocompounds following microchromatographic separations, *Anal. Chim. Acta.* **307**, 301–319 (1995) with kind permission of Elsevier Science–NL, Sara Burgerhartstraat 25, 1055 KV Amsterdam, the Netherlands.

detection were 0.2–0.5 pmol injected except for methionine, which was 2 pmol. Since methionine is a thioether, it has been conjectured that the accessibility of the sulfur atom to the surface is hindered, and that fewer electrons are transferred [1]. The product of thioether oxidations is probably sulfones (i.e., $n = 4$). The CLC–IPAD separation of dithioerythritol (DTE) and dithiane in a 60-nL

TABLE 9.1 Quantitation Parameters of Biologically Important Thiocompounds at a Au Electrode Using IPAD [1]

		Linear Range nC = $a \cdot$ (pmol) + b			Repeatability
Compound	LOD^a (pmol)	a	b	R^2	%RSD (pmol, n)
Cysteine	0.2	56.6	2.24	0.9997	4.6 (2.5, 6)
Homocysteine	0.5	33.8	−13.5	0.9984	3.6 (5, 6)
Methionine	2	15.6	−19.4	0.9999	2.5 (20, 6)
GSH	0.5	30.4	−0.14	0.9989	2.1 (5, 6)
GSSG	0.5	24.8	7.99	0.9956	2.2 (5, 6)

^{a}Calculated at S/N = 3 from injections within a S/N = 5.

injection is shown in Figure 9.3. The LODs for DTE and dithiane were determined to be 0.3 and 0.1 pmol, respectively. The decreased dispersion in the capillary chromatographic system is reflected in the lower LODs for these compounds as compared to microbore chromatography. The high selectivity of PED for thiocompounds under mildly acidic conditions reduces sample preparation and produces simpler chromatograms of complex mixtures.

The higher-frequency waveforms (i.e., >1 Hz) needed to define the narrow peaks of microchromatographic separations and the fast scan rates (i.e., >100 mV s^{-1}) of the IPAD waveform require the use of microelectrodes (i.e., ≤ 1 mm), which exhibit reduced capacitance effects. The Johnson group [2] has proposed that the upper frequency of PED waveforms is limited by the slow kinetics of

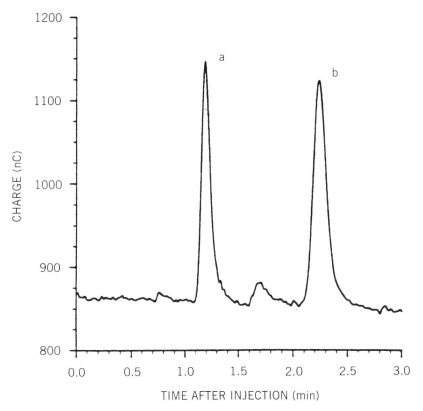

Figure 9.3. Capillary liquid chromatography–PAD separation of (*a*) DTE and (*b*) dithiane at 6 pmol (100 μM) each. Reprinted from W. R. LaCourse and G. S. Owens, Pulsed electrochemical detection of thiocompounds following microchromatographic separations, *Anal. Chim. Acta.* **307**, 301–319 (1995) with kind permission of Elsevier Science–NL, Sara Burgerhartstraat 25, 1055 KV Amsterdam, the Netherlands.

oxide formation and dissolution at Au and Pt electrodes. The upper frequencies have been theorized to be ~ 15 Hz.

The upper limits of the PED waveform will ultimately be tested by capillary electrophoresis (CE). Capillary electrophoresis is a relatively new and powerful technique for the separation of compounds of interest in complex sample matrices. As a consequence of the separation mechanism in a capillary tube (i.e., < 100 μm i.d.), CE produces very narrow electrophoretic bands and high separation efficiencies.

CE–ED has been used extensively for the determination of neurotransmitters [3] and other easily oxidized compounds at carbon electrodes. PAD has been applied following CE to the determination of carbohydrates under highly alkaline conditions [4–6]. The first application of CE–PAD was for glucose in blood (see Fig. 9.4). The level of glucose in blood was determined to be 4.25 ± 0.13 mM, which agrees well with that reported in the literature [7]. The PAD response for glucose was determined to be linear over the range of 10–1000 μM, and the mass detection limit was determined to be ~ 20 fmol. CE–PAD is ideally suited to the separation and detection of charged carbohydrates (see Fig. 9.5).

LaCourse and Owens [8] have extended the application of IPAD to the direct detection of many polar aliphatic compounds over a wide range of pH condi-

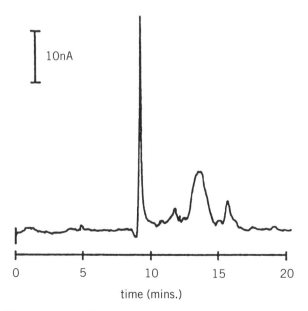

Figure 9.4. Electropherogram of human blood. Peak at ~ 9 min corresponds to 85 μM glucose. Reprinted with permission from T. J. O'Shea, S. M. Lunte, and W. R. LaCourse, *Anal. Chem.* **65**, 948–951 (1993). Copyright 1993 American Chemical Society.

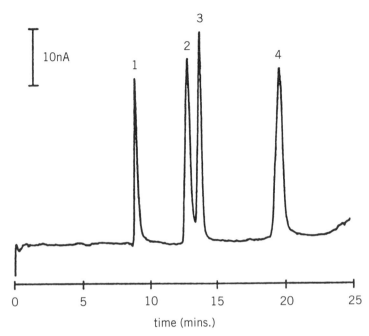

Figure 9.5. Electrophoretic separation of 1×10^{-4} M each of (1) glucosamine, (2) glucosaminic acid, (3) glucosamine 6-sulfate, and (4) glucosamine 6-phosphate. Reprinted with permission from T. J. O'Shea, S. M. Lunte, and W. R. LaCourse, *Anal. Chem.* **65**, 948–951 (1993). Copyright 1993 American Chemical Society.

tions following CE. They found that the detection of unsubstituted carbohydrates requires highly alkaline conditions, whereas amine-containing compounds (e.g., glycopeptides, peptides, aminoacids) and thiocompounds can best be detected at mildly alkaline (i.e., pH 9.0) and mildly acidic (i.e., pH 5.5) conditions, respectively. The analytical figures of merit under optimal conditions for glucose, glucosamine, and cysteine are shown in Table 9.2. Mass limits of detection are typically 10 fmol or less. Figure 9.6 shows an electropherogram for the detection of sulfur compounds in grapefruit juice. Separation efficiencies are on the order of 150,000–240,000 plates per meter. This electropherogram highlights the sensitivity and selectivity that can be obtained with PED following CE. In addition, fast-cycle PED waveforms can be used to maintain the narrow bands of CE for quantitation and the integrity of these high-efficiency separations. Figure 9.7 shows an electropherogram for the detection of (*a*) lincomycin, (*b*) cephadrine, (*c*) cephalexin, (*d*) penicillin G, and (*e*) cefazolin using CE–IPAD. It is my opinion that CE–IPAD will have a major impact on the analysis of complex biological samples for PED-active compounds, especially antibiotics.

TABLE 9.2 Quantitative Parameters of Polar Aliphatic Compounds at a Au Microelectrode in Various Operating Buffers [8]

| Buffer | Compound | Waveform | Linear Range PAD: $\mu A = a(\mu M) + b$ IPAD: $nC = a(\mu M) + b$ | | | Deviation from Linearity (μM) | LOD[a] (fmol, pg) |
			a	b	R^2		
NaOH, 16 mM pH 12	Glucose	PAD	7.97×10^{-5}	4.28×10^{-4}	0.9998	500	10, 2
		IPAD	2.13×10^{-2}	5.14×10^{-1}	0.9970	500	40, 8
	Cysteine	PAD	—	—	—	—	—
		IPAD	3.41×10^{-1}	1.79×10^{-1}	0.9956	1000	110, 10
Borate, 20 mM pH 9.3	Glucosamine	PAD	1.04×10^{-4}	2.92×10^{-4}	0.9988	500	4, 1
		IPAD	3.46×10^{-2}	7.14×10^{-1}	0.9958	1000	6, 1
	Cysteine	PAD	—	—	—	—	160, 20
		IPAD	—	—	—	—	160, 20
Acetate, 10 mM pH 5.5	Glucosamine	PAD	1.44×10^{-4}	5.24×10^{-3}	0.9936	500	90, 20
		IPAD	6.44×10^{-2}	1.74×10^{-1}	0.9978	500	30, 7
	Cysteine	PAD	3.90×10^{-4}	4.31×10^{-3}	0.9988	500	20, 2
		IPAD	2.02×10^{-1}	-2.01	0.9851	500	10, 1

[a]Calculated at S/N = 3 from injections within a S/N = 5.

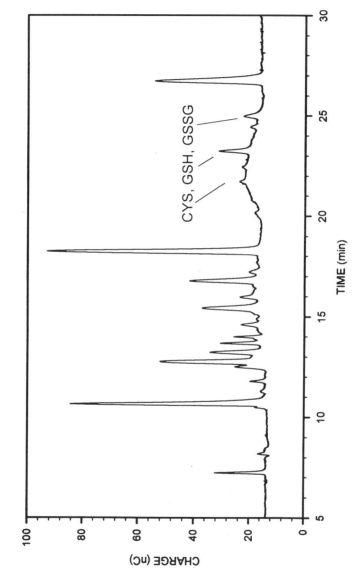

Figure 9.6. CE–IPAD of grapefruit juice highlighting bioactive sulfur-containing compounds. Conditions: 10 mM phosphate buffer, pH 7.0, 10 kV, 60 cm, 50 μm i.d.; IPAD waveform.

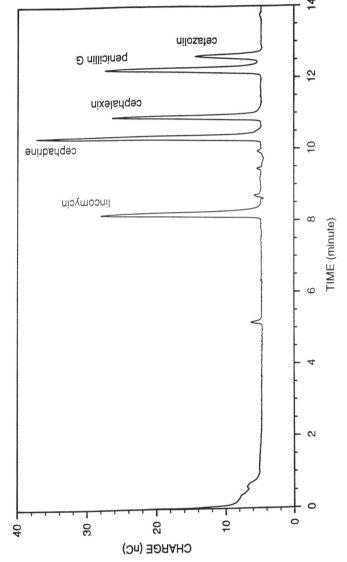

Figure 9.7. Electropherogram of a mixture of sulfur-containing antibiotics. Conditions: 10 mM phosphate buffer, pH 7.0, 10 kV. 60 cm, 50 μm i.d.; IPAD waveform. Reprinted from Ref. (23) by courtesy of Marcel Dekker, Inc.

253

INDIRECT DETECTION METHODS

The detection of electroinactive species using mode III detections was performed by Polta and Johnson [9] for the determination of chloride and cyanide ions. Suppression of the formation of PtOH accounted for negative peak formation in their flow-injection analysis system. Welch and co-workers [10–13] exploited mode III detections for the determination of penicillins. Figure 9.8 compares the output from a direct PAD waveform and an indirect PAD waveform. Note the expected negative peaks from indirect detection via mode III [10]. The separation and indirect detection of eight penicillins in milk achieved limits of detection in the range from 2 to 0.1 ppm. Welch found that the best detection limits for these compounds were accomplished using indirect detection [13]. This work was not compared with IPAD, which has been shown recently to have detection limits of 20 and 50 ppb within a milk matrix for cephapirin and ampicillin, respectively [14].

A prerequisite of PED in all its forms is adsorption of the analyte to the electrode surface prior to detection. Hence, other surface-active species that adsorb more strongly can interfere in the detection mechanism of weakly adsorbed molecules. A consequence of this phenomenon is the attenuation of the

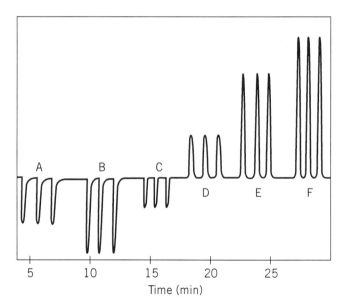

Figure 9.8. Crossover from indirect to direct PAD by variation of detection potential. Sample: 1.0 × 10^{-3} M penicillin G. E_{det} (mV): (A) 1000; (B) 1100; (C) 1200; (D) 1300; (E) 1400; (F) 1500. Reprinted with permission from L. Koprowski, E. Kirchmann, and L. E. Welch, *Electroanal.* **5**, 473–482 (1993). Copyright 1993 VCH Publishers.

analytical signal for carbohydrates when acetonitrile is added to the mobile phase. This effect is also attributed to the favored adsorption of the amine functional group over that of the alcohol group.

An alternate approach to indirect detection involves the addition of a weakly adsorbing, PAD-active reagent to the mobile phase either pre- or postcolumn [15,16]. The background signal for the reagent is then suppressed by the eluting analyte, which is preferentially adsorbed to the electrode surface. This premise is illustrated by the pulsed voltammetric response of 1 mM sorbitol obtained in the presence (-.-.-.-.-.) and absence (---------) of 0.1 mM propylamine shown in Fig. 9.9. The difference response (_____) of these plots reflects the adsorption profile of propylamine to the electrode surface. As proposed earlier, the amine group adsorbs to the electrode at potentials prior to the onset of oxidation. Note that the direct response of propylamine (.............) is less than one-tenth of the indirect response.

Figure 9.10 compares the (*A*) direct (i.e., no reagent added) and (*B*) indirect responses for (*a*) Cl⁻, (*b*) lysine, (*c*) 4-hydroxyproline, and (*d*) sucrose [16]. The detection potential is set for the oxidation of the reagent, which is detected in the

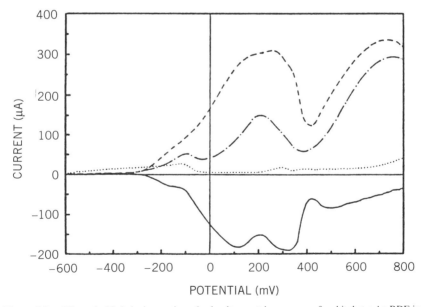

Figure 9.9. Effect of added alanine on the pulsed voltammetric response of sorbitol at a Au RDE in 0.1 M NaOH. These plots are background-corrected. Solutions: (---------) 1.0 mM sorbitol; (-.-.-.-.-.) 1.0 mM sorbitol with 0.2 mM proplyamine; (.............) 0.2 mM propylamine; and (_____) difference of sorbitol with propylamine response and sorbitol only response. Reprinteid from W. R. LaCourse, Pulsed electrochemical detection at noble metal electrodes in high performance liquid chromatography, *Analusis* **21**, 181–195 (1993) with kind permission of Elsevier Science.

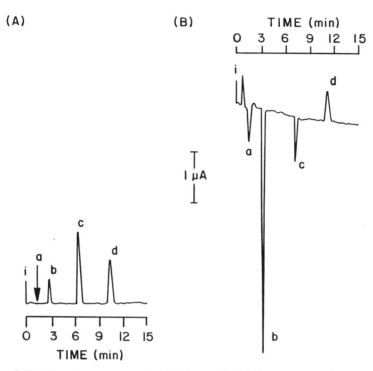

Figure 9.10. Chromatograms comparing (*A*) direct and (*B*) indirect responses for various compounds. Peaks: (*a*) Cl⁻, 14 nmol; (*b*) lysine, 14 nmol; (*c*) 4-hydroxyproline, 19 nmol; (*d*) sucrose, 7 nmol. Reprinted from W. R. LaCourse, Pulsed electrochemical detection at noble metal electrodes in high performance liquid chromatography, *Analusis* **21**, 181–195 (1993) with kind permission of Elsevier Science.

oxide-free region of the electrode. Since the mechanism of this detection is not mode III, the direct mode for Cl⁻ gives little or no response. The signal for lysine is also weak. The hydroxyl group on hydroxyproline and sucrose allows for the direct detection of these compounds. As conjectured, Cl⁻ and lysine, which strongly adsorb to the electrode surface, show strong indirect signals relative to their direct signals. Adsorption of the amine group in hydroxyproline results in an indirect signal equivalent to its direct signal. Since sucrose is also a weakly adsorbing compound, an attenuated smaller direct signal is observed in the indirect mode.

These attenuation phenomena suggest the possibility of a highly sensitive, although indirect, detection system for electroinactive analytes. It is expected that the advantages of indirect techniques will become more obvious in applications for larger molecules (e.g., polymers).

PULSED VOLTAMMETRIC DETECTION (PVD)

In dc amperometry, enhancement of selectivity is achieved through the use of multiple electrodes in various configurations (i.e., series and parallel arrangements) [17]. The multiple-electrode approach has been exploited, and has enjoyed success, using thin-layer cell designs. This approach conserves the low backgrounds (i.e., no double-layer charging currents) achieved in dc amperometry.

Greater electrochemical selectivity and analyte information can be obtained with the application of "on-the-fly" voltammetry. The generation of current versus potential plots at any time point facilitates the deconvolution of unresolved chromatographic peaks. Voltammetric detection in chromatography has been reviewed [18,19]. Double-layer charging is a major problem when performing voltammetry on-line using an analog scan. Much of this problem can be mitigated by the use of staircase voltammetry, which allows for dissipation of double-layer charging currents before sampling the signal for each incremental potential step. Voltammetric detection using pulsed waveforms, including staircase waveforms, is called *pulsed voltammetric detection* (PVD), and PVD has been applied to CLC [20].

Voltammetric detection combined with pulsed potential cleaning and reactivation has been performed by Owens et al. [21] at a Pt electrode. The measurement of small signals at high sensitivity in the presence of large background currents from oxide formation is a problem with PVD at noble-metal electrodes, and Sturrock and O'Brien [22] have overcome this problem using background compensated instrumentation with excellent results.

From a voltammetric point of view, the IPAD waveform is essentially a cyclic voltammetric experiment taking places between each set of cleaning pulses. Using staircase voltammetry during the detection step, voltammetric data can be obtained in HPLC–"IPAD," or HPLC–PVD, experiments with minimal adverse effects from double-layer charging currents. Figures 9.11A and 9.11B show the surface and contour plots, respectively, of current versus potential versus time for the separation of various biochemically active compounds (i.e., amino acids and carbohydrates) using HPLC–PVD [23]. These plots are background-corrected "on the fly." Note that the amine-containing compounds have signal at high potentials, whereas the compounds with hydroxyl groups are detectable in the oxide-free region of the electrode. At any potential a chromatogram can be extracted to afford greater selectivity, and at any time point a voltammogram can be extracted to identify or characterize the analyte or peak. In addition, the individual scans across the peak can be used to determine peak purity. Figure 9.12 shows voltammograms at (●) 4.32, (▲) 4.39, (♦) 4.44, and (■) 4.48 min of the PVD plot shown in Figure 9.11A. The purity of this peak is confirmed by the agreement in shape of the individual voltammograms. The anodic wave from -0.2 to $+0.3$ V denotes the presence of a hydroxyl group, and the "dipping" at approximately $+0.4$ V followed by an anodic peak is indicative of a compound

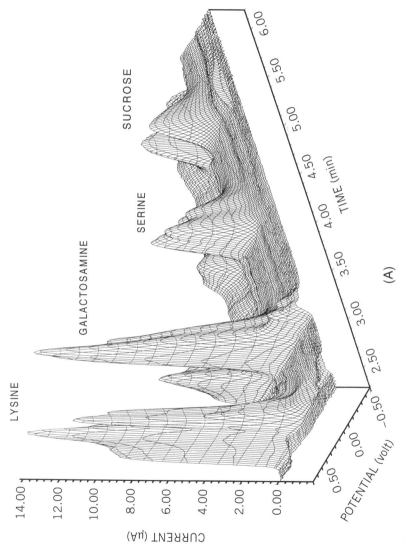

Figure 9.11. HPLC–PVD of various mixtures depicted in (*A*) surface and (*B*) contour plots for simple bioactive compounds. Note the individual signatures in the contour plot for each of the compounds.

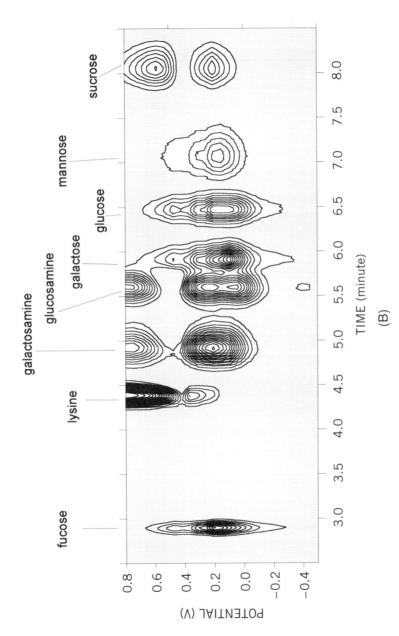

Figure 9.11. (*Continued*)

259

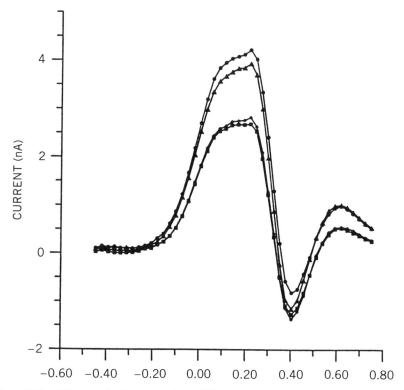

Figure 9.12. Voltammetric plots extracted from HPLC–PVD in Figure 9.11*A* for the peak labeled *serine.* The agreement in shape of the selected voltammograms confirms the purity of this peak. Voltammograms: (●) 4.32 min; (▲) 4.39 min; (♦) 4.44 min; (■) 4.48 min.

that contains an amine. This compound is *serine*. Hence, a limited degree of functional group identification is possible.

The results shown here illustrate the feasibility and doubtless importance of PVD to enhance selectivity and compound characterization, afford a limited degree of functional group identification with peak purity, and allow for quantitation of the compound of interest.

CONCLUSIONS

As illustrated throughout this book, PED is compatible with virtually all water-based separations, including ion-exchange, ion-pairing, ion-exclusion, and re-versed-phase chromatography. PED offers many advantages over alternate detection schemes for liquid chromatography. Because electrochemical detection relies on reaction at the electrode surface, detector cells can be miniaturized without sacrificing sensitivity. This advantage makes them especially suited for microbore and capillary techniques. Pulsed potential cleaning eliminates the need for daily polishing of the electrode, which renders PED more convenient experimentally than dc amperometry. The sensitivity and selectivity (e.g., sulfur-based compounds under typical reversed-phase conditions) of PED for specific functional groups on the analyte simplify the analysis of complex (e.g., biological) matrices.

Significant future developments in PED will occur for the detection of amine and sulfur compounds with an emphasis on advanced waveforms (e.g., IPAD, APAD, PVD). Applications directed toward peptides, proteins, and macromolecules are anticipated. The impressive accomplishments in HPLC–PED, thus far, have only accentuated the need for novel chromatographic separations for polar aliphatic compounds, a deeper understanding of PED and its limits, and application of this technology to real-world bioanalytical problems of critical significance.

REFERENCES

1. W. R. LaCourse and G. S. Owens, *Anal. Chim. Acta* **307**, 301–319 (1995).
2. R. E. Roberts and D. C. Johnson, *Electroanalysis* **4**, 741–749 (1992).
3. M. A. Hayes, S. D. Gilman, and A. G. Ewing, in *Capillary Electrophoresis Technology*, N. A. Guzman, ed., Marcel Dekker, New York, 1993.
4. T. J. O'Shea, S. M. Lunte, and W. R. LaCourse, *Anal. Chem.* **65**, 948–951 (1993).
5. W. Lu and R. M. Cassidy, *Anal. Chem.* **65**, 2878–2881 (1993).
6. P. L. Weber, T. Kornfelt, K. N. Lausen, and S. M. Lunte, *Anal. Biochem.* **225**, 135–42 (1995).
7. *The Merck Manual*, 15th ed., R. Berkow, ed., Merck Sharp & Dohme Research Laboratories, Rahway, NJ, 1987, p. 2413.
8. W. R. LaCourse and G. S. Owens, *Electrophoresis* **17**, 310–318 (1996).
9. J. A. Polta and D. C. Johnson, *Anal. Chem.* **57**, 1373–1376 (1985).
10. L. Koprowski, E. Kirchmann, and L. E. Welch, *Electroanalysis* **5**, 473 (1993).
11. E. Kirchmann and L. E. Welch, *J. Chromatogr.* **633**, 111 (1993).
12. E. Kirchmann, R. L. Earley, and L. E. Welch, *J. Liq. Chromatogr.* **17**, 1755 (1994).
13. S. Altunata, R. L. Earley, D. M. Mossman, and L. E. Welch, **42**, 17–25 (1995).
14. C. O. Dasenbrock, C. M. Zook, and W. R. LaCourse, paper presented at Ohio Valley Symposium, June 1996.
15. D. C. Johnson and W. R. LaCourse, *Electroanalysis* **4**, 367–380 (1992).
16. W. R. LaCourse, *Analusis* **21**, 181–195 (1993).
17. P. T. Kissinger, in *Laboratory Techniques in Electroanalytical Chemistry*, P. T. Kissinger and W. R. Heineman, eds., Marcel Dekker, New York, 1984, Chapter 22.
18. P. Jandik, P. R. Haddad, and P. E. Sturrock, *CRC Crit. Rev. Anal. Chem.* **20**, 1–74 (1988).
19. P. R. Haddad and P. Jandik, in *Ion Chromatography*, J. G. Tarter, ed., Marcel Dekker, New York, 1987.
20. R. T. Kennedy and J. W. Jorgenson, *Anal. Chem.* **61**, 436 (1989).
21. D. S. Owens, C. M. Johnson, P. E. Sturrock, and A. Jaramillo, *Anal. Chim. Acta* **197**, 249 (1987).
22. P. E. Sturrock and G. E. O'Brien, *Anal. Chim. Acta* **324**, 135 (1996).
23. W. R. LaCourse, in *Electrochemical Detection in Liquid Chromatography and Capillary Electrophoresis*, P. T. Kissinger, ed., Marcel Dekker, in press.

Appendix A Pulsed Voltammetry Program

Pulsed voltammetry (PV) is the definitive technique for the optimization of many PED waveforms, especially PAD. Since there is no commercial instrumentation available to perform this technique in its entirety, I am including a brief description of the instrumentation and, most importantly, a fundamental PV program.

In our laboratory pulsed voltammetric data are typically collected either at a rotating-disk electrode (RDE) or a flow-through electrochemical cell connected to a potentiostat to which external voltages can be inputted and the resulting output (i.e., current) collected. The potentiostat is therefore under computer control using a high-speed A/D–D/A (analog-to-digital and vice versa) expansion board. All pulsed voltammetric waveforms are generated via ASYST scientific software (ASYST Software Technologies, Inc., Rochester, NY). A schematic drawing of the overall system is shown in Figure A.1. A PV system can be put together from a wide variety of equivalent system components, and similarly, software other than ASYST can be used to control the potentiostat and perform PV.

The following program was written in ASYST Scientific Software (Version 4.0) for a Model DAS-1600 high-speed AD/DA board (Metrabyte Corp., Woburn, MA). ASYST software programing is built around user-defined 'words', or subroutines. For instance, the 'word' to alter the input values is

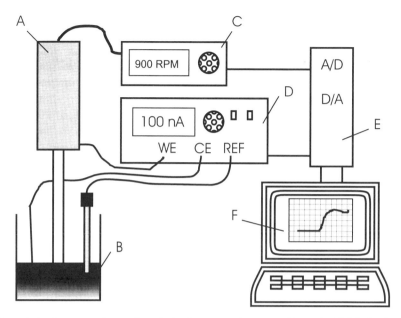

Figure A.1. Instrumental setup for pulsed voltammetry. A, rotator with attached RDE; B, electrochemical cell; C, controller for rotator with external inputs; D, potentiostat with external inputs and outputs; E, A/D–D/A interface; and F, computer system with appropriate software.

designated as AGIN. It begins with a colon, and the end is denoted with a semicolon. Once a 'word' has been defined, it can be used within other 'words' or invoked to run the program. By taking a closer look at the 'word' AGIN, we see that an infinite loop is initiated with the BEGIN command.

```
: AGIN

    BEGIN

    INPUTS

    CR ." CHANGE PARAMETERS (Y or N)?" "INPUT CHANGE ":=

    CHANGE " Y" "= IF CHANGER THEN

    CHANGE " N" "=

    UNTIL

    CR ." DO YOU WANT TO PRINT THIS SCREEN(Y OR N)? " "INPUT CHANGE ":=

    CHANGE " Y" "= IF

    SCREEN.PRINT

    THEN

;
```

The INPUTS subroutine, or 'word', is used to display all the input parameters of the program and their present values. The CR denotes a carriage return, or a line is skipped, and the phrase "CHANGE PARAMETERS (Y or N)?" is displayed on the screen. The user types in either Y or N, which is assigned to the variable CHANGE. ASYST coding uses a reverse logic, which often takes some getting used to for a programmer at any level. First, the variable CHANGE is compared with Y, which enters one into a conditional IF,THEN statement. The coding above states that if CHANGE is equivalent to Y, the subroutine CHANGER, which is used to change the input parameters, is summoned. If CHANGE does not equal Y, then CHANGE is compared with N. If CHANGE is equivalent to N, then the BEGIN loop is exited via the UNTIL command. The AGIN subroutine asks the user if the display, which shows all the input parameters and values from the INPUTS subroutine, is to be printed. An IF,THEN statement similar to the one above is used to either print the display using the SCREEN.PRINT command or exit the AGIN subroutine. A more detailed description of ASYST programming can be obtained from the software manufacturer. With a basic knowledge of programming and the annotation within the program, this coding can be converted readily to another language.

\ ********** PAD OPTIMIZATION PROGRAM ***********

\ Initialization of computer system variables. These 'word' commands, or subroutines are

\ defined by the software.

NORMAL.DISPLAY FORGET.ALL ECHO.OFF 4096 DEF.STACK

STACK.CLEAR 8 2 FIX.FORMAT

\ ********** DECLARATION OF VARIABLES **********

\ Declaration/naming of all program variables.

REAL SCALAR E1LO	REAL SCALAR E1HI	REAL SCALAR E2LO
REAL SCALAR E2HI	REAL SCALAR E3LO	REAL SCALAR E3HI
REAL SCALAR ERLO	REAL SCALAR ERHI	REAL SCALAR EDIF
REAL SCALAR ESTEP	REAL SCALAR RSTEP	REAL SCALAR RDIF
REAL SCALAR T1LO	REAL SCALAR T1HI	REAL SCALAR TDLO
REAL SCALAR TDHI	REAL SCALAR TILO	REAL SCALAR TIHI
REAL SCALAR T2LO	REAL SCALAR T2HI	REAL SCALAR T3LO
REAL SCALAR T3HI	REAL SCALAR TSTEP	REAL SCALAR ITOE
REAL SCALAR TDIF	REAL SCALAR TT	

INTEGER SCALAR E1LOD	INTEGER SCALAR E1HID	INTEGER SCALAR E2LOD
INTEGER SCALAR E2HID	INTEGER SCALAR E3LOD	INTEGER SCALAR E3HID
INTEGER SCALAR ERLOD	INTEGER SCALAR ERHID	INTEGER SCALAR RDIFD
INTEGER SCALAR EDIFD	INTEGER SCALAR ESTEPD	INTEGERSCALARRSTEPD
INTEGER SCALAR FIND	INTEGER SCALAR WAVN	INTEGER SCALAR INTN
INTEGER SCALAR DIME	INTEGER SCALAR DIMN	INTEGER SCALAR DIMT
INTEGER SCALAR DIMX	INTEGER SCALAR DIMI	INTEGER SCALAR DIMR

INTEGER SCALAR DIMALL

INTEGER SCALAR COUNT   INTEGER SCALAR COUNTA   INTEGER SCALAR COUNTB

12 STRING FILENEW 12 STRING FILENEWA 12 STRING FILENEWB   12   STRING

FILENEWC

1 STRING CHANGE

REAL DIM[24] ARRAY VAR   0 VAR :=

\ ********** INITIALIZE VARIABLES FROM PARAMETERS FILE **********

\ A parameter file is opened and the values are assigned to particular variables.

: INIT.PARAMS

FILE.OPEN PARAMETR.LST

1 SUBFILE VAR FILE>ARRAY

FILE.CLOSE

VAR [1] E1LO :=	VAR [2] E1HI :=	VAR [3] E2LO :=	VAR [4] E2HI :=
VAR [5] E3LO :=	VAR [6] E3HI :=	VAR [7] ESTEP :=	VAR [8] T1LO :=
VAR [9] T1HI :=	VAR [10] TDLO :=	VAR [11] TDHI :=	VAR [12] TILO :=
VAR [13] TIHI :=	VAR [14] T2LO :=	VAR [15] T2HI :=	VAR [16] T3LO :=
VAR [17] T3HI :=	VAR [18] TSTEP :=	VAR [19] ERLO :=	VAR [20] ERHI :=
VAR [21] RSTEP :=	VAR [22] ITOE :=	VAR [23] WAVN :=	VAR [24] INTN :=

;

```
\ ********** SCREEN OUTPUT TO SHOW VARIABLES **********

\ PV variable and their values are displayed on the screen.


: INPUTS

CR

CR

CR ."    ********** PAD WAVEFORM OPTIMIZATION PROGRAM **********    "

CR

CR

CR ."   1 Edet (-1600 to 3400 mV): LOW to HIGH " E1LO . . "  to " E1HI .

CR ."   2 Eoxd (-1600 to 3400 mV): LOW to HIGH " E2LO . . "  to " E2HI .

CR ."   3 Ered (-1600 to 3400 mV): LOW to HIGH " E3LO . . "  to " E3HI .

CR ."   4 POTENTIAL STEP (5 to 500 mV):" ESTEP .

CR ."   5 Tdet (0 to 2000 ms): LOW to HIGH " T1LO . . "  to " T1HI .

CR ."   6 Tdel (0 to 2000 ms): LOW to HIGH " TDLO . . "  to " TDHI .

CR ."   7 Tint (0 to 2000 ms): LOW to HIGH " TILO . . "  to " TIHI .

CR ."   8 Toxd (0 to 2000 ms): LOW to HIGH " T2LO . . "  to " T2HI .

CR ."   9 Tred (0 to 2000 ms): LOW to HIGH " T3LO . . "  to " T3HI .

CR ."  10 TIME STEP (1 to 50 ms): " TSTEP .

CR ."  11 ROTATION SPEED (0 to 5K RPM): LOW to HIGH " ERLO . . "  to " ERHI .

CR ."  12 ROTATION SPEED STEP (0 to 5K RPM): " RSTEP .

CR ."  13 CURRENT-to-VOLTAGE CONVERTOR (RDE, mA/V): " ITOE .

CR ."  14 # OF CYCLES PER STEP (1-1000): " WAVN .

CR ."  15 # OF INTEGRATIONS PER STEP: " INTN .

CR

;
```

\ ********** SUBROUTINE TO ALTER INPUT VARIABLES **********

\ A user-defined 'word' to change variable values.

: CHANGER

CR ." SELECT CATEGORY: # " #INPUT

CASE

1 OF ." ENTER LOW Edet (mV): " E1LO . ."   " #INPUT E1LO :=

        CR ." ENTER HIGH Edet (mV): " E1HI . ."   " #INPUT E1HI := ENDOF

2 OF ." ENTER LOW Eoxd (mV): " E2LO . ."   " #INPUT E2LO :=

        CR ." ENTER HIGH Eoxd (mV): " E2HI . ."   " #INPUT E2HI := ENDOF

3 OF ." ENTER LOW Ered (mV): " E3LO . ."   " #INPUT E3LO :=

        CR ." ENTER HIGH Ered (mV): " E3HI . ."   " #INPUT E3HI := ENDOF

4 OF ." ENTER POTENTIAL STEP (mV): " ESTEP . ."   " #INPUT ESTEP := ENDOF

5 OF ." ENTER LOW Tdet (ms): " T1LO . ."   " #INPUT T1LO :=

        CR ." ENTER HIGH Tdet (ms): " T1HI . ."   " #INPUT T1HI := ENDOF

6 OF ." ENTER LOW Tdel (ms): " TDLO . ."   " #INPUT TDLO :=

        CR ." ENTER HIGH Tdel (ms): " TDHI . ."   " #INPUT TDHI := ENDOF

7 OF ." ENTER LOW Tint (ms): " TILO . ."   " #INPUT TILO :=

        CR ." ENTER HIGH Tint (ms): " TIHI . ."   " #INPUT TIHI := ENDOF

8 OF ." ENTER LOW Toxd (ms): " T2LO . ."   " #INPUT T2LO :=

        CR ." ENTER HIGH Toxd (ms): " T2HI . ."   " #INPUT T2HI := ENDOF

9 OF ." ENTER LOW Tred (ms): " T3LO . ."   " #INPUT T3LO :=

        CR ." ENTER HIGH Tred (ms): " T3HI . ."   " #INPUT T3HI := ENDOF

10 OF ." ENTER TIME STEP (ms): " TSTEP . ."   " #INPUT TSTEP := ENDOF

11 OF ." ENTER LOW ROTATION SPEED (RPM): " ERLO . ."   " #INPUT ERLO :=

        CR ." ENTER HIGH ROTATION SPEED (RPM): " ERHI . ."   " #INPUT ERHI := ENDOF

12 OF ." ENTER ROTATION SPEED STEP (RPM): " RSTEP . ."   " #INPUT RSTEP := ENDOF

13 OF ." ENTER CURRENT-to-VOLTAGE VALUE (RDE, mA/V): " ITOE . ."   " #INPUT ITOE :=

ENDOF

14 OF ." ENTER # OF CYCLES PER STEP: " WAVN . ."   " #INPUT WAVN := ENDOF

15 OF ." ENTER # OF INTEGRATIONS PER STEP: " INTN . ."   " #INPUT INTN := ENDOF

ENDCASE

;

\ ********** SUBROUTINE TO ALTER INPUTS **********

\ A 'word' to invoke CHANGER and print the screen.

: AGIN

BEGIN

INPUTS

CR ." CHANGE PARAMETERS (Y or N)?" "INPUT CHANGE ":=

CHANGE " Y" "= IF CHANGER THEN

CHANGE " N" "=

UNTIL

CR ." DO YOU WANT TO PRINT THIS SCREEN(Y OR N)? " "INPUT CHANGE ":=

CHANGE " Y" "= IF

SCREEN.PRINT

THEN

;

\ ********** SAVE ALTERED VARIABLES **********

\ A 'word' to update the parameter file.

: SAVE.PARAMS

E1LO VAR [1] :=   E1HI VAR [2] :=   E2LO VAR [3] :=   E2HI VAR [4] :=

E3LO VAR [5] :=   E3HI VAR [6] :=   ESTEP VAR [7] :=   T1LO VAR [8] :=

T1HI VAR [9] :=   TDLO VAR [10] :=   TDHI VAR [11] :=   TILO VAR [12] :=

TIHI VAR [13] :=   T2LO VAR [14] :=   T2HI VAR [15] :=   T3LO VAR [16] :=

T3HI VAR [17] :=   TSTEP VAR [18] :=   ERLO VAR [19] :=   ERHI VAR [20] :=

RSTEP VAR [21] :=   ITOE VAR [22] :=   WAVN VAR [23] :=   INTN VAR [24] :=

REGULAR.DATAFILE   FILE.TEMPLATE   REAL DIM[24] SUBFILE END

FILE.CREATE PARAMETR.LST FILE.OPEN PARAMETR.LST 1 SUBFILE VAR ARRAY>FILE

FILE.CLOSE

\ ********** DETERMINE ARRAY DIMENSIONS **********

\ Required array dimensions are determined and assigned.

\ The CASE, or waveform parameter being studied, is determined.

: SETDIM.ARRAY

10 FIND :=

E1HI E1LO <> IF E1HI E1LO - EDIF := EDIF ESTEP / 1 + DIME := 1 FIND := THEN

E2HI E2LO <> IF E2HI E2LO - EDIF := EDIF ESTEP / 1 + DIME := 2 FIND := THEN

E3HI E3LO <> IF E3HI E3LO - EDIF := EDIF ESTEP / 1 + DIME := 3 FIND := THEN

ERHI ERLO <> IF ERHI ERLO - RDIF := RDIF RSTEP / 1 + DIMR := 4 FIND := THEN

T1HI T1LO <> IF T1HI T1LO - TDIF := TDIF TSTEP / 1 + DIMT := 5 FIND := THEN

TDHI TDLO <> IF TDHI TDLO - TDIF := TDIF TSTEP / 1 + DIMT := 6 FIND := THEN

TIHI TILO <> IF TIHI TILO - TDIF := TDIF TSTEP / 1 + DIMT := 7 FIND := THEN

T2HI T2LO <> IF T2HI T2LO - TDIF := TDIF TSTEP / 1 + DIMT := 8 FIND := THEN

T3HI T3LO <> IF T3HI T3LO - TDIF := TDIF TSTEP / 1 + DIMT := 9 FIND := THEN

   10 FIND = IF CR ." EXPERIMENT INVALID: NO HIGH/LOW VARIABLES. " QUIT THEN

FIND 4 < IF DIME DIMX := THEN

FIND 4 = IF DIMR DIMX := THEN

FIND 4 > IF DIMT DIMX := THEN

ITOE 4.884 * ITOE :=

;

\ ********** START OF PROGRAM and DECLARATION OF ARRAYS *********

INIT.PARAMS*see note at end of program*

AGIN

SAVE.PARAMS

SETDIM.ARRAY

   REAL DIM[DIMX] ARRAY POTENTIAL 0 POTENTIAL :=

   REAL DIM[DIMX] ARRAY ROTATION 0 ROTATION :=

   REAL DIM[DIMX] ARRAY RSD   0 RSD   :=

   REAL DIM[DIMX] ARRAY TIME   0 TIME   :=

   REAL DIM[DIMX] ARRAY CURRENT 0 CURRENT   :=

   REAL DIM[INTN] ARRAY CURRENTA 0 CURRENTA :=

INTEGER DIM[DIMX] ARRAY POT.OUT 0 POT.OUT :=

INTEGER DIM[DIMX] ARRAY ROT.OUT 0 ROT.OUT :=

\ ********** FILL ARRAYS and A/D VALUES **********

\ Input values are converted to digital values and corresponding arrays of

\ the waveform parameter being changed, or CASE, are filled with the appropriate values.

: SET.VARIABLES

FIND

CASE

1 OF E1HI E1LO - EDIF := EDIF ESTEP / 1 + DIME :=

    E1HI 0.819 * 1310.4 + E1HID := E1LO 0.819 * 1310.4 + E1LOD :=

    ESTEP 0.819 * ESTEPD := EDIF 0.819 * EDIFD :=

    1 COUNT := DIME 0 DO

            E1LO POTENTIAL [COUNT] :=

            E1LOD POT.OUT [COUNT] :=

            COUNT 1 + COUNT :=

            E1LO ESTEP + E1LO :=

            E1LOD ESTEPD + E1LOD :=

        LOOP

    E2LO 0.819 * 1310.4 + E2LOD :=

    E3LO 0.819 * 1310.4 + E3LOD :=

    ERLO 0.819 * ERLOD :=

    T1LO T2LO + T3LO + TT :=

    TILO 2 / DIMN :=

    DIMN INTN * DIMALL := ENDOF

2 OF E2HI E2LO - EDIF := EDIF ESTEP / 1 + DIME :=

    E2HI 0.819 * 1310.4 + E2HID := E2LO 0.819 * 1310.4 + E2LOD :=

    ESTEP 0.819 * ESTEPD := EDIF 0.819 * EDIFD :=

    1 COUNT := DIME 0 DO

            E2LO POTENTIAL [COUNT] :=

```
              E2LOD  POT.OUT [ COUNT ] :=

              COUNT 1 + COUNT :=

              E2LO ESTEP + E2LO :=

              E2LOD ESTEPD + E2LOD :=

         LOOP

   E1LO 0.819 * 1310.4 + E1LOD :=

   E3LO 0.819 * 1310.4 + E3LOD :=

   ERLO 0.819 * ERLOD :=

   T1LO T2LO + T3LO + TT :=

   TILO 2 / DIMN :=

   DIMN INTN * DIMALL := ENDOF

3 OF E3HI E3LO - EDIF := EDIF ESTEP / 1 + DIME :=

   E3HI 0.819 * 1310.4 + E3HID := E3LO 0.819 * 1310.4 + E3LOD :=

   ESTEP 0.819 * ESTEPD := EDIF 0.819 * EDIFD :=

   1 COUNT := DIME 0 DO

              E3LO POTENTIAL [ COUNT ] :=

              E3LOD  POT.OUT [ COUNT ] :=

              COUNT 1 + COUNT :=

              E3LO ESTEP + E3LO :=

              E3LOD ESTEPD + E3LOD :=

         LOOP

   E1LO 0.819 * 1310.4 + E1LOD :=

   E2LO 0.819 * 1310.4 + E2LOD :=

   ERLO 0.819 * ERLOD :=

   T1LO T2LO + T3LO + TT :=

   TILO 2 / DIMN :=

   DIMN INTN * DIMALL := ENDOF
```

```
4 OF ERHI ERLO - RDIF := RDIF RSTEP / 1 + DIMR :=

    ERHI 0.819 * ERHID := ERLO 0.819 * ERLOD :=

    RSTEP 0.819 * RSTEPD := RDIF 0.819 * RDIFD :=

    1 COUNT := DIMR 0 DO

            ERLO  ROTATION [ COUNT ] :=

            ERLOD  ROT.OUT [ COUNT ] :=

            COUNT 1 + COUNT :=

            ERLO RSTEP + ERLO :=

            ERLOD RSTEPD + ERLOD :=

        LOOP

    E1LO 0.819 * 1310.4 + E1LOD :=

    E2LO 0.819 * 1310.4 + E2LOD :=

    E3LO 0.819 * 1310.4 + E3LOD :=

    T1LO T2LO + T3LO + TT :=

    TILO 2 / DIMN :=

    DIMN INTN * DIMALL := ENDOF

5 OF E1LO 0.819 * 1310.4 + E1LOD :=

    E2LO 0.819 * 1310.4 + E2LOD :=

    E3LO 0.819 * 1310.4 + E3LOD :=

    ERLO 0.819 *        ERLOD :=

    T1HI T1LO - TDIF := TDIF TSTEP / 1 + DIMT :=

    1 COUNT := DIMT 0 DO

            T1LO TIME [ COUNT ] :=

            COUNT 1 + COUNT :=

            T1LO TSTEP + T1LO :=

        LOOP

    TILO 2 / DIMN :=
```

```
DIMN INTN * DIMALL := ENDOF
6 OF E1LO 0.819 * 1310.4 + E1LOD :=
    E2LO 0.819 * 1310.4 + E2LOD :=
    E3LO 0.819 * 1310.4 + E3LOD :=
    ERLO 0.819 *      ERLOD :=
    TDHI TDLO - TDIF := TDIF TSTEP / 1 + DIMT :=
    1 COUNT := DIMT 0 DO
              TDLO TIME [ COUNT ] :=
              COUNT 1 + COUNT :=
              TDLO TSTEP + TDLO :=
         LOOP
    T1LO T2LO + T3LO + TT :=
    TILO 2 / DIMN :=
    DIMN INTN * DIMALL := ENDOF
7 OF E1LO 0.819 * 1310.4 + E1LOD :=
    E2LO 0.819 * 1310.4 + E2LOD :=
    E3LO 0.819 * 1310.4 + E3LOD :=
    ERLO 0.819 *      ERLOD :=
    TIHI TILO - TDIF := TDIF TSTEP / 1 + DIMT :=
    1 COUNT := DIMT 0 DO
              TILO TIME [ COUNT ] :=
              COUNT 1 + COUNT :=
              TILO TSTEP + TILO :=
         LOOP
    T1LO T2LO + T3LO + TT :=
    TIHI 2 / DIMN :=
    DIMN INTN * DIMALL := ENDOF
```

8 OF E1LO 0.819 * 1310.4 + E1LOD :=

E2LO 0.819 * 1310.4 + E2LOD :=

E3LO 0.819 * 1310.4 + E3LOD :=

ERLO 0.819 *   ERLOD :=

T2HI T2LO - TDIF := TDIF TSTEP / 1 + DIMT :=

1 COUNT := DIMT 0 DO

        T2LO TIME [COUNT] :=

        COUNT 1 + COUNT :=

        T2LO TSTEP + T2LO :=

    LOOP

TILO 2 / DIMN :=

DIMN INTN * DIMALL := ENDOF

9 OF E1LO 0.819 * 1310.4 + E1LOD :=

E2LO 0.819 * 1310.4 + E2LOD :=

E3LO 0.819 * 1310.4 + E3LOD :=

ERLO 0.819 *   ERLOD :=

T3HI T3LO - TDIF := TDIF TSTEP / 1 + DIMT :=

1 COUNT := DIMT 0 DO

        T3LO TIME [COUNT] :=

        COUNT 1 + COUNT :=

        T3LO TSTEP + T3LO :=

    LOOP

TILO 2 / DIMN :=

DIMN INTN * DIMALL := ENDOF

ENDCASE

\ ********** SETUP A/D and D/A CONVERTORS **********

\ The A/D and D/A board is set-up and initialized.

SET.VARIABLES

1 1 D/A.TEMPLATE DAO1 ROT.OUT TEMPLATE.BUFFER D/A.INIT

0 0 D/A.TEMPLATE DAO0 POT.OUT TEMPLATE.BUFFER D/A.INIT

INTEGER DIM[DIMALL] ARRAY CURR.RAW 0 CURR.RAW :=

0 0 A/D.TEMPLATE ADI0 DIMN TEMPLATE.REPEAT CURR.RAW CYCLIC TEMPLATE.BUFFER

REAL DIM[DIMALL] ARRAY CURR.RAW.REAL 0 CURR.RAW.REAL := A/D.INIT

\ ********** SYSTEM STAND-BY **********

\ An appropriate waveform is applied to the electrode to equilibrate the system.

\ No current is monitored.

: HOLD.IT

FALSE

FIND

CASE

  1 OF DAO1 ERLOD D/A.OUT

    BEGIN

    DAO0 E2LOD D/A.OUT T2LO MSEC.DELAY

      E3LOD D/A.OUT T3LO MSEC.DELAY

      POT.OUT [1] D/A.OUT T1LO MSEC.DELAY

    ?KEY UNTIL ?DROP ENDOF

  2 OF DAO1 ERLOD D/A.OUT

BEGIN

DAO0 POT.OUT [1] D/A.OUT T2LO MSEC.DELAY

E3LOD D/A.OUT T3LO MSEC.DELAY

E1LOD D/A.OUT T1LO MSEC.DELAY

?KEY UNTIL ?DROP ENDOF

3 OF DAO1 ERLOD D/A.OUT

BEGIN

DAO0 E2LOD D/A.OUT T2LO MSEC.DELAY

POT.OUT [1] D/A.OUT T3LO MSEC.DELAY

E1LOD D/A.OUT T1LO MSEC.DELAY

?KEY UNTIL ?DROP ENDOF

4 OF DAO1 ROT.OUT [DIMR] D/A.OUT

BEGIN

DAO0 E2LOD D/A.OUT T2LO MSEC.DELAY

E3LOD D/A.OUT T3LO MSEC.DELAY

E1LOD D/A.OUT T1LO MSEC.DELAY

?KEY UNTIL ?DROP ENDOF

5 OF DAO1 ERLOD D/A.OUT

BEGIN

DAO0 E2LOD D/A.OUT T2LO MSEC.DELAY

E3LOD D/A.OUT T3LO MSEC.DELAY

E1LOD D/A.OUT TIME [1] MSEC.DELAY

?KEY UNTIL ?DROP ENDOF

6 OF DAO1 ERLOD D/A.OUT

BEGIN

DAO0 E2LOD D/A.OUT T2LO MSEC.DELAY

E3LOD D/A.OUT T3LO MSEC.DELAY

```
    E1LOD D/A.OUT T1LO MSEC.DELAY

    ?KEY UNTIL ?DROP ENDOF

7 OF DAO1 ERLOD D/A.OUT

    BEGIN

    DAO0 E2LOD D/A.OUT T2LO MSEC.DELAY

    E3LOD D/A.OUT T3LO MSEC.DELAY

    E1LOD D/A.OUT T1LO MSEC.DELAY

    ?KEY UNTIL ?DROP ENDOF

8 OF DAO1 ERLOD D/A.OUT

    BEGIN

    DAO0 E2LOD D/A.OUT TIME [ 1 ] MSEC.DELAY

    E3LOD D/A.OUT T3LO MSEC.DELAY

    E1LOD D/A.OUT T1LO MSEC.DELAY

    ?KEY UNTIL ?DROP ENDOF

9 OF DAO1 ERLOD D/A.OUT

    BEGIN

    DAO0 E2LOD D/A.OUT T2LO MSEC.DELAY

    E3LOD D/A.OUT TIME [ 1 ] MSEC.DELAY

    E1LOD D/A.OUT T1LO MSEC.DELAY

    ?KEY UNTIL ?DROP ENDOF

ENDCASE

;


\ ********** WAVEFORM OUTPUT and DATA COLLECTION **********

\ The actual experiment is performed. Only one CASE is executed.


: DOIT
```

```
SCREEN.CLEAR

CR

CR

CR

CR

CR

CR

CR

CR

CR

CR

CR

CR ."          EXPERIMENT IN PROGRESS. PLEASE WAIT........."

1 COUNT :=

FIND

CASE

1 OF DIME 0 DO

  TT SYNC.PERIOD

  WAVN 0 DO

  SYNCHRONIZE

  DAO0 E2LOD D/A.OUT T2LO MSEC.DELAY

      E3LOD D/A.OUT T3LO MSEC.DELAY

  POT.OUT [ COUNT ] D/A.OUT TDLO MSEC.DELAY

    ADI0 A/D.IN>ARRAY

  LOOP

    CURR.RAW CURR.RAW.REAL :=
```

```
CURR.RAW.REAL 2049 - ITOE * CURR.RAW.REAL :=

1 COUNTA :=

1 COUNTB :=

INTN 0 DO

CURR.RAW.REAL SUB[ COUNTB , DIMN ] MEAN CURRENTA [ COUNTA ] :=

COUNTA 1 + COUNTA :=

COUNTB DIMN + COUNTB :=

LOOP

CURRENTA MEAN CURRENT [ COUNT ] :=

CURRENTA VARIANCE SQRT RSD [ COUNT ] :=

COUNT 1 + COUNT :=

LOOP ENDOF

2 OF DIME 0 DO

TT SYNC.PERIOD

WAVN 0 DO

SYNCHRONIZE

DAO0 POT.OUT [ COUNT ] D/A.OUT T2LO MSEC.DELAY

  E3LOD D/A.OUT T3LO MSEC.DELAY

  E1LOD D/A.OUT TDLO MSEC.DELAY

 ADI0

 A/D.IN>ARRAY

LOOP

 CURR.RAW CURR.RAW.REAL :=

 CURR.RAW.REAL 2049 - ITOE * CURR.RAW.REAL :=

 1 COUNTA :=

 1 COUNTB :=

 INTN 0 DO
```

```
CURR.RAW.REAL SUB[ COUNTB , DIMN ] MEAN CURRENTA [ COUNTA ] :=

COUNTA 1 + COUNTA :=

COUNTB DIMN + COUNTB :=

LOOP

CURRENTA MEAN CURRENT [ COUNT ] :=

CURRENTA VARIANCE SQRT RSD [ COUNT ] :=

COUNT 1 + COUNT :=

LOOP ENDOF

3 OF DIME 0 DO

TT SYNC.PERIOD

WAVN 0 DO

SYNCHRONIZE

DAO0 E2LOD D/A.OUT T2LO MSEC.DELAY

  POT.OUT [ COUNT ] D/A.OUT T3LO MSEC.DELAY

  E1LOD D/A.OUT TDLO MSEC.DELAY

ADI0

A/D.IN>ARRAY

LOOP

  CURR.RAW CURR.RAW.REAL :=

  CURR.RAW.REAL 2049 - ITOE * CURR.RAW.REAL :=

  1 COUNTA :=

  1 COUNTB :=

  INTN 0 DO

  CURR.RAW.REAL SUB[ COUNTB , DIMN ] MEAN CURRENTA [ COUNTA ] :=

  COUNTA 1 + COUNTA :=

  COUNTB DIMN + COUNTB :=

  LOOP
```

CURRENTA MEAN CURRENT [COUNT] :=

CURRENTA VARIANCE SQRT RSD [COUNT] :=

COUNT 1 + COUNT :=

LOOP ENDOF

4 OF DIMR 0 DO

DAO1 ROT.OUT [COUNT] D/A.OUT

TT SYNC.PERIOD

WAVN 0 DO

SYNCHRONIZE

DAO0 E2LOD D/A.OUT T2LO MSEC.DELAY

  E3LOD D/A.OUT T3LO MSEC.DELAY

  E1LOD D/A.OUT TDLO MSEC.DELAY

  ADI0

  A/D.IN>ARRAY

LOOP

  CURR.RAW CURR.RAW.REAL :=

  CURR.RAW.REAL 2049 - ITOE * CURR.RAW.REAL :=

  1 COUNTA :=

  1 COUNTB :=

  INTN 0 DO

  CURR.RAW.REAL SUB[COUNTB , DIMN] MEAN CURRENTA [COUNTA] :=

  COUNTA 1 + COUNTA :=

  COUNTB DIMN + COUNTB :=

  LOOP

  CURRENTA MEAN CURRENT [COUNT] :=

  CURRENTA VARIANCE SQRT RSD [COUNT] :=

  COUNT 1 + COUNT :=

LOOP ENDOF

5 OF DIMT 0 DO

TIME [COUNT] T2LO + T3LO + TT :=

TT SYNC.PERIOD

WAVN 0 DO

SYNCHRONIZE

DAO0 E2LOD D/A.OUT T2LO MSEC.DELAY

E3LOD D/A.OUT T3LO MSEC.DELAY

E1LOD D/A.OUT TDLO MSEC.DELAY

ADI0

A/D.IN>ARRAY

LOOP

CURR.RAW CURR.RAW.REAL :=

CURR.RAW.REAL 2049 - ITOE * CURR.RAW.REAL :=

1 COUNTA :=

1 COUNTB :=

INTN 0 DO

CURR.RAW.REAL SUB[COUNTB , DIMN] MEAN CURRENTA [COUNTA] :=

COUNTA 1 + COUNTA :=

COUNTB DIMN + COUNTB :=

LOOP

CURRENTA MEAN CURRENT [COUNT] :=

CURRENTA VARIANCE SQRT RSD [COUNT] :=

COUNT 1 + COUNT :=

LOOP ENDOF

6 OF DIMT 0 DO

TT SYNC.PERIOD

```
WAVN 0 DO

SYNCHRONIZE

DAO0 E2LOD D/A.OUT T2LO MSEC.DELAY

    E3LOD D/A.OUT T3LO MSEC.DELAY

    E1LOD D/A.OUT TIME [ COUNT ] MSEC.DELAY

  ADI0

  A/D.IN>ARRAY

LOOP

  CURR.RAW CURR.RAW.REAL :=

  CURR.RAW.REAL 2049 - ITOE * CURR.RAW.REAL :=

  1 COUNTA :=

  1 COUNTB :=

  INTN 0 DO

  CURR.RAW.REAL SUB[ COUNTB , DIMN ] MEAN CURRENTA [ COUNTA ] :=

  COUNTA 1 + COUNTA :=

  COUNTB DIMN + COUNTB :=

  LOOP

  CURRENTA MEAN CURRENT [ COUNT ] :=

  CURRENTA VARIANCE SQRT RSD [ COUNT ] :=

  COUNT 1 + COUNT :=

  LOOP ENDOF

7 OF

  TT SYNC.PERIOD

  WAVN 0 DO

  SYNCHRONIZE

  DAO0 E2LOD D/A.OUT T2LO MSEC.DELAY

    E3LOD D/A.OUT T3LO MSEC.DELAY
```

```
E1LOD D/A.OUT TDLO MSEC.DELAY

ADI0

A/D.IN>ARRAY

LOOP

  CURR.RAW CURR.RAW.REAL :=

  CURR.RAW.REAL 2049 - ITOE * CURR.RAW.REAL :=

  DIMT 0 DO

    TIME [ COUNT ] 2 / DIMI :=

    1 COUNTA :=

    1 COUNTB :=

    INTN 0 DO

    CURR.RAW.REAL SUB[ COUNTB , DIMI ] MEAN CURRENTA [ COUNTA ] :=

    COUNTA 1 + COUNTA :=

    COUNTB DIMN + COUNTB :=

    LOOP

    CURRENTA MEAN CURRENT [ COUNT ] :=

    CURRENTA VARIANCE SQRT RSD [ COUNT ] :=

    COUNT 1 + COUNT :=

    LOOP ENDOF

8 OF DIMT 0 DO

  T1LO TIME [ COUNT ] + T3LO + TT :=

  TT SYNC.PERIOD

  WAVN 0 DO

  SYNCHRONIZE

  DAO0 E2LOD D/A.OUT TIME [ COUNT ] MSEC.DELAY

    E3LOD D/A.OUT T3LO MSEC.DELAY

    E1LOD D/A.OUT TDLO MSEC.DELAY
```

```
ADI0

A/D.IN>ARRAY

LOOP

  CURR.RAW CURR.RAW.REAL :=

  CURR.RAW.REAL 2049 - ITOE * CURR.RAW.REAL :=

  1 COUNTA :=

  1 COUNTB :=

  INTN 0 DO

  CURR.RAW.REAL SUB[ COUNTB , DIMN ] MEAN CURRENTA [ COUNTA ] :=

  COUNTA 1 + COUNTA :=

  COUNTB DIMN + COUNTB :=

  LOOP

  CURRENTA MEAN CURRENT [ COUNT ] :=

  CURRENTA VARIANCE SQRT RSD [ COUNT ] :=

  COUNT 1 + COUNT :=

  LOOP ENDOF

9 OF DIMT 0 DO

  T1LO T2LO + TIME [ COUNT ] + TT :=

  TT SYNC.PERIOD

  WAVN 0 DO

  SYNCHRONIZE

  DAO0 E2LOD D/A.OUT T2LO MSEC.DELAY

    E3LOD D/A.OUT TIME [ COUNT ] MSEC.DELAY

    E1LOD D/A.OUT TDLO MSEC.DELAY

  ADI0

  A/D.IN>ARRAY

  LOOP
```

```
CURR.RAW CURR.RAW.REAL :=

CURR.RAW.REAL 2049 - ITOE * CURR.RAW.REAL :=

1 COUNTA :=

1 COUNTB :=

INTN 0 DO

CURR.RAW.REAL SUB[ COUNTB , DIMN ] MEAN CURRENTA [ COUNTA ] :=

COUNTA 1 + COUNTA :=

COUNTB DIMN + COUNTB :=

LOOP

CURRENTA MEAN CURRENT [ COUNT ] :=

CURRENTA VARIANCE SQRT RSD [ COUNT ] :=

COUNT 1 + COUNT :=

LOOP ENDOF

ENDCASE

;


\ ********** PLOT DATA ON SCREEN **********
\ The collected data is plotted onscreen for immediate scrutiny.


: PLOT.IT


FIND 4 < IF 1 FIND := THEN

FIND 4 = IF 2 FIND := THEN

FIND 4 > IF 3 FIND := THEN


FIND

CASE
```

```
1 OF POTENTIAL CURRENT XY.AUTO.PLOT ENDOF

2 OF ROTATION  CURRENT XY.AUTO.PLOT ENDOF

3 OF TIME    CURRENT XY.AUTO.PLOT ENDOF

ENDCASE

"INPUT

CR

;
```

```
\ *********** SAVE DATA*************

\ A 'word' to use to save the data. The data is saved as a potential file (?E.dat) or

\ time file (?T.DAT) or  rotation speed file (?R.dat), a current file (?I.DAT),

\ and a standard deviation file (?V.DAT).
```

```
: SAVIT

CR ." DO YOU WANT TO SAVE THE DATA (Y or N)? " "INPUT CHANGE ":=

CHANGE " Y" "=

IF CR ." INPUT NAME OF FILE TO TRANSFER (7 CHRS): " "INPUT FILENEW ":=

FIND

CASE

 1 OF FILENEW " E.DAT" "CAT "DUP CR "TYPE FILENEWA ":=

        CR

        1 COUNT :=

        FILENEWA "TYPE

        FILENEWA DEFER> OUT>FILE

        CONSOLE.OFF

        CR

        CR
```

```
DIME 0 DO

POTENTIAL [ COUNT ] . CR

COUNT 1 + COUNT :=

LOOP

OUT>FILE.CLOSE

CR ENDOF

2 OF FILENEW " R.DAT" "CAT "DUP CR "TYPE FILENEWA ":=

CR

1 COUNT :=

FILENEWA "TYPE

FILENEWA DEFER> OUT>FILE

CONSOLE.OFF

CR

CR

DIMR 0 DO

ROTATION [ COUNT ] . CR

COUNT 1 + COUNT :=

LOOP

OUT>FILE.CLOSE

CR ENDOF

3 OF FILENEW " T.DAT" "CAT "DUP CR "TYPE FILENEWA ":=

CR

1 COUNT :=

FILENEWA "TYPE

FILENEWA DEFER> OUT>FILE

CONSOLE.OFF

CR
```

```
        CR

        DIMT 0 DO

        TIME [ COUNT ] . CR

        COUNT 1 + COUNT :=

        LOOP

        OUT>FILE.CLOSE

        CR ENDOF

ENDCASE

    FILENEW " I.DAT" "CAT "DUP CR "TYPE FILENEWB ":=

        CR

        1 COUNT :=

        FILENEWB "TYPE

        FILENEWB DEFER> OUT>FILE

        CONSOLE.OFF

        CR

        CR

        DIMX 0 DO

        CURRENT [ COUNT ] . CR

        COUNT 1 + COUNT :=

        LOOP

        OUT>FILE.CLOSE

        CR

    FILENEW " V.DAT" "CAT "DUP CR "TYPE FILENEWC ":=

        CR

        1 COUNT :=

        FILENEWC "TYPE

        FILENEWC DEFER> OUT>FILE
```

```
        CONSOLE.OFF

        CR

        CR

        DIMX 0 DO

        RSD [ COUNT ] . CR

        COUNT 1 + COUNT :=

        LOOP

        OUT>FILE.CLOSE

        CR

THEN

CR

CR ." ANOTHER  EXPERIMENT (Y OR N)? " "INPUT CHANGE ":=

CHANGE " Y" "=

IF LOAD PADOPTA.PGM

THEN QUIT

;


\ ********* THE END OF THE PROGRAM*********

\ The portion of the program initiates the hold cycle, performs the experiment,

\ outputs a tone when the experiment is over, plots the data, and saves it.


CR ." HIT ENTER-KEY WHEN READY."

HOLD.IT

DOIT

1000 440 TUNE

PLOT.IT

SAVIT
```

Note: The first time this program is run, the INIT.PARAMS statement should be left out of the program. This will allow the program to create the file PARAMTR.LST, which contains the last set of input values. After this file has been created, the INIT.PARAMS statement can be added back into the program, and the last set of input values used will be the default set of values.

Appendix B Tables of Applications

TABLE B.1 **Applications Related to Carbohydrates and Alditols**

Author(s)	Title of Article	Ref.
Hughes and Johnson	Amperometric detection of simple carbohydrates at platinum electrodes in alkaline solutions by application of a triple-pulse potential waveform.	A.1
Hughes and Johnson	High-performance liquid chromatographic separation with triple-pulse amperometric detection of carbohydrates in beverages.	A.2
Buchberger, Winsauer, and Breitwieser	Studies on the electrochemical detection of sugars.	A.3
Edwards and Haak	A pulsed amperometric detector for ion chromatography.	A.4
Rocklin and Pohl	Determination of carbohydrates by anion exchange chromatography with pulsed amperometric detection.	A.5
Hughes and Johnson	Triple-pulse amperometric detection of carbohydrates after chromatographic separation.	A.6
Ohsawa, Yoshimura, Watanabe, Tanaka, Yokota, Tamura, and Imaeda	Determination of xylitol in the human serum and saliva by ion chromatography with pulsed amperometric detection.	A.7
Olechno, Carter, Edwards, and Gillen	Developments in the chromatographic determination of carbohydrates.	A.8
Edwards, Pohl, and Rubin	Determination of carbohydrates using pulsed amperometric detection combined with anion exchange separations.	A.9

(continued)

TABLE B.1   (*Continued*)

Author(s)	Title of Article	Ref.
Neuburger and Johnson	Pulsed amperometric detection of carbohydrates at gold electrodes with a two-step potential waveform.	A.10
Neuburger and Johnson	Comparison of the pulsed amperometric detection of carbohydrates at gold and platinum electrodes for flow injection and liquid chromatographic systems.	A.11
Neuburger and Johnson	Pulsed coulometric detection of carbohydrates at a constant detection potential at gold electrodes in alkaline media.	A.12
Koizumi, Kubuta, Tanimoto, and Okada	Determination of cyclic glucans by anion-exchange chromatography with pulsed amperometric detection.	A.13
Townsend, Hardy, Hindsgaul, and Lee	High-performance anion-exchange chromatography of oligosaccharides using pellicular resins and pulsed amperometric detection.	A.14
Haginaka and Nomura	Liquid chromatographic determination of carbohydrates with pulsed amperometric detection and a membrane reactor.	A.15
Chen, Yet, and Shao	New methods for rapid separation and detection of oligosaccharides from glycoproteins.	A.16
Hardy and Townsend	Separation of positional isomers of oligosaccharides and glycopeptides by high-performance anion-exchange chromatography with pulsed amperometric detection.	A.17
Larew, Mead, and Johnson	Flow-injection determination of starch and total carbohydrates with an immobilized glucoamylase reactor and pulsed amperometric detection.	A.18
Larew and Johnson	Quantitation of chromatographically separated malto-oligosaccharides with a single calibration curve using a post-column enzyme reactor and pulsed amperometric detection.	A.19
Welch, Mead, and Johnson	A comparison of pulsed amperometric detection and conductivity detection for carbohydrates.	A.20
Hardy, Townsend, and Lee	Monosaccharide analysis of glycoconjugates by anion exchange chromatography with pulsed amperometric detection.	A.21
Larew and Johnson	Transient generation of diffusion layer alkalinity for the pulsed amperometric detection of glucose in low capacity buffers having neutral and acidic pH values.	A.22
Wang and Zopf	Liquid ion-exchange chromatography under pressure of milk oligosaccharides using a pulsed amperometric detection.	A.23
Townsend, Hardy, Cummings, Carver, and Bendiak	Separation of branched sialylated of oligosaccharides using high-pH anion-exchange chromatography with pulsed amperometric detection.	A.24
Olechno, Carter, Edwards, Gillen, Townsend, Lee, and Hardy	Chromatographic analysis of glycoprotein derived carbo-hydrates: pellicular ion exchange resins and pulsed electrochemical detection.	A.25
Bindra and Wilson	Pulsed amperometric detection of glucose in biological fluids at a surface-modified gold electrode.	A.26

(*continued*)

TABLE B.1 (*Continued*)

Author(s)	Title of Article	Ref.
Garleb, Bourquin, and Fahey	Neutral monosaccharide composition of various fibrous substrates: a comparison of hydrolytic procedures and use of anion-exchange high-performance liquid chromatography with pulsed amperometric detection of monosaccharides.	A.27
Haginaka, Nishimura, Wakai, Yusuda, Koizumi, and Nomura	Determination of cyclodextrins and branched cyclodextrins by reversed-phase chromatography with pulsed amperometric detection and a membrane reactor.	A.28
Hardy and Townsend	Separation of fucosylated oligosaccharides using high pH anion-exchange chromatography with pulsed amperometric detection.	A.29
Chatterton, Harrison, Thornley, and Bennett	Purification and quantification of ketoses (fructosylsucroses) by gel permeation and anion exchange chromatography.	A.30
Hardy	Liquid chromatographic analysis of the carbohydrates of glycoproteins.	A.31
Jansen	HPLC of oligosaccharides and monosaccharides from glycoproteins.	A.32
Pollman	Ion chromatographic determination of lactose, galactose, and dextrose in grated cheese using pulsed amperometric detection.	A.33
Koizumi, Kubota, Tanimoto, and Okada	High-performance anion-exchange chromatography of homogeneous D-gluco-oligosaccharides and -polysaccharides (polymerization degree $\geq$ 50) with pulsed amperometric detection.	A.34
Townsend, Hardy, and Lee	Separation of oligosaccharides using high-performance anion-exchange chromatography with pulsed amperometric detection.	A.35
Hardy	Monosaccharide analysis of glycoconjugates by high-performance anion-exchange chromatography with pulsed amperometric detection.	A.36
Liu, Chiu, and Lo	A direct, sensitive analysis of monosaccharides after hydrolysis of glycoproteins by anion exchange chromatography coupled with pulsed amperometric detection.	A.37
Honda, Suzuki, Zaiki, and Kakehi	Analysis of N- and O-glycosidically bound sialooligosaccharides in glycoproteins by high-performance liquid chromatography with pulsed amperometric detection.	A.38
Martens and Frankenberger	Determination of glycuronic acids by high-performance anion chromatography with pulsed amperometric detection.	A.39
Zhu and Xu	Determination of oligosaccharides by anion chromatography with pulsed amperometric detection.	A.40
Blom, Andersson, Krasnewich, and Gahl	Pulsed amperometric detection of carbohydrates in lysosomal storage disease fibroplasts: a new screening technique for carbohydrate storage diseases.	A.41
Martens and Frankenberger	Determination of glycuronic acids by high-performance anion chromatography with pulsed amperometric detection.	A.42

(*continued*)

TABLE B.1 (*Continued*)

Author(s)	Title of Article	Ref.
Gunasingham, Tan, and Ng	Pulsed amperometric detection of glucose using a mediated enzyme electrode.	A.43
White and Widmer	Application of high-performance anion-exchange chromatography with pulsed amperometric detection to sugar analysis in citrus juices.	A.44
Andrews and King	Selection of potentials for pulsed amperometric detection of carbohydrates at gold electrodes.	A.45
Swallow and Low	Analysis and quantitation of the carbohydrates in honey using high-performance liquid chromatography.	A.46
Manzi, Diaz, and Varki	High-pressure liquid chromatography of sialic acids on a pellicular resin anion-exchange column with pulsed amperometric detection; a comparison with six other systems.	A.47
Fleming, Kapembwa, Laker, Levin, and Griffin	Rapid and simultaneous determination of lactulose and mannitol in urine, by HPLC with pulsed amperometric detection, for use in studies of intestinal permeability.	A.48
Martens and Frankenberger	Determination of saccharides by high-performance anion-exchange chromatography with pulsed amperometric detection.	A.49
LaCourse, Mead, and Johnson	Anion-exchange separation of carbohydrates with pulsed amperometric detection using a pH-selective reference electrode.	A.50
Hotchkiss and Hicks	Analysis of oligogalacturonic acids with 50 or fewer residues by high-performance anion-exchange chromatography and pulsed amperometric detection.	A.51
Weitzhandler and Hardy	Sensitive blotting assay for the detection of glycopeptides in peptide maps.	A.52
Townsend	Assessment of protein glycosylation using high-pH anion-exchange chromatography with pulsed electrochemical detection.	A.53
Bowers	A new analytical cell for carbohydrate analysis with a maintenance-free reference electrode.	A.54
Sun, Mou, Lin, and Lu	Analysis of monosaccharides in lipopolysaccharides by anion-exchange chromatography with pulsed amperometric detection.	A.55
Ruo, Wang, Dordal, and Atkinson	Assay of inulin in biological fluids by high-performance liquid chromatography with pulsed amperometric detection.	A.56
Krull and Cote	Determination of gulose and/or guluronic acid by ion chromatography and pulsed amperometric detection.	A.57
Peschet and Giacalone	A new concept in sugar analysis: pulsed amperometric coupled ionic chromatography.	A.58
Sun, Mou, and Lu	Analysis of oligosaccharides and monosaccharides by anion exchange chromatography with pulsed amperometric detection.	A.59
Koizumi, Fukuda, and Hizukuri	Estimation of the distributions of chain length of amylopectins by high-performance liquid chromatography with pulsed amperometric detection.	A.60

(*continued*)

TABLE B.1 (*Continued*)

Author(s)	Title of Article	Ref.
Clarke, Sarabia, Keenleyside, MacLachlan, and Whitfield	The compositional analysis of bacterial extracellular polysaccharides by high-performance anion-exchange chromatography.	A.61
Lampio and Finne	Sugar analysis of glycoproteins and glycolipids after methanolysis by high-performance liquid chromatography with pulsed amperometric detection.	A.62
van Riel and Olieman	Selectivity control in the anion-exchange chromatographic determination of saccharides in dairy products using pulsed amperometric detection.	A.63
Ammeraal, Delgado, Tenbarge, and Friedman	High-performance anion-exchange chromatography with pulsed amperometric detection of linear and branched glucose oligosaccharides.	A.64
LaCourse and Johnson	Optimization of waveforms for pulsed amperometric detection (p.a.d.) of carbohydrates following separation by liquid chromatography.	A.65
Dekker, Van der Meer, and Olieman	Sensitive pulsed amperometric detection of free and conjugated bile acids in combination with gradient reversed-phase HPLC.	A.66
Martens and Frankenberger	Determination of saccharides in biological materials by high-performance anion-exchange chromatography with pulsed amperometric detection.	A.67
Mou, Sun, and Lu	Determination of xylose oligomers and monosaccharides by anion-exchange chromatography with pulsed amperometric detection.	A.68
Garleb, Bourquin, and Fahey	Galacturonate in pectic substances from fruits and vegetables: comparison of anion exchange HPLC with pulsed amperometric detection to standard colorimetric procedure.	A.69
Swallow, Low, and Petrus	Detection of orange juice adulteration with beet medium invert sugar using anion-exchange liquid chromatography with pulsed amperometric detection.	A.70
Anumula and Taylor	Rapid characterization of asparagine-linked oligosaccharides isolated from glycoproteins using a carbohydrate analyzer.	A.71
Feste and Khan	Separation of glucooligosaccharides and polysaccharide hydrolyzates by gradient elution hydrophilic interaction chromatography with pulsed amperometric detection.	A.72
Sawert	Oligosaccharide mapping. Anion-exchange chromatography with pulsed amperometric detection (HPAE-PAD).	A.73
Solbrig-Lebuhn	Carbohydrate determination by anion-exchange chromatography with pulsed electrochemical detection (HPAE-PAD).	A.74
Soga, Inoue, and Yamaguchi	Determination of carbohydrates by hydrophilic interaction chromatography with pulsed amperometric detection using postcolumn pH adjustment.	A.75
Stumm and Baltes	Determination of the synthetic carbohydrate polydextrose in foods by ion chromatography with pulsed amperometric detection.	A.76

(*continued*)

TABLE B.1 *(Continued)*

Author(s)	Title of Article	Ref.
Hermentin, Witzel, Doenges, Bauer, Haupt, Patel, Parekh, and Brazel	The mapping by high-pH anion-exchange chromatography with pulsed amperometric detection and capillary electrophoresis of the carbohydrate moieties of human plasma α-1-acid glycoprotein.	A.77
Kragten, Kamerling, and Vliegenthart	Composition analysis of carboxymethylcellulose by high-pH anion-exchange chromatography with pulsed amperometric detection.	A.78
Quigley and Englyst	Determination of neutral sugars and hexosamines by high-performance liquid chromatography with pulsed amperometric detection.	A.79
Kragten, Kamerling, Vliegenhart, Botter, and Batelaan	Composition analysis of sulfoethylcellulose by high-pH anion-exchange chromatography with pulsed amperometric detection.	A.80
Tomiya, Suzuki, Awaya, Mizuno, Matsubara, Nakano, and Kurono	Determination of monosaccharides and sugar alcohols in tissues from diabetic rats by high-performance liquid chromatography with pulsed amperometric detection.	A.81
Golbrig-Lebuhn	Carbohydrate analysis. Anion-exchange chromatography with pulsed electrochemical detection (HPAE-PAD).	A.82
Feste and Khan	Separation of glucose polymers by hydrophilic interaction chromatography on aqueous size-exclusion columns using gradient elution with pulsed amperometric detection.	A.83
Koizumi, Kubota, Ozaki, Shigenobu, Fukuda, and Tanimoto	Analyses of isomeric mono-O-methyl-D-glucoses, D-glucobioses and D-glucose monophosphates by high-performance anion-exchange chromatography with pulsed amperometric detection.	A.84
Hermentin, Witzel, Vliegenthart, Kamerling, Nimtz, and Condrat	A strategy for the mapping of N-glycans by high-pH anion-exchange chromatography with pulsed amperometric detection.	A.85
Prodolliet, Blanc, Bruelhart, Obert, and Parchet	Determination of carbohydrates in soluble coffee by high-performance anion-exchange chromatography with PAD.	A.86
Tabata and Dohi	An assay for oligo-glucantransferase activity in the glycogen debranching enzyme system by using HPLC with a pulsed amperometric detector.	A.87
White and Cancalon	Detection of beet sugar adulteration of orange juice by liquid chromatography/pulsed amperometric detection with column switching.	A.88
Tanaka, Nakamori, Akanuma, and Yabuuchi	High performance liquid chromatographic determination of 1,5-anhydroglucitol in human plasma for diagnosis of diabetes mellitus.	A.89
Kubota, Fukuda, Ohtsuji, and Koizumi	Microanalysis of β-cyclodextrin and glucosyl-β-cyclodextrin in human plasma by high-performance liquid chromatography with pulsed amperometric detection.	A.90
Mopper, Schultz, Chevolot, Germain, Revuelta, and Dawson	Determination of sugars in unconcentrated seawater and other natural waters by liquid chromatography with pulsed amperometric detection.	A.91

(continued)

TABLE B.1 (*Continued*)

Author(s)	Title of Article	Ref.
Haginaka, Nishimura, and Yasuda	Determination of cyclodextrins in serum by reversed-phase chromatography with pulsed amperometric detection and a membrane reactor.	A.92
Corradini, Cristalli, and Corradini	High performance anion-exchange chromatography with pulsed amperometric detection of nutritionally significant carbohydrates.	A.93
Spiro, Kates, Koller, O'Neill, Albersheim, and Darvill	Purification and characterization of biologically active 1,4-linked α-D-oligogalacturonides after partial digestion of polygalacturonic acid with endopolygalacturonase.	A.94
Hotchkiss and Hicks	Analysis of pectate lyase-generated oligogalacturonic acids by high-performance anion-exchange chromatography with pulsed amperometric detection.	A.95
Adams, Zegeer, Bohnert, and Jensen	Anion exchange separation and pulsed amperometric detection of inositols from flower petals.	A.96
Fleming, Kynaston, Laker, Pearson, Kapembwa, and Griffin	Analysis of multiple sugar probes in urine and plasma by high-performance anion-exchange chromatography with pulsed electrochemical detection. Application in the assessment of intestinal permeability in human immunodeficiency virus infection.	A.97
Van Nifterik, Xu, Laurent, Mathieu, and Rakoto	Analysis of cellulose and kraft pulp ozonolysis products by anion-exchange chromatography with pulsed amperometric detection.	A.98
Taha and Deits	Detection of glycinamide ribonucleotide by HPLC with pulsed amperometry: application to the assay or glutamine:5-phosphoribosyl-1-pyrophosphate amidotransferase.	A.99
Rohrer, Cooper, and Townsend	Identification, quantification, and characterization of glycopeptides in reversed-phase HPLC separations of glycoprotein proteolytic digests.	A.100
Lu and Cassidy	Pulsed amperometric detection of carbohydrates separated by capillary electrophoresis.	A.101
Fukuda, Kubota, Ikuta, Hasegawa, and Koizimi	Microanalysis of β-cyclodextrin and glucosyl-β-cyclodextrin in biological matrixes by high-performance liquid chromatography with pulsed amperometric detection.	A.102
Pastore, Lavagnini, and Versini	Ion chromatographic determination of monosaccharides from trace amounts of glycosides isolated from grape musts.	A.103
Sanghi, Kok, Koomen, and Hoek	Determination of inert sugars in urine by liquid chromatography with pulsed amperometric detection.	A.104
Low and Wudrich	Detection of inexpensive sweetener addition to grapefruit juice by HPLC-PAD.	A.105
Kynaston, Fleming, Laker, and Pearson	Simultaneous quantification of mannitol, 3-*o*-methyl glucose, and lactulose in urine by HPLC with pulsed electrochemical detection, for use in studies of intestinal permeability.	A.106
Stefansson and Lu	Ion-pair LC of sugars at alkaline pH and pulsed electrochemical detection.	A.107

(*continued*)

TABLE B.1   (*Continued*)

Author(s)	Title of Article	Ref.
O'Shea, Lunte, and LaCourse	Detection of carbohydrates by capillary electrophoresis with pulsed amperometric detection.	A.108
LaCourse and Johnson	Optimization of waveforms for pulsed amperometric detection of carbohydrates based on pulsed voltammetry.	A.109
Trotzer, Hofsommer, and Rubach	Carbohydrate analysis with anion-exchange chromatography and pulsed amperometric detection with the Dionex DX-300 chromatography module.	A.110
Corradini	Recent advances in carbohydrate analysis by high-performance anion-exchange chromatography coupled with pulsed amperometric detection (HPAEC-PAD).	A.111
Yu and El Rassi	Preparation of amino-zirconia bonded phases and their evaluation in hydrophilic interaction chromatography of carbohydrates with pulsed amperometric detection.	A.112
Kuroda, Taka, Kajikawa, Niimi, Ishida, and Kawanishi	Determination of sugars and polyols in red cells by high-performance liquid chromatography with pulsed amperometric detection; 1. red cell sorbitol.	A.113
Corradini, Corradini, Huber, and Bonn	Synthesis of a polymeric-based stationary phase for carbohydrate separation by high-pH anion-exchange chromatography with pulsed amperometric detection.	A.114
Corradini, Cristalli, and Corradini	Determination of carbohydrates in fruit-based beverages by high-performance anion-exchange chromatography with pulsed amperometric detection (HPAEC-PAD).	A.115
Andersen, Goochee, Cooper, and Weitzhandler	Monosaccharide and oligosaccharide analysis of isoelectric focusing-separated and blotted granulocyte colony-stimulating factor glycoforms using high-pH anion-exchange chromatography with pulsed amperometric detection.	A.116
Herber and Robinett	Determination of carbon sources in fermentation media using high-performance anion-exchange liquid chromatography and pulsed amperometric detection.	A.117
Tsai, Gu, and Byrd	Quantification of polysaccharides in Haemophilus influenzae type b conjugate and polysaccharide vaccines by high-performance anion-exchange chromatography with pulsed amperometric detection.	A.118
Timmermans, van Leeuwen, Tournois, de Wit, and Vliegenthart	Quantitative analysis of the molecular weight distribution of inulin by means of anion-exchange HPLC with pulsed amperometric detection.	A.119
Quigley and Englyst	Determination of the uronic acid constituents of non-starch polysaccharides by high-performance liquid chromatography with pulsed amperometric detection.	A.120
Roberts and Johnson	Fast-pulsed electrochemical detection at noble metal electrodes: the frequency-dependent response at gold electrodes for chromatographically separated carbohydrates.	A.121
Sullivan and Douek	Determination of carbohydrates in wood, pulp and process liquor samples by high-performance anion-exchange chromatography with pulsed amperometric detection.	A.122

(*continued*)

TABLE B.1 *(Continued)*

Author(s)	Title of Article	Ref.
Swallow and Low	Determination of honey authenticity by anion-exchange liquid chromatography.	A.123
Slaughter and Livingston	Separation of fructan isomers by high performance anion exchange chromatography.	A.124
Marko-Varga, Buttler, Gorton, Olsson, Durand, and Barcelo	Qualitative and quantitative carbohydrate analysis of fermentation substrates and broths by liquid chromatographic techniques.	A.125
Friedman, Levin, and McDonald	α-Tomatine determination in tomatoes by HPLC using pulsed amperometric detection.	A.126
Park, Park, Lee, Kim, and Park	Microanalysis of ginseng saponins by ion chromatography with pulsed amperometric detection.	A.127
Friedman and Levin	α-Tomatine content in tomato and tomato products determined by HPLC with pulsed amporometric detection	A.128
Kuroda, Tada, Kajikawa, Niimi, Ishida, and Kawanishi	Determination of sugars and polyols in red blood cells by high-performance liquid chromatography with pulsed amperometric detection; 2. red blood cell myo-inositol.	A.129
Wilson, Cataldo, and Andersen	Determination of total nonstructural carbohydrates in tree species by high-performance anion-exchange chromatography with pulsed amperometric detection.	A.130
Zook and LaCourse	Pulsed amperometric detection of carbohydrates in fruit juices following high-performance anion-exchange chromatography.	A.131
Corradini, Corradini, Huber, and Bonn	High-performance anion-exchange chromatography of carbohydrates using a new resin and pulsed amperometric detection.	A.132
Jorge and Abdul-Wajid	Sailyl-Tn-KLH, glycoconjugate analysis and stability by high-pH anion-exchange chromatography with PAD.	A.133
Roberts and Johnson	Variations in PED response at a gold microelectrode as a function of waveform parameters when applied to alditols and carbohydrates separated by capillary electrophoresis.	A.134
Kerherve, Charriere, and Gadel	Determination of marine monosaccharides by high-pH anion-exchange chromatography with pulsed amperometric detection.	A.135
Mou and Li	Progress in carbohydrate analysis by anion exchange separation-pulsed amperometric detection.	A.136
Stuckel and Low	Maple syrup authenticity analysis by anion-exchange liquid chromatography with pulsed amperometric detection.	A.137
Kelly, Kimball, and Johnston	Quantitation of digitoxin, digoxin, and their metabolites by high-performance liquid chromatography using pulsed amperometric detection.	A.138
Torto, Buttler, Gorton, Marko-Varga, Stalbrand, and Tjerneld	Monitoring of enzymatic hydrolysis of ivory nut mannan using on-line microdialysis sampling and anion-exchange chromatography with integrated pulsed amperometric detection.	A.139

(continued)

TABLE B.1   (*Continued*)

Author(s)	Title of Article	Ref.
Rohrer	Separation of asparagine-linked oligosaccharides by high-pH anion-exchange chromatography with pulsed amperometric detection: empirical relationships between oligosaccharide structure and chromatographic retention.	A.140
Dunkel and Amato	Analysis of endo-(1,5)-α-L-arabinose degradation patterns of linear (1,5)-α-L-arabinose-oligosaccharides by high-performance anion-exchange chromatography with pulsed amperometric detection.	A.141
Rohrer, Thayer, Avdalovic, and Weitzhandler	HPAEC-PAD analysis of monoclonal antibody glycosylation.	A.142
Prodolliet, Bugner, and Feinberg	Determination of carbohydrates in soluble coffee by anion-exchange chromatography with pulsed amperometric detection: interlaboratory study.	A.143
Cooper and Rohrer	Separation of neutral asparagine-linked oligosaccharides by high-pH anion-exchange chromatography with pulsed amperometric detection.	A.144
Chaplin	Analysis of bile acids and their conjugates using high-pH anion-exchange chromatography with pulsed amperometric detection.	A.145
Weber, Kornfelt, Lausen, and Lunte	Characterization of glycopeptides from recombinant coagulation factor VIIa by high-performance liquid chromatography and capillary zone electrophoresis using ultraviolet and pulsed electrochemical detection.	A.146
Wong and Jane	Effects of pushing agents on the separation and detection of debranched amylopectin by high-performance anion-exchange chromatography with pulsed amperometric detection.	A.147
Hikima, Kakizaki, and Hasebe	Enzyme sensor for L-lactate using differential pulse amperometric detection.	A.148
Kimura, Shirota, Suzuki, and Ohtsu	High-sensitivity carbohydrate analysis by semi-microcolumn LC with pulsed amperometric detector.	A.149
Elliott, Olsen, and Tallman	Detection of glucose at a rotating gold composite electrode by pulsed amperometric detection.	A.150
Low	Food authenticity analysis by anion-exchange liquid chromatography.	A.151
Hotchkiss, El-Bahtimy, and Fishman	Analysis of pectin structure by HPAEC-PAD.	A.152
Zook, Patel, LaCourse, and Ralapati	Characterization of tobacco products by high-performance anion exchange chromatography-pulsed amperometric detection.	A.153
Gough, Luke, Beeley, and Geddes	Human salivary glucose analysis by high-performance ion-exchange chromatography and pulsed amperometric detection.	A.154
Ichiki, Semma, Sekiguchi, Nakamura, and Ito	Determination of sucralose in food by anion-exchange chromatography with pulsed amperometric detection.	A.155

(*continued*)

TABLE B.1 (*Continued*)

Author(s)	Title of Article	Ref.
Mayer, Ruffner, and Rast	Assessment of the mode of action of chitinases with anion exchange chromatography/pulsed amperometric detection.	A.156
Torto, Marko-Varga, Gorton, Staalbrand, and Tjerneld	Online quantitation of enzymatic mannan hydrolyzates in small-volume bioreactors by microdialysis sampling and column liquid chromatography-integrated pulsed electrochemical detection.	A.157
Tyagarajan, Forte, and Townsend	Exoglycosidase purity and linkage specificity: assessment using oligosaccharide substrates and high-pH anion-exchange chromatography with pulsed amperometric detection.	A.158
Weber and Lunte	Capillary electrophoresis with pulsed amperometric detection of carbohydrates and glycopeptides.	A.159
Madigan, McMurrough, and Smyth	Application of gradient ion chromatography with pulsed electrochemical detection to the analysis of carbohydrates in brewing.	A.160
Bernarl, Del Nozal, Toribio, and Del Alamo	HPLC analysis of carbohydrates in wines and instant coffees using anion-exchange chromatography coupled to pulsed amperometric detection.	A.161
Frias, Price, Fenwick, Hedley, Sorensen, and Vidal-Valverde	Improved method for the analysis of α-galactoside in pea seeds by capillary zone electrophoresis. Comparison with high-performance liquid chromatography-triple-pulsed amperometric detection.	A.162
Wang, Klegerman, and Groves	Analysis of antineoplastic polysaccharides from Mycobacterium bovis BCG vaccine by high-performance anion-exchange chromatography with pulsed amperometric detection.	A.163
Weitzhandler, Slingsby, Jagodzinski, Pohl, Narayanan, and Avdalovic	Eliminating monosaccharide peak tailing in high pH anion-exchange chromatography with pulsed amperometric detection.	A.164
Sigvardson, Eliason, and Herbranson	Determination of raffinose and lactobionic acid in ViaSpan by anion exchange chromatography with pulsed amperometric detection.	A.165

TABLE B.2   Applications Related to Aliphatic Alcohols

Author(s)	Title of Article	Ref.
Hughes, Meschi, and Johnson	Amperometric detection of simple alcohols in aqueous solutions by application of a triple-pulse potential waveform at platinum electrodes.	B.1
LaCourse, Johnson, Rey, and Slingsby	Pulsed amperometric detection of aliphatic alcohols in liquid chromatography.	B.2
Le Fur, Etievant, and Meunier	Interest of pulsed electrochemical detection for the analysis of flavor-active alcohols separated by liquid chromatography.	B.3
Schiavon, Comisso, Toniolo, and Bontempelli	Pulsed amperometric detection of ethanol in breath by gold electrodes supported on ion exchange membranes (solid polymer electrolytes).	B.4
Lozano, Chirat, Graille, and Pioch	Measurement of free glycerol in biofuels.	B.5

TABLE B.3   Applications Related to Aminoalcohols, Aminosugars, and Aminoglycosides

Author(s)	Title of Article	Ref.
Polta, Johnson, and Merkel	Liquid chromatographic separation of aminoglycosides with pulsed amperometric detection.	C.1
LaCourse, Jackson, and Johnson	Pulsed amperometric detection of alkanolamines following ion-pair chromatography.	C.2
Statler	Determination of tobramycin using high-performance liquid chromatography with pulsed amperometric detection.	C.3
Campbell, Carson, and Van Bramer	Improved detection of alkanolamines by liquid chromatography with electrochemical detection.	C.4
Roston and Rhinebarger	Evaluation of HPLC with pulsed-amperometric detection for analysis of an aminosugar drug substance.	C.5
Martens and Frankenberger	Determination of aminosaccharides by high-performance anion-exchange chromatography with pulsed amperometric detection.	C.6
McLaughlin and Henion	Determination of aminoglycoside antibiotics by reversed-phase ion-pair high-performance liquid chromatography coupled wtih pulsed amperometry and ion spray mass spectrometry.	C.7
Phillips and Simmonds	Determination of spectinomycin using cation-exchange chromatography with pulsed amperometric detection.	C.8
Kaine and Wolnik	Forensic investigation of gentamicin sulfates by anion-exchange ion chromatography with pulsed electrochemical detection.	C.9
Dobberpuhl and Johnson	Pulsed electrochemical detection of alkanolamines separated by multimodal high-performance liquid chromatography.	C.10
Adams, Schepers, Roets, and Hoogmartens	Determination of neomycin sulfate by liquid chromatography with pulsed electrochemical detection.	C.11

TABLE B.4 Applications Related to Amines, Aminoacids, Peptides, and Proteins

Author(s)	Title of Article	Ref.
Polta and Johnson	The direct electrochemical detection of amino acids at a platinum electrode in alkaline chromatographic effluent.	D.1
Welch, LaCourse, Mead, Johnson, and Hu	Comparison of pulsed coulometric detection and potential-sweep-pulsed coulometric detection for underivatized amino acids in liquid chromatography.	D.2
Welch, LaCourse, Mead, and Johnson	A comparison of pulsed amperometric detection and conductivity detection of underivatized amino acids in liquid chromatography.	D.3
Donaldson and Adlard	Analysis of δ-L-α-aminoadipyl-L-cysteinyl-D-valine by ion chromatography and pulsed amperometric detection.	D.4
Donaldson, Broby, Adlard, and Bucke	High-pressure liquid chromatography and pulsed amperometric detection of castanospermine and related alkaloids.	D.5
Welch and Johnson	Liquid chromatographic separation with electrochemical detection of the phenylhydantoin and methylthiohydantoin derivatives of amino acids.	D.6
Martens and Frankenberger	Pulsed amperometric detection of amino acids separated by anion exchange chromatography.	D.7
Draisci, Cavalli, Lucentini, and Stacchini	Ion-exchange separation and pulsed amperometric detection for determination of biogenic amines in fish products.	D.8
Sadik and Wallace	Pulsed amperometric detection of proteins using antibody containing conducting polymers.	D.9
Sadik, John, Wallace, Barnett, Clarke, and Laing	Pulsed amperometric detection of thaumatin using antibody-containing poly(pyrrole) electrodes.	D.10
Dobberpuhl and Johnson	Pulsed electrochemical detection at the ring of a ring-disk electrode applied to a study of amine adsorption at gold electrodes.	D.11
Sadik and Wallace	Pulsed amperometric detection of proteins using antibody containing conducting polymers (Erratum).	D.12
van Riel and Olieman	Selective detection in RP-HPLC of Tyr-, Trp-, and sulfur-containing peptides by pulsed amperometry at platinum.	D.13
Lu, Zhao, and Wallace	Pulsed electrochemical detection of proteins using conducting polymer based sensors.	D.14
Dobberpuhl, Hoekstra, and Johnson	Pulsed electrochemical detection at gold electrodes applied to monoamines and diamines following their chromatographic separation.	D.15
Weitzhandler, Pohl, Rohrer, Narayanan, Slingsby, and Avdalovic	Eliminating amino acid and peptide interference in high-performance anion-exchange pulsed amperometric detection glycoprotein monosaccharide analysis.	D.16
Dobberpulh and Johnson	A study of ethylamine at the gold rotating ring-disk electrode using pulsed electrochemical detection at the ring.	D.17

TABLE B.5   Applications Related to Sulfur-Containing Compounds

Author(s)	Title of Article	Ref.
Polta and Johnson	Pulsed amperometric detection of sulfur compounds. Part I. Initial studies at platinum electrodes in alkaline solutions.	E.1
Polta, Johnson, and Luecke	Pulsed amperometric detection of sulfur compounds. Part II. Dependence of response on adsorption time.	E.2
Ngoviwatchai and Johnson	Pulsed amperometric detection of sulfur-containing pesticides in reversed-phase liquid chromatography.	E.3
Neuburger and Johnson	Pulsed coulometric detection with automatic rejection of background signal in surface-oxide-catalyzed anodic detections at gold electrodes in flow-through cells.	E.4
Doerge and Yee	Liquid chromatographic determination of ethylenethiourea using pulsed amperometric detection.	E.5
Vandeberg, Kowagoe, and Johnson	Pulsed amperometric detection of sulfur compounds: thiourea at gold electrodes.	E.6
Vandeberg and Johnson	Pulsed electrochemical detection of cysteine, cystine, methionine, and glutathione at gold electrodes following their separation by liquid chromatography.	E.7
Koprowski, Kirchmann, and Welch	The electrochemical oxidation of penicillins of gold electrodes.	E.8
Kirchmann and Welch	High-performance liquid chromatographic separation and electrochemical detection of penicillins.	E.9
Kirchmann, Earley, and Welch	The electrochemical detection of penicillins in milk.	E.10
Vandeberg and Johnson	Comparison of pulsed amperometric detection and integrated voltammetric detection for organic sulfur compounds in liquid chromatography.	E.11
Altunata, Earley, Mossman, and Welch	Pulsed electrochemical detection of penicillins using three and four step waveforms.	E.12
Tudos and Johnson	Dissolution of gold electrodes in alkaline media containing cysteine.	E.13
LaCourse and Owens	Pulsed electrochemical detection of thiocompounds following microchromatographic separations.	E.14
Liu, Li, and Wang	Deteration of cysteine and reduced glutathione in human plasma by liquid chromatography with pulsed amperometric electrochemical detection using a platinum-particles modified glassy carbon electrode.	E.15
Owens and LaCourse	Pulsed electrochemical detection of sulfur-containing compounds following microbore liquid chromatography.	E.16

TABLE B.6 Inorganic, Electroinactive, and Miscellaneous Applications

Author(s)	Title of Article	Ref.
Stastny and Volf	A pulse voltammetric monitor for liquid chromatography.	F.1
Polta, Yeo, and Johnson	Flow-injection system for the rapid and sensitive assay of concentrated aqueous solutions of strong acids and bases.	F.2
Polta and Johnson	Pulsed amperometric detection of elctroinactive adsorbates at platinum electrodes in a flow injection system.	F.3
Rocklin	Ion chromatography with pulsed amperometric detection. Simultaneous determination of formic acid, formaldehyde, acetaldehyde, propionaldehyde, and butyraldehyde.	F.4
Thomas and Sturrock	Determination of carbamates by high-performance liquid chromatography with electrochemical detection using pulsed-potential cleaning.	F.5
Whang and Tsai	Analysis of triorganotin compounds by high-performance liquid chromatography with reverse-pulse amperometric detection.	F.6
Hara, Saeki, and Nomura	New pulse-amperometric detection in ion-exchange chromatography of heavy metal ions.	F.7
Stojanovic, Bond, and Butler	A comparative study of the cylindrical wire, thin-layer, and wall-jet detector cells for the determination of inorganic arsenic by ion exclusion chromatography with constant and pulsed amperometric detection.	F.8
Williams and Johnson	Pulsed voltammetric detection of arsenic(III) at platinum electrodes in acidic media.	F.9
Wagner and McGarrity	Determination of sulfite in beer using ion-exclusion chromatography and pulsed amperometric detection.	F.10
Ianniello	Determination of Suttocide A in cosmetic formulations by single column ion chromatography with pulsed amperometric detection (SCIC-PAD).	F.11
Nair	Determination of trace levels of cyanamide in a novel potassium channel activator bulk drug by pulsed electrochemical detection.	F.12
Hsi and Tsai	Liquid chromatography with pulsed amperometric detection for speciation of mercury.	F.13
Taha and Deits	Detection of metabolites of the Entner-Doudoroff pathway by HPLC with pulsed amperometry: application to assays for pathway enzymes.	F.14
Le Fur, Meunier, and Etievant	Comparative investigation of pulsed electrochemical and ultraviolet detection in the determination of flavor-active aldehydes separated by HPLC.	F.15
Wagner	Stabilization of sulfite for automated analysis using ion exclusion chromatography combined with pulsed amperometric detection.	F.16
Dilleen, Lawrence, and Slater	Determination of 2-furaldehyde in transformer oil using flow injection with pulsed amperometric detection.	F.17
Park, Hong, and You	Optimization of wave forms for pulsed amperometric detection of cyanide and sulfite with silver-working electrode.	F.18

(*continued*)

TABLE B.6   (*Continued*)

Author(s)	Title of Article	Ref.
Wen and Cassidy	Anodic and cathodic pulse amperometric detection of metal ions separated by capillary electrophoresis.	F.19
Shi and Johnson	Determination of formaldehyde in air by ion-excluision and ion-exchange chromatography with pulsed amperometric detection.	F.20

TABLE B.7   Instrumental Developments and PED Reviews

Author(s)	Title of Article or Chapter	Ref.
Pentari and Efstathiou	Construction and applications of a microcomputer controlled pulsed amperometric detector system.	G.1
Weiss	Ion chromatography with pulsed amperometric detection—theory and applications.	G.2
Ramstad and Milner	Instrumental methods for electrosorptive detection in liquid chromatography.	G.3
Austin-Harrison and Johnson	Pulsed amperometric detection based on direct and indirect anodic reactions: a review.	G.4
Cochrane	Ion exchange separations in conjunction with electrochemical detection.	G.5
Rocklin, Henshall, and Rubin	A multimode electrochemical detector for non-UV-absorbing molecules.	G.6
Johnson and LaCourse	Liquid chromatography with pulsed electrochemical detection at gold and platinum electrodes.	G.7
Lee	Carbohydrate analysis with high performance anion exchange chromatography using pulsed amperometric detector.	G.8
Soga	New method for sugar analysis using HPLC.	G.9
Roberts and Johnson	Fast pulsed amperometric detection at noble metal electrodes: a study of oxide formation and dissolution kinetics at gold in 0.1M sodium hydroxide.	G.10
Johnson and LaCourse	Pulsed electrochemical detection at noble metal electrodes in liquid chromatography.	G.11
Johnson, Dobberpul, Roberts, and Vandeberg	Pulsed amperometric detection of carbohydrates, amines and sulfur species in ion chromatography. The current state of research.	G.12
LaCourse	Pulsed electrochemical detection at noble metal electrodes in high-performance liquid chromatography.	G.13
Rocklin, Tullsen, and Marucco	Maximizing signal-to-noise ratio in direct current and pulsed amperometric detection.	G.14
Roberts and Johnson	Fast-pulsed amperometric detection at noble metal electrodes: a study of oxide formation and dissolution at platinum in 0.1M NaOH.	G.15
Johnson and LaCourse	Pulsed electrochemical detection of carbohydrates at gold electrodes following liquid chromatographic separation.	G.16
LaCourse and Owens	Pulsed electrochemical detection of nonchromophoric compounds following capillary electrophoresis.	G.17
LaCourse and Dasenbrock	High performance liquid chromatography-pulsed electrochemical detection for the analysis of antibiotics.	G.18
LaCourse	Pulsed electrochemical detection.	G.19

REFERENCES

Carbohydrates and Alditols

A.1. S. Hughes and D. C. Johnson, *Anal. Chim. Acta* **132**, 11–22 (1981).

A.2. S. Hughes and D. C. Johnson, *J. Agric. Food Chem.* **30**(4), 712–714 (1982).

A.3. W. Buchberger, K. Winsauer, and C. Breitwieser, *Fresenius' Z. Anal. Chem.* **311:5**, 517 (1982).

A.4. P. Edwards and K. K. Haak, **15**(4), 78–87 (1983).

A.5. R. D. Rocklin and C. A. Pohl, *J. Liq. Chromatogr.* **6**(9), 1577–1590 (1983).

A.6. S. Hughes and D. C. Johnson, *Anal. Chim. Acta* **149**, 1–10 (1983).

A.7. K. Ohsawa, Y. Yoshimura, S. Watanabe, H. Tanaka, A. Yokota, K. Tamura, and K. Imaeda, *Anal. Sci.* **2**(2), 165–168 (1986).

A.8. J. D. Olechno, S. R. Carter, W. T. Edwards, and D. G. Gillen, *Am. Biotechnol. Lab.* **5**(5), 38–50 (1987).

A.9. W. T. Edwards, C. A. Pohl, and R. Rubin, *Tappi J.* **70**(6), 138–140 (1987).

A.10. G. G. Neuburger and D. C. Johnson, *Anal. Chem.* **59**(1), 150–154 (1987).

A.11. G. G. Neuburger and D. C. Johnson, *Anal. Chem.* **59**(1), 203–204 (1987).

A.12. G. G. Neuburger and D. C. Johnson, *Anal. Chim. Acta* **192**(2), 205–213 (1987).

A.13. K. Koizumi, Y. Kubuta, T. Tanimoto, and Y. Okada, *J. Chromatogr.* **454**, 303–310 (1988).

A.14. R. R. Townsend, M. R. Hardy, O. Hindsgaul, and Y. C. Lee, *Anal. Biochem.* **174**(2), 459–470 (1988).

A.15. J. Haginaka and T. Nomura, *J. Chromatogr.* **447**(1), 268–271 (1988).

A.16. L. Chen, M. G. Yet, and M. Shao, *FASEB J.* **2**(12), 2819–2824 (1988).

A.17. M. R. Hardy and R. R. Townsend, *Proc. Natl. Acad. Sci.* (USA) **85**(10), 3289–3293 (1988).

A.18. L. A. Larew, D. A. Mead Jr., and D. C. Johnson, *Anal. Chim. Acta* **204**(1–2), 43–51 (1988).

A.19. L. A. Larew and D. C. Johnson, *Anal. Chem.* **60**(18), 1867–1872 (1988).

A.20. L. E. Welch, D. A. Mead, Jr., and D. C. Johnson, *Anal. Chim. Acta.* **204**(1–2), 323–327 (1988).

A.21. M. R. Hardy, R. R. Townsend, and Y. C. Lee, *Anal. Biochem.* **170**(1), 54–62 (1988).

A.22. L. A. Larew and D. C. Johnson, *J. Electroanal. Chem. Interfacial Electrochem.* **264**(1–2), 131–147 (1989).

A.23. W. T. Wang and D. Zopf, *Carbohydr. Res.* **189**, 1–11 (1989).

A.24. R. R. Townsend, M. R. Hardy, D. A. Cummings, J. P. Carver, and B. Bendiak, *Anal. Biochem.* **182**(1), 1–8 (1989).

A.25. J. D. Olechno, S. R. Carter, W. T. Edwards, D. G. Gillen, R. R. Townsend, Y. C. Lee, and M. R. Hardy, *Tech. Protein Chem.* 364–376 (1989).

A.26. D. S. Bindra and G. S. Wilson, *Anal. Chem.* **61**(22), 2566–2570 (1989).

A.27. K. A. Garleb, L. D. Bourquin, and G. C. Fahey, Jr., *J. Agric. Food Chem.* **37**(5), 1287–1293 (1989).

A.28. J. Haginaka, Y. Nishimura, J. Wakai, H. Yasuda, K., Koizumi, and T. Nomura, *Anal. Biochem.* **179**(2), 336–340 (1989).

A.29. M. R. Hardy and R. R. Townsend, *Carbohydr. Res.* **188**, 1–7 (1989).

A.30. J. N. Chatterton, P. A. Harrison, W. R. Thornley, and J. H. Bennett, *Plant Physiol. Biochem.* **27**(2), 289–295 (1989).

A.31. M. R. Hardy, *LC-GC* **7**(3), 242–246 (1989).

A.32. C. Jansen, *Chem. Labor. Betr.* **40**(3), 248–252 (1989).

A.33. R. M. Pollman, *J. Assoc. Off. Anal. Chem.* **72**(3), 425–428 (1989).

A.34. K. Koizumi, Y. Kubota, T. Tanimoto, and Y. Okada, *J. Chromatogr.* **464**(2), 365–373 (1989).

A.35. R. R. Townsend, M. R. Hardy, and Y. C. Lee, *Meth. Enzymol.*, Vol. 179: *Complex Carbohydr.*, Part F, 65–76 (1989).

A.36. M. R. Hardy, *Meth. Enzymol.*, Vol. 179, *Complex Carbohydr.* Part F, 76–82 (1989).

A.37. C. S. Liu, E. Chiu, and T. B. Lo, *Huaxue* **48**(4), 295–302 (1990).

A.38. S. Honda, S. Suzuki, S. Zaiki, and K. Kakehi, *J. Chromatogr.* **523**, 189–200 (1990).

A.39. D. A. Martens and W. T. Frankenberger, Jr., *Chromatographia* **30**(11–12), 651–656 (1990).

A.40. Y. Zhu and S. Xu, *Sepu* **8**(5), 315–316 (1990).

A.41. H. J. Blom, H. C. Andersson, D. M. Krasnewich, and W. A. Gahl, *J. Chromatogr.* **533**, 11–21 (1990).

A.42. D. A. Martens and W. T. Frankenberger, Jr., *Chromatographia* **30**(5–6), 249–254 (1990).

A.43. H. Gunasingham, C. H. Tan, and H-M. Ng, *J. Electroanal. Chem. Interfacial Electrochem.* **287**(2), 349–362 (1990).

A.44. D. R. White, Jr. and W. W. Widmer, *J. Agric. Food Chem.* **38**(10), 1918–1921 (1990).

A.45. R. W. Andrews and R. M. King, *Anal. Chem.* **62**(19), 2130–2134 (1990).

A.46. K. W. Swallow and N. H. Low, *J. Agric. Food Chem.* **38**(9), 1828–1832 (1990).

A.47. A. E. Manzi, S. Diaz, and A. Varki, *Anal. Biochem.* **188**(1), 20–32 (1990).

A.48. S. C. Fleming, M. S. Kapembwa, M. F. Laker, G. E. Levin, and G. E. Griffin, *Clin. Chem.* **36**(5), 797–799 (1990).

A.49. D. A. Martens and W. T. Frankenberger, Jr., *Chromatographia* **29**(1–2), 7–12 (1990).

A.50. W. R. LaCourse, D. A. Mead, Jr., and D. C. Johnson, *Anal. Chem.* **62**(2), 220–224 (1990).

A.51. A. T. Hotchkiss, Jr. and K. B. Hicks, *Anal. Biochem.* **184**(2), 200–206 (1990).

A.52. M. Weitzhandler and M. Hardy, *J. Chromatogr.* **510**, 225–232 (1990).

A.53. R. R. Townsend, GBF Monogr. **15**: *Protein Glycosylation: Cell., Biotechnol. Anal. Aspects* 147–160 (1991).

A.54. M. L. Bowers, *J. Pharm. Biomed. Anal.* **9**(10–12), 1133–1137 (1991).

A.55. Q. Sun, S. Mou, J. Lin, and D. Lu, *Anal. Sci.* **7** (Suppl.), *Proc. Int. Congr. Anal. Sci.* Part 1, 157–160 (1991).

A.56. T. I. Ruo, Z. Wang, M. S. Dordal, and A. J. Atkinson, Jr., *Clin. Chim. Acta* **204**(1–3), 217–222 (1991).

A.57. L. H. Krull and G. L. Cote, *Carbohydr. Polym.* **17**(3), 205–207 (1991).

A.58. J. L. Peschet and A. Giacalone, *Ind. Aliment. Agric.* **108**(7–8), 583–586 (1991).

A.59. Q. Sun, S. Mou, and D Lu, *Huaxue Tongbao* **8**, 39–41 (1991).

A.60. K. Koizumi, M. Fukuda, and S. Hizukuri, *J. Chromatogr.* **585**(2), 233–238 (1991).

A.61. A. J. Clarke, V. Sarabia, W. Keenleyside, P. R. MacLachlan, and C. Whitfield, *Anal. Biochem.* **199**(1), 68–74 (1991).

A.62. A. Lampio and J. Finne, *Anal. Biochem.* **197**(1), 132–136 (1991).

A.63. J. van Riel and C. Olieman, *Carbohydr. Res.* **215**(1), 39–46 (1991).

A.64. R. N. Ammeraal, G. A. Delgado, F. L. Tenbarge, and R. B. Friedman, *Carbohydr. Res.* **215**(1), 179–192 (1991).

A.65. W. R. LaCourse and D. C. Johnson, *Carbohydr. Res.* **215**(1), 159–178 (1991).

A.66. R. Dekker, R. Van der Meer, and C. Olieman, *Chromatographia* **31**(11–12), 548–553 (1991).

A.67. D. A. Martens and W. T. Frankenberger, Jr., *J. Chromatogr.* **564**(1–2), 297–309 (1991).

A.68. S. Mou, Q. Sun, and D. Lu, *J. Chromatogr.* **564**(1–2), 289–295 (1991).

A.69. K. A. Garleb, L. D. Bourquin, and G. C. Fahey, Jr., *J. Food Sci.* **56**(2), 423–426 (1991).

A.70. K. W. Swallow, N. H. Low, and D. R. Petrus, *J. Assoc. Off. Anal. Chem.* **74**(2), 341–345 (1991).

A.71. K. R. Anumula and P. B. Taylor, *Eur. J. Biochem.* **195**(1), 269–280 (1991).

A.72. A. S. Feste and I. Khan, *J. Chromatogr.* **630**(1–2), 129–139 (1992).

A.73. A. Sawert, *GIT Fachz. Lab.* **36**(9), 875–881 (1992).

A.74. H. Solbrig-Lebuhn, *Zuckerindustrie* (Berlin) **117**(12), 979–983 (1992).

A.75. T. Soga, Y. Inoue, and K. Yamaguchi, *J. Chromatogr.* **625**, 151–155 (1992).

A.76. I. Stumm and W. Baltes, *Z. Lebensm.-Unters. Forsch.* **195**(3), 246–249 (1992).

A.77. P. Hermentin, R. Witzel, R. Doenges, R. Bauer, H. Haupt, T. Patel, R. B. Parekh, and D. Brazel, *Anal. Biochem.* **206**(2), 419–429 (1992).

A.78. E. A. Kragten, J. P. Kamerling, and J. F. G. Vliegenthart, *J. Chromatogr.* **623**(1), 49–53 (1992).

A.79. M. E. Quigley and H. N. Englyst, *Analyst* **117**(11), 1715–1716 (1992).

A.80. E. A. Kragten, J. P. Kamerling, J. F. G. Vliegenhart, H. Botter, and J. G. Batelaan, *Carbohydr. Res.* **233**, 81–86 (1992).

A.81. N. Tomiya, T. Suzuki, J. Awaya, K. Mizuno, A. Matsubara, K. Nakano, and M. Kurono, *Anal. Biochem.* **206**(1), 98–104 (1992).

A.82. H. Solbrig-Lebuhn, *LaborPraxis* **16**(8), 786–789 (1992).

A.83. A. S. Feste and I. Khan, *J. Chromatogr.* **607**, 7–18 (1992).

A.84. K. Koizumi, Y. Kubota, H. Ozaki, K. Shigenobu, M. Fukuda, and T. Tanimoto, *J. Chromatogr.* **595**(1–2), 340–345 (1992).

A.85. P. Hermentin, R. Witzel, J. F. G. Vliegenthart, J. P. Kamerling, M. Nimtz, and H. S. Condrat, *Anal. Biochem.* **203**(2), 281–289 (1992).

A.86. J. Prodolliet, M. B. Blanc, M. Bruelhart, L. Obert, and J. M. Parchet, *Colloq. Sci. Int. Cafe* **14**, 211–219 (1992).

A.87. S. Tabata and Y. Dohi, *Carbohyr. Res.* **230**(1), 179–183 (1992).

A.88. D. R. White, Jr. and P. F. Cancalon, *J. AOAC Int.* **75**(3), 584–587 (1992).

A.89. S. Tanaka, K. Nakamori, H. Akanuma, and M. Yabuuchi, *Biomed. Chromatogr.* **6**(2), 63–66 (1992).

A.90. Y. Kubota, M. Fukuda, K. Ohtsuji, and K. Koizumi, *Anal. Biochem.* **201**(1), 99–102 (1992).

A.91. K. Mopper, C. A. Schultz, L. Chevolot, C. Germain, R. Revuelta, and R. Dawson, *Environ. Sci. Technol.* **26**(1), 133–138 (1992).

A.92. J. Haginaka, Y. Nishimura, and H. Yasuda, *J. Pharm. Biomed. Anal.* **11**(10), 1023–1026 (1993).

A.93. C. Corradini, A. Cristalli, and D. Corradini, *J. Liq. Chromatogr.* **16**(16), 3471–3485 (1993).

A.94. M. D. Spiro, K. A. Kates, A. L. Koller, M. A. O'Neill, P. Albersheim, and A. G. Darvill, *Carbohyr. Res.* **247**, 9–20 (1993).

A.95. A. T. Hotchkiss, Jr. and K. B. Hicks, *Carbohyr. Res.* **247**, 1–7 (1993).

A.96. P. Adams, A. Zegeer, H. J. Bohnert, and R. G. Jensen, *Anal. Biochem.* **214**(1), 321–324 (1993).

A.97. S. C. Fleming, J. A. Kynaston, M. F. Laker, A. D. J. Pearson, M. S. Kapembwa, and G. E. Griffin, *J. Chromatogr.* **640**(1–2), 293–297 (1993).

A.98. L. Van Nifterik, J. Xu, J. L. Laurent, J. Mathieu, and C. Rakoto, *J. Chromatogr.* **640**(1–2), 335–343 (1993).

A.99. T. S. Taha and T. L. Deits, *Anal. Biochem.* **213**(2), 323–328 (1993).

A.100. J. S. Rohrer, G. A. Cooper, and R. R. Townsend, *Anal. Biochem.* **212**(1), 7–16 (1993).

A.101. W. Lu and R. M. Cassidy, *Anal. Chem.* **65**(20), 2878–2881 (1993).

A.102. M. Fukuda, Y. Kubota, A. Ikuta, K. Hasegawa, and K. Koizimi, *Anal. Biochem.* **212**(1), 289–291 (1993).

A.103. P. Pastore, I. Lavagnini, and G. Versini, *J. Chromatogr.* **634**(1), 47–56 (1993).

A.104. S. K. Sanghi, W. T. Kok, G. C. M. Koomen, and F. J. Hoek, *Anal. Chim. Acta* **273**(1–2), 443–447 (1993).

A.105. N. H. Low and G. G. Wudrich, *J. Agric. Food Chem.* **41**(6), 902–909 (1993).

A.106. J. A. Kynaston, S. C. Fleming, M. F. Laker, and A. D. J. Pearson, *Clin. Chem.* **39**(3), 453–456 (1993).

A.107. M. Stefansson and B. Lu, *Chromatographia* **35**(1–2), 61–66 (1993).

A.108. T. J. O'Shea, S. M. Lunte, and W. R. LaCourse, *Anal. Chem.* **65**(7), 948–951 (1993).

A.109. W. R. LaCourse and D. C. Johnson, *Anal. Chem.* **65**(1), 50–55 (1993).

A.110. A. Trotzer, H.-J. Hofsommer, and K. Rubach, *Fluess. Obst.* **61**(12), 581–589 (1994).

A.111. C. Corradini, *Ann. Chim.* (Rome) **84**(9–10), 385–396 (1994).

A.112. J. Yu and Z. El Rassi, *J. High Resolut. Chromatogr.* **17**(11), 773–778 (1994).

A.113. N. Kuroda, S. Taka, T. Kajikawa, M. Niimi, T. Ishida, and K. Kawanishi, *Tonyobyo* (Tokyo) **37**(9), 695–698 (1994).

A.114. C. Corradini, D. Corradini, C. G. Huber, and G. K. Bonn, *J. Chromatogr.* **685**(2), 213–220 (1994).

A.115. C. Corradini, A. Cristalli, and D. Corradini, *Ital. J. Food Sci.* **6**(1), 103–111 (1994).

A.116. D. C. Andersen, C. F. Goochee, G. Cooper, and M. Weitzhandler, *Glycobiology* **4**(4), 459–467 (1994).

A.117. W. K. Herber and R. S. Robinett, *J. Chromatogr.* **676**(2), 287–295 (1994).

A.118. C.-M. Tsai, X.-X. Gu, and R. A. Byrd, *Vaccine* **12**(8), 700–706 (1994).

A.119. J. W. Timmermans, M. B. van Leeuwen, H. Tournois, D. de Wit, and J. F. G. Vliegenthart, *J. Carbohyr. Chem.* **13**(6), 881–888 (1994).

A.120. M. E. Quigley and H. N. Englyst, *Analyst* **119**(7), 1511–1518 (1994).

A.121. R. E. Roberts and D. C. Johnson, *Electroanalysis* **6**(4), 269–273 (1994).

A.122. J. Sullivan and M. Douek, *J. Chromatogr.* **67**(1–2), 339–350 (1994).

A.123. K. W. Swallow and N. H. Low, *J. AOAC Int.* **77**(3), 695–702 (1994).

A.124. L. H. Slaughter and D. P. Livingston III, *Carbohydr. Res.* **253**, 287–291 (1994).

A.125. G. Marko-Varga, T. Buttler, L. Gorton, L. Olsson, G. Durand, and D. Barcelo, *J. Chromatogr.* **665**, 317–332 (1994).

A.126. M. Friedman, C. E. Levin, and G. M. McDonald, *J. Agric. Food Chem.* **42**(9), 1959–1964 (1994).

A.127. M. K. Park, J. H. Park, M. Y. Lee, S. J. Kim, and I. J. Park, *J. Liq. Chromatogr.* **17**(5), 1171–1182 (1994).

A.128. M. Friedman and C. E. Levin, *J. Agric. Food Chem.* **43**(6), 1507–1511 (1995).

A.129. N. Kuroda, S. Tada, T. Kajikawa, M. Niimi, T. Ishida, and K. Kawanishi, *Tonyobyo* (Tokyo) **38**(12), 979–983 (1995).

A.130. R. Wilson, A. Cataldo, and C. P. Andersen, *Can. J. For. Res.* **25**(12), 2022–2028 (1995).

A.131. C. M. Zook and W. R. LaCourse, *Curr. Seps.* **14**(2), 48–52 (1995).

A.132. C. Corradini, D. Corradini, C. G. Huber, and G. K. Bonn, *Chromatographia* **41**(9/10), 511–515 (1995).

A.133. P. Jorge and A. Abdul-Wajid, *Glycobiology* **5**(8), 759–764 (1995).

A.134. R. E. Roberts and D. C. Johnson, *Electroanalysis* **7**(11), 1015–1019 (1995).

A.135. P. Kerherve, B. Charriere, and F. Gadel, *J. Chromatogr.* **718**(2), 283–289 (1995).

A.136. S. Mou and Z. Li, Sepu, **13**(5), 320–324 (1995).

A.137. J. G. Stuckel and N. H. Low, *J. Agric. Food Chem.* **43**(2), 3046–3051 (1995).

A.138. K. L. Kelly, B. A. Kimball, and J. J. Johnston, *J. Chromatogr.* **711**(2), 289–295 (1995).

A.139. N. Torto, T. Buttler, L. Gorton, g. Marko-Varga, H. Stalbrand, and F. Tjerneld, *Anal. Chim. Acta* **313**(1–2), 15–24 (1995).

A.140. J. S. Rohrer, *Glycobiology* **5**(4), 359–360 (1995).

A.141. M. P. H. Dunkel and R. Amato, *Carbohydr. Res.* **268**(1), 151–158 (1995).

A.142. J. Rohrer, J. Thayer, N. Avdalovic, and M. Weitzhandler, *Tech. Protein Chem.* **VI**, 65–73 (1995).

A.143. J. Prodolliet, E. Bugner, and M. Feinberg, *J. AOAC Int.* **78**(3), 768–782 (1995).

A.144. G. A. Cooper and J. S. Rohrer, *Anal. Biochem.* **226**(1), 182–184 (1995).

A.145. M. F. Chaplin, *J. Chromatogr.* **664**(2), 431–434 (1995).

A.146. P. L. Weber, T. Kornfelt, K. N. Lausen, and S. M. Lunte, *Anal. Biochem.* **225**(1), 135–142 (1995).

A.147. K. S. Wong and J. Jane, *J. Liq. Chromatogr.* **18**(1), 63–80 (1995).

A.148. S. Hikima, T. Kakizaki, and K. Hasebe, *Fresenius' J. Anal. Chem.* **351**(2–3), 237–240 (1995).

A.149. T. Kimura, O. Shirota, A. Suzuki, and Y. Ohtsu, *Kuromatogurafi* **17**(2), 156–157 (1996).

A.150. P. T. Elliott, S. A. Olsen, and D. E. Tallman, *Electroanalysis* **8**(5), 443–446 (1996).

A.151. N. H. Low, *Am. Lab.* **28**(6), 35M–35W (1996).

A.152. A. T. Hotchkiss Jr., K. El-Bahtimy, and M. L. Fishman, *Mod. Meth. Plant Anal.* **17**, *Plant Cell Wall Analysis* 129–146 (1996).

A.153. C. M. Zook, P. M. Patel, W. R. LaCourse, and S. Ralapati, *J. Agric. Food Chem.* **44**(7), 1773–1779 (1996).

A.154. H. Gough, G. A. Luke, J. A. Beeley, and D. A. M. Geddes, *Arch. Oral Biol.* **41**(2), 141–145 (1996).

A.155. H. Ichiki, M. Semma, Y. Sekiguchi, M. Nakamura, and Y. Ito, *Nippon Shokuhin Kagaku Gakkaishi* **2**(2), 119–121 (1996).

A.156. C. Mayer, H. P. Ruffner, and D. M. Rast, *Adv. Chitin Sci.* **1**, 1308–113 (1996).

A.157. N. Torto, G. Marko-Varga, L. Gorton, H. Staalbrand, and F. Tjerneld, *J. Chromatogr.* **725**(1), 165–175 (1996).

A.158. K. Tyagarajan, J. G. Forte, and R. R. Townsend, *Glycobiology* **6**(1), 83–93 (1996).

A.159. P. L. Weber and S. M. Lunte, *Electrophoresis* **17**(2), 302–309 (1996).

A.160. D. Madigan, I. McMurrough, and M. R. Smyth, *J. Am. Soc. Brew. Chem.* **54**(1), 45–49 (1996).

A.161. J. L. Bernarl, M. J. Del Nozal, L. Toribio, and M. Del Alamo, *J. Agric. Food Chem.* **44**(2), 507–511 (1996).

A.162. J. Frias, K. R. Price, G. R. Fenwick, C. L. Hedley, H. Sorensen, and C. Vidal-Valverde, *J. Chromatogr.* **719**, 213–219 (1996).

A.163. R. Wang, M. E. Klegerman, and M. J. Groves, *J. Liq. Chromatogr.* **19**(16), 2691–2698 (1996).

A.164. M. Weitzhandler, R. Slingsby, J. Jagodzinski, C. Pohl, L. Narayanan, and N. Avdalovic, *Anal. Biochem.* **241**(1), 135–136 (1996).

A.165. K. W. Sigvardson, M. S. Eliason, and D. E. Herbranson, *J. Pharm. Biomed. Anal.* **15**(2), 227–231 (1996).

Alcohols

B.1. S. Hughes, P. L. Meschi, and D. C. Johnson, *Anal. Chim. Acta* **132**, 1–10 (1981).

B.2. W. R. LaCourse, D. C. Johnson, M. A. Rey, and R. W. Slingsby, *Anal. Chem.* **63**(2), 134–139 (1991).

B.3. E. Le Fur, P. X. Etievant, and J. M. Meunier, *J. Agric. Food Chem.* **42**(2), 320–326 (1994).

B.4. G. Schiavon, N. Comisso, R. Toniolo, and G. Bontempelli, *Electroanalysis* **8**(6), 544–548 (1996).

B.5. P. Lozano, N. Chirat, J. Graille, and D. Pioch, *Fresenius' J. Anal. Chem.* **354**(3), 319–322 (1996).

Aminoalcohols, Aminosugars, and Aminoglycosides

C.1. J. A. Polta, D. C. Johnson, and K. E. Merkel, *J. Chromatogr.* **324**(2), 407–414 (1985).

C.2. W. R. LaCourse, W. A. Jackson, and D. C. Johnson, *Anal. Chem.* **61**(22), 2466–2471 (1989).

C.3. J. A. Statler, *J. Chromatogr.* **527**(1), 244–246 (1990).

C.4. D. L. Campbell, S. Carson, and D. Van Bramer, *J. Chromatogr.* **546**(1–2), 381–385 (1991).

C.5. D. A. Roston and R. R. Rhinebarger, *J. Liq. Chromatogr.* **14**(3), 539–556 (1991).

C.6. D. A. Martens and W. T. Frankenberger, Jr., *Talanta* **38**(3), 245–251 (1991).

C.7. L. G. McLaughlin and J. D. Henion, *J. Chromatogr.* **591**, 195–206 (1992).

C.8. J. G. Phillips and C. Simmonds, *J. Chromatogr.* **675**(1–2), 123–128 (1994).

C.9. L. A. Kaine and K. A. Wolnik, *J. Chromatogr.* **674**(1–2), 255–261 (1994).

C.10. D. A. Dobberpuhl and D. C. Johnson, *J. Chromatogr.* **694**(1), 391–398 (1995).

C.11. E. Adams, R. Schepers, E. Roets, and J. Hoogmartens, *J. Chromatogr.* **741**(2), 233–240 (1996).

Amines, Aminoacids, Peptides, and Proteins

D.1. J. A. Polta and D. C. Johnson, *J. Liq. Chromatogr.* **6**(10), 1727–1743 (1983).

D.2. L. E. Welch, W. R. LaCourse, D. A. Mead Jr., D. C. Johnson, and T. Hu, *Anal. Chem.* **61**(6), 555–559 (1989).

D.3. L. E. Welch, W. R. LaCourse, D. A. Mead, Jr., and D. C. Johnson, *Talanta* **37**(4), 377–380 (1990).

D.4. M. J. Donaldson and M. W. Adlard, *J. Chromatogr.* **509**(2), 347–356 (1990).

D.5. M. J. Donaldson, H. Broby, M. W. Adlard, and C. Bucke, *Phytochem. Anal.* **1**(1), 18–21 (1990).

D.6. L. E. Welch and D. C. Johnson, *J. Liq. Chromatogr.* **13**(7), 1387–1409 (1990).

D.7. D. A. Martens and W. T. Frankenberger, Jr., *J. Liq. Chromatogr.* **15**(3), 423–439 (1992).

D.8. R. Draisci, S. Cavalli, L. Lucentini, and A. Stacchini, *Chromatographia* **35**(9–12), 584–590 (1993).

D.9. O. A. Sadik and G. G. Wallace, *Anal. Chim. Acta* **279**(2), 209–212 (1993).

D.10. O. A. Sadik, M. J. John, G. G. Wallace, D. Barnett, C. Clarke, and D. G. Laing, *Analyst* **199**(9), 1997–2000 (1994).

D.11. D. A. Dobberpuhl and D. C. Johnson, *Anal. Chem.* **67**(7), 1254–1258 (1995).

D.12. O. A. Sadik and G. G. Wallace, *Anal. Chim. Acta* **302**(1), 131 (1995).

D.13. J. A. M. van Riel and C. Olieman, *Anal. Chem.* **67**, 3911–3915 (1995).

D.14. W. Lu, H. Zhao, and G. G. Wallace, *Anal. Chim. Acta* **315**(1–2), 27–32 (1995).

D.15. D. A. Dobberpuhl, J. C. Hoekstra, and D. C. Johnson, *Anal. Chim. Acta* **322**(1–2), 55–62 (1996).

D.16. M. Weitzhandler, C. Pohl, J. Rohrer, L. Narayanan, R. Slingsby, and N. Avdalovic, *Anal. Biochem.* **241**(1), 128–134 (1996).

D.17. D. A. Dobberpuhl and D. C. Johnson, *Electroanalysis* **8**(8–9), 726–731 (1996).

Sulfur-Containing Compounds

E.1. T. Z. Polta and D. C. Johnson, *J. Electroanal. Chem. Interfacial Electrochem.* **209**(1), 159–169 (1986).

E.2. T. Z. Polta, D. C. Johnson, and G. R. Luecke, *J. Electroanal. Chem. Interfacial Electrochem.* **209**(1), 171–181 (1986).

E.3. A. Ngoviwatchai and D. C. Johnson, *Anal. Chim. Acta* **215**(1–2), 1–12 (1988).

E.4. G. G. Neuburger and D. C. Johnson, *Anal. Chem.* **60**(20), 2288–2293 (1988).

E.5. D. R. Doerge and A. B. K. Yee, *J. Chromatogr.* **586**(1), 158–160 (1991).

E.6. P. J. Vandeberg, J. L. Kowagoe, and D. C. Johnson, *Anal. Chim. Acta.* **260**(1), 1–11 (1992).

E.7. P. J. Vandeberg and D. C. Johnson, *Anal. Chem.* **65**(20), 2713–2718 (1993).

E.8. L. Koprowski, E. Kirchmann, and L. E. Welch, *Electroanalysis* **5**, 473 (1993).

E.9. E. Kirchmann and L. E. Welch, *J. Chromatogr.* **633**, 111 (1993).

E.10. E. Kirchmann, R. L. Earley, and L. E. Welch, *J. Liq. Chromatogr.* **17**, 1755 (1994).

E.11. P. J. Vandeberg and D. C. Johnson, *Anal. Chim. Acta* **290**(3), 317–327 (1994).

E.12. S. Altunata, R. L. Earley, D. M. Mossman, and L. E. Welch, *Talanta* **42**(1), 17–25 (1995).

E.13. A. J. Tudos and D. C. Johnson **67**(3), 557–560 (1995).

E.14. W. R. LaCourse and G. S. Owens, *Anal. Chim. Acta* **307**(2–3), 301–319 (1995).

E.15. A. Liu, T. Li, and E. Wang, *Anal. Sci.* **11**(4), 597–603 (1995).

E.16. G. S. Owens and W. R. LaCourse, *Curr. Seps.* **14**(3/4), 82–88 (1996).

Inorganic, Electroinactive, and Miscellaneous Applications

F.1. M. Stastny and R. Volf, *Sb. Vys. Sk. Chem.-Technol. Praze, Anal. Chem.* **H18**, 73–80 (1983).

F.2. J. A. Polta, I. H. Yeo, and D. C. Johnson, *Anal. Chem.* **57**(2), 563–564 (1985).

F.3. J. A. Polta and D. C. Johnson, *Anal. Chem.* **57**(7), 1373–1376 (1985).

F.4. R. D. Rocklin, *Adv. Chem. Ser.* **210**, *Formaldehyde*, 13–21 (1985).

F.5. M. B. Thomas and P. E. Sturrock, *J. Chromatogr.* **357**(2), 318–324 (1986).

F.6. C. W. Whang and W. L. Tsai, *J. Chin. Chem. Soc.* **36**(3), 179–186 (1989).

F.7. M. Hara, Y. Saeki, and N. Nomura, *Toyama Daigaku Kyoikugakubu Kiyo, B* **40**, 43–47 (1992).

F.8. R. S. Stojanovic, A. M. Bond, and E. C. V. Bulter, *Electroanalysis* **4**(4), 453–461 (1992).

F.9. D. G. Williams and D. C. Johnson, *Anal. Chem.* **64**(17), 1785–1789 (1992).

F.10. H. P. Wagner and M. J. McGarrity, *J. Am. Soc. Brew. Chem.* **50**(1), 1–3 (1992).

F.11. R. M. Ianniello, *J. Liq. Chromatogr.* **15**(17), 3045–3063 (1992).

F.12. J. B. Nair, *J. Chromatogr.* **671**(1–2), 367–374 (1994).

F.13. T. S. Hsi and J. S. Tsai, *J. Clin. Chem. Soc.* (Taipei) **41**(3), 315–322 (1994).

F.14. T. S. Taha and T. L. Deits, *Anal. Biochem.* **219**(1), 115–120 (1994).

F.15. E. Le Fur, J.-M. Meunier, and P. X. Etievant, *J. Agric. Food Chem.* **42**(12), 2760–2765 (1994).

F.16. H. P. Wagner, *J. Am. Soc. Brew. Chem.* **53**(2), 82–84 (1995).

F.17. J. W. Dilleen, C. M. Lawrence, and J. M. Slater, *Analyst* **121**(6), 755–759 (1996).

F.18. S.-W. Park, S.-W. Hong, and J.-H. You, *Bull. Korean Chem. Soc.* **17**(2), 143–147 (1996).

F.19. J. Wen and R. M. Cassidy, *Anal. Chem.* **68**(6), 1047–1053 (1996).

F.20. Y. Shi and B. J. Johnson, *Analyst* **121**(10), 1507–1510 (1996).

Instrumental Development and PED Reviews

G.1. J. G. Pentari and C. E. Efstathiou, *Anal. Instrum.* **15**(4), 329–345 (1986).

G.2. J. Weiss, *GIT-Suppl.* **3**, 65–71 (1986).

G.3. T. Ramstad and D. Milner, *Anal. Instrum.* **18**(2), 147–176 (1989).

G.4. D. S. Austin-Harrison and D. C. Johnson, *Electroanalysis* **1**(3), 189–197 (1989).

G.5. R. Cochrane, in *Proceedings of International Conference on Ion Exchange Processes*, P. A. Williams and M. J. Hudson, eds., Elsevier, New York, 1990, pp. 219–229.

G.6. R. D. Rocklin, A. Henshall, and R. B. Rubin, *Am. Lab.* **22**(5), 34–49 (1990).

G.7. D. C. Johnson and W. R. LaCourse, *Anal. Chem.* **62**(10), 589A–597A (1990).

G.8. Y. C. Lee, *Kagaku to Kogyo* **43**(6), 953–957 (1990).

G.9. T. Soga, *Gekkan Fudo Kemikaru* **7**(12), 44–48 (1991).

G.10. R. Roberts and D. C. Johnson, *Electroanalysis* **4**(8), 741–749 (1992).

G.11. D. C. Johnson and W. R. LaCourse, *Electroanalysis* **4**(4), 367–380 (1992).

G.12. D. C. Johnson, D. Dobberpuhl, R. Roberts, and P. Vandeberg, *J. Chromatogr.* **640**(1–2), 79–96 (1993).

G.13. W. R. LaCourse, *Analusis* **21**(4), 181–195 (1993).

G.14. R. D. Rocklin, T. R. Tullsen, and M. G. Marucco, *J. Chromatogr.* **671**(1–2), 109–114 (1994).

G.15. R. E. Roberts and D. C. Johnson, *Electroanalysis* **6**(3), 193–199 (1994).

G.16. D. C. Johnson and W. R. LaCourse, *J. Chromatogr. Libr.* **58** 391–429 (1995).

G.17. W. R. LaCourse and G. S. Owens, *Electrophoresis* **17**(2), 310–318 (1996).

G.18. W. R. LaCourse and C. O. Dasenbrock, in *Advances in Chromatography*, P. R. Brown and E. Gruska, eds., Marcel Dekker, New York, in press.

G.19. W. R. LaCourse, in *Electrochemical Detection in Liquid Chromatography and Capillary Electrophoresis*, P. T. Kissinger, ed., Marcel Dekker, in press.

INDEX

Activated pulsed amperometric detection (APAD), 8, 135
    waveform optimization, 176
Alcohols, 198–205, 304
    polyalcohols, 205
    quantitation, 205
Alkanolamines, 206
Amines, 213–215, 305
Aminoacids, 138, 141, 215–216
Aminoalcohols, 206–208, 304
Aminoglycosides, 209–213, 304
    hygromycin B, 211
    nebramycin factors, 209
    neomycin A, B, and C, 212
    tobramycin, 211
    spectinomycin, 211
    streptomycin, 211–212
Aminosugars 209, 304
Ampere (amp), 14
Amperometry, DC, 2, 61–63
    applications, 82–85
Anode, 17
Antibiotics, 144
    aminoglycosides, 209–213, 304
    cephalosporins, 220–222
    CE–IPAD, 250
    multicycle waveforms, 144
    penicillins, 220–222
        indirect detection, 254
APAD, *see* activated pulsed amperometric detection
Applications, 207
    alcohols, 198–205, 304
        acidic conditions, 199–200
        alkaline conditions, 201–203
        polyalcohols, 205
    alditols, 293
    amines, 213–215, 305
    aminoalcohols, 206–208, 304

aminosugars 209, 304
aminoglycosides, 209–213, 304
    hygromycin B, 211
    nebramycin factors, 209
    neomycin A, B, and C, 212
    nobramycin, 211
    spectinomycin, 211
    streptomycin, 211–212
    tobramycin, 211
antibiotics in milk, 222
carbohydrates, 182, 193, 293
    complex, 191, 196
    glycoconjugates, 191
    homopolymers, 191
    monosaccharides, 185, 194
        phosphorylated, 193
    oligosaccharides, 186
    starch hydrolyzates, 190, 192
cephalosporin, 204
distribution, 182
electroinactive compounds, 307
glycoconjugates, 191
inorganic compounds, 223, 307
monosaccharides, 185, 194
    phosphorylated, 193
oligosaccharides, 186
peptides, 216, 305
proteins, 216, 305
serial UV and PAD, 203
starch hydrolyzates, 190, 192
sulfite, 224
sulfur-containing compounds, 218–223, 306
thiols, reduced/oxidized, 246
Auxiliary electrode, 239

Baseline
    drift, 243

Baseline (*Continued*)
  noise, 243
Batteries, 17
Bioapplications, *see also* Applications
  carbohydrates, 193
  thiocompounds, 219
Bulk solution, 41

Capillary electrophoresis (CE), 249
  electrochemical detection, 249
  integrated pulsed amperometric
    detection, 249–253
  antibiotics, 250
  carbohydrates, 249–250
  pulsed amperometric detection, 10,
    249–253
  pulsed electrochemical detection,
    249–253
Capillary liquid chromatography
  (CLC), 247
Carbohydrates, 9. *See also* Applications
  dissociation constants, 184
  quantitation, 100, 157, 193–197
  waveform optimization, 155–165
Catalytic stabilization, 2
Cathode, 17
CE–IPAD, *see* Capillary electrophoresis–integrated pulsed
  amperometric detection
CE–PAD, *see* Capillary electro-
  phoresis–pulsed amperometric
  detection
Cephalosporins, 220–222
Charge, 13
Charge-transfer polarization, 38
Charging current, 35
Chloride
  indirect detection, 255–256
Chronoamperometry, 63
Conductivity detection, 61
Conductors, 14

Cottrell equation, 45
Coulomb, 13
Coulometry, 34
Current, 14, 33–35, 41, 43, 74
  faradaic, 34
  nonfaradaic, 34
Current sampling, 117
Cyclic voltammetry, 50–52
Cysteine, 219

DC amperometry, 2, 61–63
  applications, 82–85
  comparison with PAD, 107
Diffusion layer, 40, 44–46
Dissolved O_2, 241

E_{act}, 135
E_{det}, 112, 155–156
E_{oxd}, 112, 159–160
E_{red}, 112, 150–164
ED, *see* Electrochemical detection
Electrical double-layer, 35
Electrical potential, 14
Electrocatalysis, 90
Electrocatalytic detection, 108
Electrochemical cell(s), 15, 17,
    239–241
  galvanic, 17
  neutrality, 17
  porous electrodes, 68
  potential, 19
  shorthand notation, 18
  thin-layer, 64
  tubular electrode(s), 67
  Uni-Jet electrode, 69
  wall-Jet electrode, 66
Electrochemical detection (ED), 1
  amperometry, 61
  applied potential
    optimization, 75
    selectivity, 78

coulometry, 61
electrolytic, 17
indirect with PAD, 110
mobile phase limitations, 71–74
potential limits, 73
quantitation, 82
selectivity, multiple electrodes,
79–81
stabilization
surface adsorption, 89–90
π-resonance, 86–89
working electrode, 69–70
Electrochemical mechanism
mass-transport control, 57
surface-control, 58
Electrode(s), 15
auxiliary, 47, 239
electrical double layer, 35
hydrodynamic, 52–58
indicator, 30
inert, 16
noble metal, 90
polarity, 17
polarization, 35–39
potential, 16, 19, 21
effect of activity, 24
effect of current, 32–33
formal, 25
galvanic, 35–36
ohmic potential loss, 36
reference, 28–31, 238
Ag/AgCl, 30
calomel electrode, 28–29
response, 16
standard potential, 19
working, 27, 236–237
Electrode interface, 39
Electrode potential, 16, 19, 21
effect of activity, 24
effect of current, 32–33
formal, 25
galvanic, 35–36
ohmic potential loss, 36

Electrode pretreatment, 4–6
pulsed potential cleaning, 6
pulsed potential cleaning, histori-
cal, 6
Electrode reactivation, 4
Electrolysis, 32
Electromotive force, 14
Electron transfer, 41
activation energy barrier, 41–42
Electron, 13
Electron-transfer kinetics, 71
Electroplating, 17
Equilibrium constant, 22

Faraday, 14
Faraday's law, 34
Flow-through cell designs, 63–69

Glutathione, 219–220
oxidized (GSSG), 219–220
reduced (GSH), 219–220
Glycosequencing, 10
Gouy–Chapman diffuse layer, 40

Half-cell, 15
Half-reactions, 15
High frequency waveforms, 164, 248
High-performance anion exchange
chromatography (HPAEC), 7,
184
baseline shifting, 187
carbohydrates, effect of hydroxide
on separation, 184
effect of pusher ion, 186
hydroxide gradients, 187
separation of carbohydrates, mechanism, 184–185
High-performance liquid chromatography (HPLC), 1, 61
pulsed electrochemical detection

High-performance liquid chromatography (HPLC) (*Continued*)
  auxiliary electrode, 239
  carbohydrate reactions, 185, 187
  dissolved O_2, 241
  electrochemical cell, 239–241
  microbore, 246
  reference electrodes, 238
  storage, 238
  system problems, 241
  working electrode
    maintenance, 236–237
    polishing, 237
  pulsed voltammetric detection
    contour plot, 259
    purity, 257, 260
    surface plot, 258
Homocysteine, 219
HPAEC, *see* High-performance anion
  exchange chromatography
HPLC, *see* High-performance liquid
  chromatography (HPLC)
Hydrodynamic voltammetry, 53,
  76–78
  mass-transfer, 53
  voltammogram(s), 76–78

Indirect detection, 110, 254–256
  chloride, 255–256
  penicillins, 254
  post-column reagent addition, 255
Inner Helmholtz plane (IHP), 39
Inorganic compounds, 223, 307
Instrumental
  design, 308
  developments, 308
  setup, 230–233
Insulators, 14
Integrated pulsed amperometric detection (IPAD), 7, 135, 137
  amino acids, 138, 141
  guidelines, 176
  thiocompounds, 142

  waveform, 135
  waveform optimization, 176–177
Integrated voltammetric detection
  (IVD), 137
IPAD, *see* Integrated pulsed amperometric detection
iR-drop, 36
IVD, *see* Integrated voltammetric
  detection

Junction potential, 25

Kinetic polarization, 37
Kinetics
  electron-transfer, 71
  oxide dissolutions, 129–130
  oxide-formation, 124–129

Levich equation, 55
Limiting current equation, 45
Linear sweep voltammetry (LSV),
  49
Liquid junction potential, 25–26
LSV, *see* Linear sweep voltammetry

Maintenance
  working electrode, 236–237
Mass transfer, 37
Mass-transport, 42–44
  convection, 43
  diffusion, 43
  migration, 43
Mechanisms
  mode I, 108, 122
  mode II, 110, 122
  mode III, 110
Methionine, 219
MHDV, *see* Modulated hydrodynamic
  voltammetry
Microelectrode, 245

Mode I, 108, 122
Mode II, 110, 122
Mode III, 110
Modulated hydrodynamic voltamme-
    try (MHDV), 132
Monosaccharides, 185, 194
Multicycle waveforms, 144
    antibiotics, 144
Multiplex-PAD (MPAD), 145

Nernst equation, 22–24
Normal hydrogen electrode (NHE),
    20

Ohm's law, 14
Oligosaccharide
    enzyme reaction, 197
    quantitation, 197
Optimization
    carbohydrates, 155–165
    effect of ionic strength, 178
    effect of organic solvents, 179
    effect of pH, 177–178
    effect of temperature, 179
    system consideration, 177–179
    waveform, 149
Optimized waveform
    pulsed amperometric detection
        carbohydrates, 108
        glucose, 108
        lysine, 166
        thiocompounds, 169
Outer Helmholtz plane (OHP), 39
Overpotential, 36–39
    charge-transfer, 38
    concentration, mass transfer, 37
    kinetic, 37
Oxidant, 14
Oxidation, 14
Oxide dissolution kinetics, 129–130
Oxide formation kinetics, 124–129
Oxide-catalyzed detections, 110, 122

Oxide-free detections, 108
Oxidizing agent, 14

PAD, *see* Pulsed amperometric detec-
    tion
PCD, *see* Pulsed coulometric detec-
    tion
PED, *see* Pulsed electrochemical
    detection
Penicillins, 220–222
    indirect detection, 254
Peptides, 216, 305
Phosphorylated monosaccharides, 193
Polarography, 48–49
    differential normal pulse, 49
    differential pulse, 49
    double differential pulse, 49
    normal pulse, 49
    reverse pulse, 49
Polishing, 237
Porous electrodes, 68
Post-column reagent addition,
    233–236
    indirect detection, 255
Post-peak dipping, 116
Potential sweep-pulsed coulometric
    detection (PS-PCD), 7
Potentiometry, 27–32
Potentiostat, 47
Proteins, 216, 305
PS–PCD, *see* Potential sweep-pulsed
    coulometric detection
Pulsed amperometric detection
    (PAD), 7, 86
    beneficial effects of adsorption,
        206
    comparison with DC amperometry,
        107
    current sampling, 117
    effect on surface oxide formation,
        127
    origin of peaks
        mode I (oxide-free), 113

Pulsed amperometric detection (PAD)
    (*Continued*)
    mode II (oxide-catalyzed), 114
    oxide-catalyzed detections
        limitations, 115
    oxide-free detections, 113
    response
        carbohydrates and analogues, 157
        mobile phase effects, 119
    waveform, 112
        high frequency, 164
    waveform optimization
        adsorption effects
            amines, 167
            thiocompounds, 170
        amines and aminoacids, 165–
            168
        carbohydrates, 155–165
        E_{det}, 155–156
        E_{oxd} and t_{oxd}, 159–160
        E_{red} and t_{red}, 160–164
        multiple parameter, 173
        thiocompounds, 168–171
        t_{del}, 157–159
        t_{int}, 157–159
Pulsed coulometric detection (PCD),
    7, 118
Pulsed electrochemical detection
    (PED), 7
    antibiotics, 210–213
    concept, 105
    glucose, 99
    historical, 7
    instrumental setup, 230–233
    mechanisms, 108, 110, 130–132
        mode I, 108
        mode II, 110
        mode III, 110
    microelectrode, 245
        advantages, 245
    pH effect(s), 139, 188
    pH reference electrode, 189
    post-column reagent addition,
        233–236

reviews, 308
selectivity, 145–147
waveforms
    high-frequency, 248
waveform optimization,
    effect of ionic strength, 178
    effect of organic solvents, 179
    effect of pH, 177–178
    effect of temperature, 179
    system considerations, 177–179
Pulsed voltammetry (PV), 8, 153,
    262
    instrumentation, 154
    mechanistic tool, 181
    program, 265–292
        Asyst, word, 262
        description, 262
    quantitative, 180
    response, 177
    S/N, 155
Pulsed voltammetric detection (PVD),
    8, 153, 257–260
    contour plot, 257, 259
    purity, 257, 260
    surface plot, 257–258
PV, *see* Pulsed voltammetry
PVD, *see* Pulsed voltammetric detec-
    tion

Quantitation
    alcohols, 205
    carbohydrates, 193–197
    IPAD, 247
    oligosaccharide, 197
        enzyme reaction, 197
    thiocompounds, 247

Randles–Sevick equation, 49
Reduced analyte signal, 243
Reducing agent, 14
Reductant, 14
Reduction, 14

Reference electrodes, 238
storage, 238
Resistance (R), 14
Resonance stabilization, 2
Reverse pulsed amperometric detection (RPAD), 8, 132–134
waveform optimization, 176
Reversible reaction, 15
Rotating disk voltammetry, 55–58
RPAD, *see* Reverse pulsed amperometric detection

Salt bridge, 17
Saturated calomel electrode (SCE), 28
Selectivity, 145–147
applied potential, 78
multiple electrodes, 79–81
PED, 145–147
Standard hydrogen electrode (SHE), 20
Sulfite, 224
Sulfur-containing pesticides, 218
Sulfur-containing compounds, 218–223, 306
Supporting electrolyte, 47
Surface oxide formation, 124–126

t_{int}, 112, 157–159
t_{det}, 112, 157–159
t_{int}, 112, 157–159
t_{oxd}, 112, 159–160
t_{red}, 112, 160–164
Thin-layer cell, 64, 142
Thiocompounds, 219
quantitation, 247
Thiols, reduced/oxidized, 246
Troubleshooting
baseline
drift, 243
noise, 243

HPLC–PED, 242–244
reduced analyte signal, 243
Tubular electrodes, 67

Uni-jet cell, 69

Voltage, 14
Voltammetric resolution, 111
Voltammetric wave, 50
Voltammetry, 46
adsorption effects, 101
alcohols and glycols, 95–96
amines, 101–103
Au, effect of pH, 93–95
Au in NaOH, 92–93, 123
caffeic acid, 87
catechol, 86
carbohydrates, 96–98
1,2 cyclohexanediol, 86
glucose, mechanism of detection, 99
glucose in NaOH, 105
irreversibility, 52
Pd in NaOH, 93
Pt in Acid, 123
Pt in NaOH, 90–92
square-wave, 49
stationary electrode, 49–52
thiocompounds, 103–105
voltammogram(s), 49

Wall-jet electrode, 66
Waveform(s,) 47–48
cyclic voltammetry, 152
DC amperometry, 63
high-frequency, 164
optimization, 149
amines, 165–168
carbohydrates, 155–165
thiocompounds, 168–175
pulsed amperometric detection, 108, 112

Waveform optimization, 176–177
    effect of ionic strength, 178
    effect of organic solvents, 179
    effect of pH, 177–178
    effect of temperature, 179
    multiple parameters, 173
    pulsed electrochemical detection,
        149–181
        activated pulsed amperometric
            detection, 176

integrated pulsed amperometric
    detection, 176–178
pulsed amperometric detection,
    155–175
reverse pulsed amperometric
    detection, 176
system considerations, 177–179
Working electrode, 236–237
    polishing, 237